Lecture Notes in Applied and Computational Mechanics

Volume 78

Series editors

Friedrich Pfeiffer, Technische Universität München, Garching, Germany
e-mail: pfeiffer@amm.mw.tum.de

Peter Wriggers, Universität Hannover, Hannover, Germany
e-mail: wriggers@ikm.uni-hannover.de

About this Series

This series aims to report new developments in applied and computational mechanics - quickly, informally and at a high level. This includes the fields of fluid, solid and structural mechanics, dynamics and control, and related disciplines. The applied methods can be of analytical, numerical and computational nature.

More information about this series at http://www.springer.com/series/4623

Sergio Conti · Klaus Hackl
Editors

Analysis and Computation of Microstructure in Finite Plasticity

Springer

Editors
Sergio Conti
Inst. für Angewandte Mathematik
Universität Bonn
Bonn
Germany

Klaus Hackl
Lehrstuhl für Mechanik - Materialtheorie
Ruhr-Universität Bochum
Bochum
Germany

ISSN 1613-7736 ISSN 1860-0816 (electronic)
Lecture Notes in Applied and Computational Mechanics
ISBN 978-3-319-38443-6 ISBN 978-3-319-18242-1 (eBook)
DOI 10.1007/978-3-319-18242-1

Springer Cham Heidelberg New York Dordrecht London

Softcover reprint of the hardcover 1st edition 2015

Printed on acid-free paper

Springer International Publishing AG Switzerland is part of Springer Science+Business Media (www.springer.com)

Preface

The present volume presents research carried out within the research group FOR 797 "Analysis and computation of microstructure in finite plasticity" (*MICROPLAST*), financed by the German Science Foundation (DFG). The research group was established in 2007 and came to an end in 2015. During this period more that thirty researchers contributed actively to this endeavor, organized in eight sub-projects.

The aim of *MICROPLAST* was to understand and model dislocation based microstructure formation and evolution in materials, and develop tools to compute it effectively. This is important because plasticity determines a number of industrial and natural phenomena, ranging from the deformation of the earth crust to forming of metals. Most plastic processes are strongly influenced by the formation and evolution of dislocation patterns and other plastic microstructures. Classical plasticity models were able to make useful predictions without accounting for the microstructure, and focusing on a phenomenological understanding of the macroscopic material response. These models have, however, strong limitations, concerning in particular the transferability of the results, the applicability to large deformations, to small samples, and the treatment of ageing and fatigue.

The foundation of *MICROPLAST* was inspired both by new mathematical and modeling developments and by new experimental techniques. Around the turn of the millennium novel variational concepts became available allowing to formulate finite plasticity and associated microstructure formation within a rigorous and powerful mathematical framework. The systematic variational formulation of the evolution of internal variables via dissipation potentials and dissipation distances made it possible to apply the concepts of relaxation theory and Γ-convergence, that were originally developed for static problems, to the evolution of inelastic materials.

At the same time experimental techniques had advanced to a stage where it became possible to perform in situ measurements of dislocation microstructures and their evolution, thus giving valuable information for the development of the corresponding models. For this reason, *MICROPLAST* was established as an interdisciplinary cooperation of scientists from the fields of mathematics, mechanics and materials sciences, working closely together on a common goal.

This volume contains reports on the various achievements reached within *MICROPLAST*. These range from experimental evidence on the mechanisms of microstructure formation to micromechanical and multiscale models of these processes. They include derivations of relaxed envelopes of non-convex energies using novel developments of variational calculus, as well as full mathematical analysis of the models established, and the development of suitable and fast numerical schemes.

It is our hope that we succeeded in giving an insight into a new developing field, and that the readers will find the material in this volume interesting and beneficial for their research.

The contributions in this volume have undergone an internal peer-review. We are convinced that this process led to significant improvements in the papers contained in this book.

Finally, we would like to thank the German Science Foundation for their generous support and the always very pleasant and professional cooperation.

April 2015

Sergio Conti
Klaus Hackl

MICRO*PLAST*

Contents

Chapter 1
Numerical Algorithms for the Simulation of Finite Plasticity with Microstructures

Carsten Carstensen, Dietmar Gallistl, and Boris Krämer

Abstract. This article reports on recent developments in the analysis of finite element methods for nonlinear PDEs with enforced microstructures. The first part studies the convergence of an adaptive finite element scheme for the two-well problem in elasticity. The analysis is based on the relaxation of the classical model energy by its quasiconvex envelope. The second part aims at the computation of *guaranteed lower energy bounds* for the two-well problem with nonconforming finite element methods that involve a stabilization for the discrete linear Green strain tensor. The third part of the paper investigates an adaptive discontinuous Galerkin method for a degenerate convex problem from topology optimization and establishes some equivalence to nonconforming finite element schemes.

1.1 Introduction

Mathematical models in the framework of nonlinear elasticity for phase transformations in solids [BJ87, CK88, BJ92] and for elastoplastic deformations [CHM02] lead to variational problems for which the existence of minimizers cannot be obtained by the direct method in the calculus of variations, for further applications and references see [CDK11, CDK13, CDK15] and the literature quoted therein.

Carsten Carstensen
Humboldt-Universität zu Berlin, Institut für Mathematik,
Unter den Linden 6, D-10099 Berlin, Germany
e-mail: cc@math.hu-berlin.de

Dietmar Gallistl
Universität Bonn, Institut für Numerische Simulation,
Wegelerstraße 6, D-53115 Bonn, Germany
e-mail: gallistl@ins.uni-bonn.de

Boris Krämer
Humboldt-Universität zu Berlin, Institut für Mathematik,
Unter den Linden 6, D-10099 Berlin, Germany
e-mail: kraemer@math.hu-berlin.de

S. Conti and K. Hackl (eds.), *Analysis and Computation of Microstructure in Finite Plasticity*,
Lecture Notes in Applied and Computational Mechanics 78, DOI: 10.1007/978-3-319-18242-1_1

The infimization of the energy enforces oscillations of the infimizing sequence on finer and finer scales called microstructures and converge only weakly but not strongly. Typically the weak limit is not a minimizer of the original minimization problem and has to be replaced by a generalized minimizer which involves the gradient Young measure associated to the sequence of deformation gradients [KP91, MŠ99].

The numerical simulation of problems of this kind is an onging challenging task and a direct minimization of the nonconvex energy in a finite element space leads to strongly mesh-dependent effects [Lus96, Chi00, Car01]. An alternative approach is based on a minimization of the relaxed variational problem [Dac08] obtained by replacing the energy density W by its quasiconvex relaxation W^{qc}, that is, one minimizes

$$I^{qc}(v) := \int_\Omega W^{qc}(\varepsilon(v))\,dx - \int_\Omega f\cdot v\,dx \quad \text{among all } v \text{ in } \mathscr{A} := u_D + H_0^1(\Omega;\mathbb{R}^2). \tag{1.1}$$

Here the function $u_D \in H^1(\Omega;\mathbb{R}^2)$ defines the Dirichlet boundary conditions for the problem and the linear Green strain $\varepsilon(v)$ is the symmetric part of the deformation gradient Du. Since the energy density in the relaxed minimization problem satisfies the necessary convexity conditions in the vector-valued calculus of variations, problem (1.1) allows for a minimizer in $\mathscr{A}$. Moreover, any minimizer u characterizes a macroscopic deformation of the original problem in the sense that there exists a sequence $(u_j)_{j\in\mathbb{N}}$ which infimizes the energy of the original variational problem and converges weakly to u. If this convergence is also strong in $H^1(\Omega;\mathbb{R}^2)$, then the minimum of the energy is attained and u is a classical minimizer of the original problem.

This approach is extremely appealing, if an explicit formula for W^{qc} is known. In this case one can construct for a given deformation gradient F, a corresponding gradient Young measure ν with center of mass F which realizes the relaxed energy, $W^{qc}(F) := \langle \nu, W\rangle := \int_{\mathbb{R}^{2\times 2}} W(F)\,d\nu(F)$, and provides at the same time a representation for the stress variable $\sigma(F) = DW^{qc}(F) := \langle \nu, DW\rangle := \int_{\mathbb{R}^{2\times 2}} DW(F)\,d\nu(F)$; see [BKK00] and [CM02] for a discussion of the regularity of the stress variable. In this way one obtains the associated stresses which are of fundamental importance in engineering applications. A successful example of this approach in the numerical analysis of a relaxed problem can be found in [CP97, CP01].

A posteriori error estimation for relaxed nonconvex problems or degenerate convex problems typically encounters the reliability-efficiency gap [CJ03]. This means that reliable a posteriori error estimators converge with a worse rate compared to the true error and, hence, is not efficient: The efficiency index even diverges towards ∞.

The motivation of effective numerical simulations of microstructures in finite plasticity arose in [CHM02], where it is shown that a typical time-step in finite plasticity leads to a non convex minimization problem and that shear bands may be seen as microstructures in the corresponding minimization process. The relaxation models in the post-modern calculus of variations [Dac08] appears as the only feasible approach for a computer simulation in [Car01] and this provoked massive research on relaxation models of single- and multiple-slip systems and their time evolution in this research group. The numerical treatment needs an extra justification and, even if some convergence analysis arises naturally in [CP97] for a class of convexified model problems, there remain severe open questions in the adaptive mesh-refining [Car08] and in efficient and reliable error control [CJ03]. The mathematical understanding of the performance of the related discretization schemes could not follow the lines of an implicit function theorem [CD04] because of the too restrictive smallness and uniform polyconvexity assumptions. The latter at least seemingly contradicts the concept of a related hull in finite plasticity. The research of this project therefore started at the understanding of the convexified model problems of [CM02, CP97, Car08] and their generalization to polyconvex problems with standard [CP97, Car08] and nonstandard discretization [CGR12] and a focus on adaptive mesh-refining with a complete a priori and posterior error analysis.

In the first part of this paper, we outline the convergence analysis for the relaxation of the classical model energy in a two-dimensional setting for which the relaxation was obtained in [Koh91, LC88, Pip91]; see (1.6) below for the precise formula. From the point of view of numerical analysis, one striking advantage of the relaxed minimization problem is that the macroscopic deformation u can, in principle, be computed with a strongly convergent sequence of minimizers in suitable finite element spaces. The reliability-efficiency gap does not prevent the convergence proof of the associated stresses for a large class of variational problems with energy densities that fail to be strictly convex [Car08].

The second part of this paper is devoted to the computation of *guaranteed lower energy bounds* for the two-well problem of [Koh91, LC88, Pip91]. The nonconforming finite element method serves as the main tool for deriving those lower bounds.

In the third part of the paper, we investigate a discontinuous Galerkin method for a degenerate convex problem from topology optimization. The reliability-efficiency gap motivated stabilized finite element methods (FEMs) [BC10, BC14] for degenerate convex minimization problems. The recent developments of [CL15] could improve the reliability-efficiency gap with duality methods and nonconforming FEMs. The discontinuous Galerkin method here appears as a natural choice of a stabilized discontinuous method and this paper succeeds in establishing an equivalence to a nonconforming finite element method. The conclusion in Section 1.6 connects the research in this project with the achievements of this research group and contains various open questions for future research.

1.2 Preliminaries and Notation

Let $\Omega \subseteq \mathbb{R}^2$ be a bounded polygonal Lipschitz domain with outer unit normal ν along the boundary $\partial\Omega$. Let $\mathscr{T}$ be a regular triangulation of Ω into triangles in the sense of Ciarlet, with edges $\mathscr{F}$ and vertices $\mathscr{N}$. The interior (resp. boundary) edges are denoted by $\mathscr{F}(\Omega)$ (resp. $\mathscr{F}(\partial\Omega)$). Analogously let $\mathscr{N}(\Omega)$ denote the interior vertices and $\mathscr{N}(\partial\Omega)$ denote the vertices on the boundary. The set of edges of a triangle $T \in \mathscr{T}$ reads $\mathscr{F}(T)$, the set of vertices of T is denoted by $\mathscr{N}(T)$. For any

$T \in \mathcal{T}$ let $h_T = \operatorname{diam}(T)$ and define the piecewise constant mesh-size function $h_{\mathcal{T}}$ by $h_{\mathcal{T}}|_T := h_T$. The length of an edge $F \in \mathcal{F}$ is denoted by h_F. For any interior edge $F \in \mathcal{F}(\Omega)$, there exist two adjacent triangles T_+ and T_- such that $F = \partial T_+ \cap \partial T_-$. Let $\nu_F = (\nu_F(1); \nu_F(2))$ denote the fixed normal vector of F that points from T_+ to T_-. For $F \in \mathcal{F}(\partial\Omega)$, let ν_F denote the outward unit normal vector of Ω. The tangential vector of an edge F is denoted by $\tau_F := (-\nu_F(2); \nu_F(1))$. Given any (possibly vector-valued) function v, define the jump and the average of v of across $F \in \mathcal{F}(\Omega)$ by

$$[v]_F := v|_{T_+} - v|_{T_-} \qquad \text{and} \qquad \langle v \rangle_F := (v|_{T_+} + v|_{T_-})/2 \qquad \text{along } F.$$

For a boundary edge $F \in \mathcal{F}(\partial\Omega) \cap \mathcal{F}(T_+)$, define $[v]_F := v|_F - u_D|_F$ and $\langle v \rangle_F := (v|_F + u_D|_F)/2$ for the prescribed Dirichlet data u_D.

For any $T \in \mathcal{T}$, the space of polynomial functions of degree at most k is denoted by $P_k(T)$. The space of piecewise polynomials reads

$$P_k(\mathcal{T}) = \{v \in L^2(\Omega) \mid \forall T \in \mathcal{T}, v|_T \in P_k(T)\}.$$

The piecewise action of the derivative D is denoted by D_{NC}. The symmetric part of the gradient reads $\varepsilon := \operatorname{sym} D$ and its piecewise action reads $\varepsilon_{\mathrm{NC}}$. The L^2 projection onto piecewise constants with respect to $\mathcal{T}$ is denoted by Π_0.

Standard notation on Lebesgue and Sobolev spaces applies throughout this paper; $H^{-1}(\Omega)$ denotes the dual spaces of $H^1_0(\Omega)$. The space of smooth functions with compact support in Ω is denoted by $\mathcal{D}(\Omega)$. The L^2 norm over the domain Ω is abbreviated as $\|\cdot\| := \|\cdot\|_{L^2(\Omega)}$. The L^2 inner product reads $(\cdot,\cdot)_{L^2(\Omega)}$. The integral mean is denoted by $\fint$. The space of real 2×2 matrices reads $\mathbb{M} \equiv \mathbb{R}^{2\times 2}$. The symmetric part of a matrix $A \in \mathbb{M}$ reads $\operatorname{sym} A$. The space of symmetric 2×2 matrices reads $\mathbb{S} := \operatorname{sym}\mathbb{M}$. The dot denotes the product of two one-dimensional lists of the same length while the colon denotes the Euclidean product of matrices, e.g., $a \cdot b = a^\top b \in \mathbb{R}$ for $a, b \in \mathbb{R}^2$ and $A : B = \sum_{j,k=1}^2 A_{jk} B_{jk}$ for 2×2 matrices A, B. The measure $|\cdot|$ is context-sensitive and refers to the number of elements of some finite set or the length of an edge or the area of some domain and not just the modulus of a real number or the Euclidean length of a vector.

1.3 Convergent Adaptive Finite Element Method for the Two-Well Problem in Elasticity

In this section we outline the convergence analysis for the relaxation of the classical model energy

$$W(E) = \min\{\frac{1}{2}\langle \mathbb{C}(E - A_1), E - A_1\rangle + w_1, \frac{1}{2}\langle \mathbb{C}(E - A_2), E - A_2\rangle + w_2\} \quad (1.2)$$

in a two-dimensional setting for which the relaxation was obtained in [Koh91, LC88, Pip91]. see (1.6) below for the precise formula with given symmetric matrices A_1

and A_2 called the wells. It turns out that the quasiconvex relaxation is in fact the convex relaxation if and only if the two preferred strains A_1 and A_2 are compatible, see [Koh91, Lemma 4.1] for necessary and sufficient conditions for compatibility. The case of compatible wells was analyzed in [Car08] and therefore we focus on the incompatible case in this paper. Moreover, we assume that the matrix $A_1 - A_2$ is not proportional to the identity matrix since in this case the uniqueness of minimizers may be lost [Ser96, Remark 2.2]. Hence we assume that the eigenvalues η_1 and η_2 of the matrix $A_1 - A_2$ satisfy

$$0 < \eta_1 < \eta_2. \tag{1.3}$$

We refer to the problem as nonconvex since for incompatible wells the relaxation is not convex but quasiconvex.

1.3.1 Review of the Model Problem

In this subsection, we recall the model two-well problem following the discussion in [CD14]. The starting point is the nonconvex energy density W for a two-dimensional model in linear elasticity with linear kinematics for a phase transforming material with two preferred elastic strains A_1 and $A_2 \in \mathbb{S}$ and elasticity tensor $\mathbb{C}$ for which

$$W(E) := \min\{W_1(E), W_2(E)\} \quad \text{for all } E \in \mathbb{S} \tag{1.4}$$

with suitable constants $w_j \in \mathbb{R}$ and

$$W_j(E) := \frac{1}{2}\langle \mathbb{C}(E - A_j), E - A_j\rangle + w_j \quad \text{for } j = 1, 2. \tag{1.5}$$

The focus lies on the classical case of an isotropic Hooke's law with bulk modulus $\kappa > 0$ and shear modulus $\mu > 0$, i.e.,

$$\mathbb{C}E := \kappa(\mathrm{tr}E)1_{2\times 2} + 2\mu\,\mathrm{dev}E \quad \text{for any } E \in \mathrm{sym}\mathbb{M} \equiv \mathbb{S}.$$

Since A_1 and A_2 are symmetric matrices, we may relabel the matrices in such a way that the eigenvalues η_1 and η_2 of $A_1 - A_2$ satisfy $\eta_1 \geq |\eta_2|$ and, after a suitable change of coordinates, we may suppose that the eigenvectors are parallel to the coordinate axes, i.e., $A_1 - A_2 = \mathrm{diag}(\eta_1, \eta_2)$. It is well-established (see, e.g., Lemma 4.1 in [Koh91]) that A_1 and A_2 are incompatible as linear elastic strains if and only if $\eta_2 > 0$. The relaxed energy density W^{qc} was computed by Kohn [Koh91], Lurie and Cherkaev [LC88], and Pipkin [Pip91]. As mentioned, e.g., in [Koh91], Section 4, the relaxation is piecewise quadratic and globally C^1, and in the notation of this reference given by the expression below. Define $\nu := (\kappa - \mu)/\mu$ as well as (for $j = 1, 2$)

$$\gamma_j := (\kappa - \mu)\mathrm{tr}(A_1 - A_2) + 2\mu\eta_j \quad \text{and} \quad g := \frac{\gamma_1^2}{\kappa + \mu} = \frac{\gamma_1^2}{\mu(\mu+2)}.$$

Let

$$\begin{aligned}
\mathscr{P}_1 &= \{E \in \mathbb{S} \,|\, W_1(E) - W_2(E) + g/2 \le 0\},\\
\mathscr{P}_2 &= \{E \in \mathbb{S} \,|\, W_1(E) - W_2(E) - g/2 \ge 0\},\\
\mathscr{P}_{\mathrm{rel}} &= \{E \in \mathbb{S} \,|\, |W_1(E) - W_2(E)| \le g/2\}.
\end{aligned}$$

The quasiconvex envelope of W [Koh91, Pip91, LC88] reads

$$W^{\mathrm{qc}}(E) = \begin{cases} W_1(E) & \text{if } E \in \mathscr{P}_1,\\ W_2(E) & \text{if } E \in \mathscr{P}_2,\\ W_2(E) - \dfrac{1}{2g}(W_2(E) - W_1(E) + g/2)^2 & \text{if } E \in \mathscr{P}_{\mathrm{rel}} \end{cases} \tag{1.6}$$

for any $E \in \mathbb{S}$. This gives rise to the macroscopic energy

$$I^{\mathrm{qc}}(v) := \int_\Omega W^{\mathrm{qc}}(\varepsilon(v))\,dx - \int_\Omega f \cdot v\,dx. \tag{1.7}$$

Let

$$\gamma := \mu(\widetilde{\nu} - (\widetilde{\nu} + 2)\gamma_2/\gamma_1) \quad \text{for} \quad \widetilde{\nu} := (\kappa - \mu)/\mu. \tag{1.8}$$

The translated energy utilizes the shifted energy density

$$\Phi(X) = W^{\mathrm{qc}}(\mathrm{sym}X) - \gamma \det X \quad \text{for any } X \in \mathbb{M} \tag{1.9}$$

and amounts to

$$E(v) := \int_\Omega \Phi(Dv)\,dx - \int_\Omega f \cdot v\,dx \quad \text{for any } v \in \mathscr{A}.$$

It turns out [CD14] that this convex functional has (possibly non-unique) minimizers in $u \in \mathscr{A}$ which lead to a unique pseudostress $\tau := D\Phi(Du)$.

1.3.2 Adaptive Algorithm

This section describes an algorithm for adaptive mesh-refining. The following paragraphs are (in parts) a repetition of material published in [CD14].

Given an initial shape-regular triangulation $\mathscr{T}_0$, this scheme generates a sequence of triangulations $\mathscr{T}_\ell$ and corresponding finite element spaces $V^{(\ell)}$ which are shape-regular depending on the initial configuration. In particular, all constants are independent of ℓ.

1.3.2.1 INPUT

The input required by the numerical scheme is a shape-regular triangulation $\mathscr{T}_0$ of the bounded domain $\Omega \subset \mathbb{R}^2$, the associated finite element space $V^{(0)} = V(\mathscr{T}_0)$ of continuous functions which are on all elements affine polynomials with values in $\mathbb{R}^2$, and a fixed parameter Θ with $0 < \Theta < 1$ for the marking strategy. For simplicity, we assume that the Dirichlet condition u_D is contained in $V^{(0)}$.

1.3.2.2 SOLVE and the Discrete Minimization Problems

Given the triangulation $\mathscr{T}_\ell$, $\ell \in \mathbb{N}_0$, with the corresponding discrete spaces $V^{(\ell)} = V(\mathscr{T}_\ell)$ and $V_0^{(\ell)} = V_0(\mathscr{T}_\ell)$ on the level ℓ, compute the discrete solution $u_\ell \in u_D + V_0^{(\ell)} \equiv \mathscr{A}_\ell$ as the unique minimizer of the energy functional I^{qc} on $\mathscr{A}_\ell$. For simplicity, we suppose that the discrete solution is computed exactly. Then, the discrete stress is given by

$$\sigma_\ell = DW^{\mathrm{qc}}(\varepsilon(u_\ell)) \in L^2(\mathscr{T}_\ell;\mathbb{S}).$$

Note that DW^{qc} is piecewise affine and globally continuous and hence globally Lipschitz continuous. Since $\varepsilon(u_\ell) \in P_0(\mathscr{T}_\ell;\mathbb{S})$ is piecewise constant, so is $\sigma_\ell \in L^2(\mathscr{T}_\ell;\mathbb{S})$.

1.3.2.3 ESTIMATE

Suppose that T_+ and T_- are two distinct triangles in $\mathscr{T}_\ell$ with a common edge $F = \partial T_+ \cap \partial T_- \in \mathscr{F}_\ell(\Omega)$ of length $|F|$. The unit normal vector

$$\nu_F = \nu_{T_+}|_F = -\nu_{T_-}|_F \quad \text{along } F$$

is defined up to the orientation which we fix as the orientation of the outer normal ν_{T_+} of T_+ along F. Given the discrete stress $\sigma_\ell = DW^{\mathrm{qc}}(\varepsilon(u_\ell)) \in L^2(\mathscr{T}_\ell;\mathbb{S})$ of the previous subsection, the jump of σ_ℓ across the edge is defined as

$$[\sigma_\ell]_F \nu_F = \sigma_\ell|_{T_+} \nu_{T_+} + \sigma_\ell|_{T_-} \nu_{T_-} = \left(\sigma_\ell|_{T_+} - \sigma_\ell|_{T_-}\right) \nu_F \quad \text{along } F.$$

Let $\mathscr{F}(T)$ denote the set of the three edges of a triangle $T \in \mathscr{T}_\ell$ and $\mathscr{F}_{int}(T) = \mathscr{F}(T) \setminus \mathscr{F}_\ell(\partial\Omega)$ the subset of interior edges. To each triangle $T \in \mathscr{T}_\ell$ with area $|T|$ we associate the error estimator contribution $\eta_\ell(T)$ given by

$$\eta_\ell^2(T) = |T|\,\|f + \operatorname{div}\sigma_\ell\|^2_{L^2(T)} + |T|^{1/2} \sum_{F \in \mathscr{F}_{int}(T)} \|[\sigma_\ell]_F \nu_F\|^2_{L^2(F)}.$$

The sum

$$\eta_\ell^2 = \sum_{T \in \mathscr{T}_\ell} \eta_\ell^2(T)$$

is indeed an error estimator for the accompanying pseudo-stress approximations from the translated energy minimization problem, see [CD14]. However, the upper bound η_ℓ of the pseudo-stress error is not sharp, the reliable error estimator η_ℓ is not efficient. This dramatic difficulty in the a posteriori error control is called reliability-efficiency gap in [CJ03] and is caused by the degenerate convexity typically encountered in relaxed variational problems in the effective modelling of microstructures.

1.3.2.4 MARK and REFINE

Suppose that all element contributions $(\eta_\ell^2(T) : T \in \mathscr{T}_\ell)$ defined in the previous subsection are known on the current level ℓ with triangulation $\mathscr{T}_\ell$. Given the input parameter $\Theta \in (0,1)$ select a subset $\mathscr{M}_\ell$ of $\mathscr{T}_\ell$ (of minimal cardinality) with

$$\Theta \eta_\ell^2 \leq \sum_{T \in \mathscr{M}_\ell} \eta_\ell^2(T) =: \eta_\ell^2(\mathscr{M}_\ell). \tag{1.10}$$

This selection condition is also called *bulk criterion* or Dörfler marking [Dör96, MNS02] and is easily arranged with some greedy algorithm.

Any marked element is bisected according to the rules in Figure 1.1 and further mesh refinements may be necessary (e.g., via newest vertex bisection) such that $\mathscr{T}_{\ell+1}$ is a refinement of $\mathscr{T}_\ell$ with $\mathscr{M}_\ell \subset \mathscr{T}_\ell \setminus \mathscr{T}_{\ell+1}$.

Theorem 1.3.2 does not need the refinement with five bisections to obtain the interior node property and may focus on green-blue-red or green-blue refinement strategies.

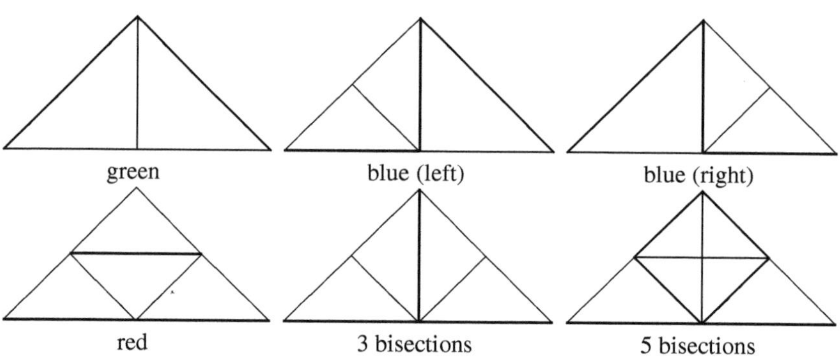

Fig. 1.1 Possible refinements of a triangle (up to rotations)

1.3.2.5 OUTPUT and Convergence Result

For a given triangulation $\mathscr{T}_\ell$ the adaptive scheme generates the triangulation at the next level $\mathscr{T}_{\ell+1}$ by a successive completion of the subroutines

$$\texttt{SOLVE} \rightarrow \texttt{ESTIMATE} \rightarrow \texttt{MARK} \rightarrow \texttt{REFINE} \tag{1.11}$$

Based on the input triangulation $\mathscr{T}_0$, this scheme defines a sequence of meshes $\mathscr{T}_0, \mathscr{T}_1, \mathscr{T}_2, \dots$ and associated discrete subspaces

$$V^{(0)} \subsetneqq V^{(1)} \subsetneqq \cdots \subsetneqq V^{(\ell)} \subsetneqq V^{(\ell+1)} \subsetneqq \cdots \subsetneqq V = H^1(\Omega;\mathbb{R}^2) \tag{1.12}$$

with discrete minimizers $u_\ell \in u_D + V_0^{(\ell)}$, $\ell \in \mathbb{N}_0$. The main properties of this sequence of solutions are formulated in Theorem 1.3.1 and Theorem 1.3.2.

1.3.3 Convergence for the Deformation Gradient

The first main result of [CD14] shows strong convergence for three out of four components in the deformation gradient. The fact that the last component cannot be controlled is related to the degenerate convexity of the relaxed energy.

Theorem 1.3.1. *Let $u \in \mathscr{A}$ be a minimizer of I^{qc}, and let u_h be a minimizer of I^{qc} in a finite element space $u_D + V_{h,0}$ with $u_D \in V_h$ and Courant finite element method with respect to some shape-regular triangulation $\mathscr{T}_h$. Then there exist constants C_1 and C_2 which depend on the triangulation only through the shape-regularity such that, in a suitable coordinate system with $A_1 - A_2 = \operatorname{diag}(\eta_1, \eta_2)$,*

$$\begin{aligned}\|\partial_1(u-u_h)_1\|_{H^{-1}(\Omega)} + \sum_{j,k=1,2;(j,k)\neq(1,1)} \|\partial_k(u-u_h)_j\| \\ \leq C_1 \min_{v_h \in u_D + h,0} \left(I^{qc}(v_h) - I^{qc}(u)\right).\end{aligned}$$

If $u \in H^2(\Omega;\mathbb{R}^2)$ then

$$\min_{v_h \in u_D + V_{h,0}} \left(I^{qc}(v_h) - I^{qc}(u)\right) \leq C_2 h \|D^2 u\|.$$

The second main result of [CD14] guarantees the convergence of the adaptive mesh-refining process of 1.3.2.

Theorem 1.3.2. *Suppose that the assumptions in Theorem 1.3.1 hold. Then the sequence $(u_\ell)_{\ell \in \mathbb{N}}$ with $u_\ell \in u_D + V_0^{(\ell)}$, $\ell \in \mathbb{N}_0$, computed by the adaptive scheme converges with respect to the weak topology of $H^1(\Omega;\mathbb{R}^2)$ to the unique minimizer u of the variational integral I^{qc} in the class of admissible functions $\mathscr{A}$. Moreover, the energies $I^{qc}(u_\ell)$ converge, i.e.,*

$$\lim_{\ell \to \infty} I^{qc}(u_\ell) = I^{qc}(u) = \min_{v \in u_D + H_0^1(\Omega;\mathbb{R}^2)} I^{qc}(v),$$

and, in a suitable coordinate system with $A_1 - A_2 = \operatorname{diag}(\eta_1, \eta_2)$, all components of the deformation gradient except the $(1,1)$-component converge strongly $L^2(\Omega)$, i.e.,

$$\|\partial_1(u-u_\ell)_1\|_{H^{-1}(\Omega)} + \sum_{j,k=1,2;(j,k)\neq(1,1)} \|\partial_k(u-u_\ell)_j\| \to 0 \text{ as } \ell \to \infty.$$

One key ingredient in the proof is the observation [Koh91] that the relaxation of the energy (1.4) can be written as the sum of a convex and a polyaffine function which in the case at hand is a multiple of the determinant. This special structure has, e.g., been used in [Ser96, Ser98] to obtain uniqueness results and regularity of phase boundaries while our approach is in the spirit of the translation method which has been widely used in homogenization theory to separate nonconvex terms with special structure, usually polyaffine functions, from others terms. The key lemma of [CD14] reads as follows.

Lemma 1.3.3 (convexity control). *There exists some matrix* $\mathbf{D}$ *with the maximal eigenvalue* $\rho(\mathbf{D})$ *such that the constant* $\lambda_1 := \max\{1/(4(\gamma_1^2+\gamma_2^2)), 4\rho(\mathbf{D})\}$ *satisfies*

$$\lambda_1 |D\Phi(A) - D\Phi(B)|^2 \leq \Phi(A) - \Phi(B) - D\Phi(B) : (A-B) \quad \textit{for all } A, B \in \mathbb{M}.$$

Proof. The proof in [CD14, Theorem 4.1] leads to the existence of the constant λ_1. The matrix $\mathbf{D}$ and the new explicit expression of λ_1 are derived in the appendix of this paper. □

1.4 Guaranteed Lower Energy Bounds for the Two-Well Problem

The conforming finite element method from the previous section leads to upper energy bounds. This section discusses the computation of guaranteed lower energy bounds with nonconforming finite elements.

1.4.1 Nonconforming FEM and Discrete Energy Functional

The nonconforming P_1 finite element space (also named after Couzeix-Raviart) is defined by

$$\mathrm{CR}^1(\mathscr{T}) := \{v \in P_1(\mathscr{T}) \mid v \text{ is continuous in the interior edges' midpoints}\}.$$

The space of nonconforming finite element functions that vanish in the boundary edges' midpoints is denoted by

$$\mathrm{CR}^1_0(\mathscr{T}) := \{v \in \mathrm{CR}^1(\mathscr{T}) \mid v \text{ vanishes in the boundary edges' midpoints}\}.$$

Set

$$\mathrm{CR}^1(\mathscr{T};\mathbb{R}^2) := [\mathrm{CR}^1(\mathscr{T})]^2 \quad \text{and} \quad \mathrm{CR}^1_0(\mathscr{T};\mathbb{R}^2) := [\mathrm{CR}^1_0(\mathscr{T})]^2.$$

For a triangle T, the Crouzeix-Raviart interpolation $I_{\mathrm{CR}} : H^1(T) \to P_1(T)$ acts on $v \in H^1(T)$ through

$$I_{\mathrm{CR}} v(\mathrm{mid}(F)) = \fint_F v\, ds \quad \text{for all } F \in \mathscr{F}(T)$$

and enjoys the integral mean property of the gradient

$$\nabla I_{\mathrm{CR}} v = \fint_T \nabla v\, dx. \tag{1.13}$$

The following approximation property of I_{CR} with the constant $\varkappa := \sqrt{1/48 + j_{1,1}^{-2}} = 0.298234942888$ is proven in [CG14].

Proposition 1.4.1 (Thm. 4 of [CG14]). *For any $v \in H^1(T)$ on a triangle T the Crouzeix-Raviart interpolation operator satisfies*

$$\|v - I_{\mathrm{CR}} v\|_{L^2(T)} \le \varkappa h_T \|\nabla(v - I_{\mathrm{CR}} v)\|_{L^2(T)}. \qquad \square$$

It is well-known that the classical nonconforming P_1 FEM may be unstable for problems involving the linearized Green strain tensor due to the lack of a discrete Korn inequality. Indeed, the nonconforming FEM allows for configurations with piecewise (infinitesimal) rigid body motions $v_h \in \mathrm{CR}_0^1(\mathscr{T})$ such that $D_{\mathrm{NC}} v_h$ is a nonzero skew-symmetric matrix field. The technique from [HL03] employs, with some positive parameter α, the stabilization term

$$\alpha \sum_{F \in \mathscr{F}} \fint_F |[v_h]_F|^2\, ds.$$

The following discrete Korn inequality is proven in [HL03].

Proposition 1.4.2 (Proposition 2.2 of [HL03]). *For any $\alpha > 0$ there exists a positive constant $C(\alpha)$ which only depends on the shape-regularity in $\mathscr{T}$ such that any $v_h \in \mathrm{CR}^1(\mathscr{T})$ satisfies*

$$C(\alpha)^{-1} \|D_{\mathrm{NC}} v_h\|^2 \le \|\varepsilon_{\mathrm{NC}}(v_h)\|^2 + \alpha \sum_{F \in \mathscr{F}} \fint_F |[v_h]_F|^2\, ds. \qquad \square$$

The discrete convex energy functional reads (for all $v_h \in \mathrm{CR}^1(\mathscr{T}; \mathbb{R}^2)$)

$$E_{\mathrm{NC}}(v_h) := \int_\Omega \Phi(D_{\mathrm{NC}} v_h)\, dx + \alpha \sum_{F \in \mathscr{F}} \fint_F |[v_h]_F|^2\, dx - \int_\Omega f \cdot v_h\, ds.$$

The discrete set of admissible functions reads

$$\mathscr{A}_{\mathrm{CR}} := I_{\mathrm{CR}} u_D + \mathrm{CR}_0^1(\mathscr{T}; \mathbb{R}^2)$$

and gives rise to the discrete problem

$$u_h \in \arg\min_{v_h \in \mathscr{A}_{\mathrm{NC}}} E_{\mathrm{NC}}(v_h).$$

Note that super-linear growth of E_{NC} (and thus well-posedness of the minimization problem) follows with Lemma 1.4.3 below provided $\alpha > 0$ is sufficiently large (dependent on γ). This restriction on α is required because the term in (1.9) involves the determinant of quadratic growth.

1.4.2 Lower Energy Bounds

Let $C_{\mathrm{ell}} = 2\min\{\mu, \kappa\}$ denote the ellipticity constant that satisfies

$$C_{\mathrm{ell}}|E|^2 \le |E|^2_{\mathbb{C}} := E : \mathbb{C}E \quad \text{for any } E \in \mathbb{S}. \tag{1.14}$$

Lemma 1.4.3 (growth condition). *The constants* $C_1 := 2/C_{\mathrm{ell}} = \max\{1/\mu, 1/\kappa\}$ *and* $C_2 := C_1\,(\max\{|A_1|_{\mathbb{C}}, |A_2|_{\mathbb{C}}\} - \min\{w_1, w_2 - g/2\})$ *satisfy, for any* $E \in \mathbb{S}$, *that*

$$|E|^2 \le C_1 W^{\mathrm{qc}}(E) + C_2.$$

Proof. Let $E \in \mathbb{S}$. In case that $E \in \mathscr{P}_{\mathrm{rel}}$, the definition of $\mathscr{P}_{\mathrm{rel}}$ yields $|W_1(E) - W_2(E)| \le g/2$ and, hence,

$$\frac{1}{2g}(W_2(E) - W_1(E) + g/2)^2 \le g/2. \tag{1.15}$$

For any $j = 1, 2$, the Young inequality reads

$$\frac{1}{2}|E|^2_{\mathbb{C}} - |A_j|^2_{\mathbb{C}} \le |E - A_j|^2_{\mathbb{C}}. \tag{1.16}$$

The combination of (1.15)–(1.16) with the definition of W^{qc} from (1.6) results in

$$\frac{1}{2}|E|^2_{\mathbb{C}} - \max\{|A_1|_{\mathbb{C}}, |A_2|_{\mathbb{C}}\} + \min\{w_1, w_2 - g/2\} \le W^{\mathrm{qc}}(E).$$

The ellipticity (1.14) and elementary algebra conclude the proof. □

Lemma 1.4.4 (Korn-type inequality). *Any* $v \in \mathscr{A}$ *satisfies*

$$\|Dv\|^2 \le 4\|\varepsilon(v)\|^2 + 5\|Du_D\|^2. \tag{1.17}$$

Proof. The Korn inequality

$$\|Dv\|^2 \le 2\|\varepsilon(v)\|^2 \quad \text{for any } v \in H^1_0(\Omega; \mathbb{R}^2)$$

is an elementary consequence of the integration by parts formula. For a general function $v \in A$, the split $v = v_0 + v_D$ into $v_0 \in H^1_0(\Omega; \mathbb{R}^2)$ and the harmonic extension $v_D \in \mathscr{A}$ of the Dirichlet data $u_D|_{\partial\Omega}$ leads to

$$\|Dv\|^2 \leq 2\|\varepsilon(v_0)\|^2 + \|Dv_D\|^2.$$

The Young inequality $4ab \leq a^2 + 4b^2$ for any $(a,b) \in \mathbb{R}^2$ implies

$$0 \leq \|\varepsilon(v_0)\|^2 + 4\|\varepsilon(v_D)\|^2 + 4(\varepsilon(v_0), \varepsilon(v_D))_{L^2(\Omega)}.$$

Therefore,

$$\frac{1}{2}\|\varepsilon(v_0)\|^2 \leq \|\varepsilon(v_0)\|^2 + 2\|\varepsilon(v_D)\|^2 + 2(\varepsilon(v_0), \varepsilon(v_D))_{L^2(\Omega)} = \|\varepsilon(v)\|^2 + \|\varepsilon(v_D)\|^2.$$

Hence, the elementary estimate $\|\varepsilon(v_D)\| \leq \|D(v_D)\|$ leads to

$$\|Dv\|^2 \leq 4\|\varepsilon(v)\|^2 + 5\|Dv_D\|^2.$$

Since the harmonic extension v_D minimizes the H^1 seminorm subject to the boundary conditions $u_D|_{\partial\Omega}$, any other extension $u_D \in \mathscr{A}$ provides the upper bound $\|Dv_D\| \leq \|Du_D\|$. This proves (1.17). □

Lemma 1.4.5. *With the Friedrichs constant C_F, the constants C_1, C_2 from Lemma 1.4.3, and γ from (1.8), any $v \in \mathscr{A}$ satisfies*

$$\|Dv\|^2 \leq 8C_1 E(v) + 8\gamma C_1 \int_\Omega \det Dv\, dx + 8|\Omega| C_2 + 16 C_1^2 C_F^2 \|f\|^2 + 5\|u_D\|^2.$$

Proof. The Korn-type estimate (1.17), Lemma 1.4.3 and the definition of E imply

$$\begin{aligned}\|\varepsilon(v)\|^2 &\leq C_1 \int_\Omega W^{\mathrm{qc}}(\varepsilon(v))\, dx + \int_\Omega C_2\, dx \\ &= C_1 E(v) + C_1 \int_\Omega f \cdot v\, dx + C_1 \gamma \int_\Omega \det Du\, dx + \int_\Omega C_2\, dx.\end{aligned}$$

The Friedrichs inequality with constant C_F and the Young inequality prove

$$\begin{aligned}C_1 \int_\Omega f \cdot v\, dx &\leq C_1 \|f\|\, \|v\| \\ &\leq C_1 C_F \|f\|\, \|Dv\| \leq 2C_1^2 C_F^2 \|f\|^2 + \frac{1}{8}\|Dv\|^2.\end{aligned}$$

The combination of the foregoing displayed formulas proves the result. □

Lemma 1.4.6. *It holds*

$$\begin{aligned}&\lambda_1 \|D\Phi(Du) - D\Phi(D_{\mathrm{NC}} I_{\mathrm{CR}} u)\|^2 + E_{\mathrm{NC}}(u_h) \\ &\qquad \leq E(u) + \varkappa \|h_{\mathscr{T}} f\|\, \|(1 - \Pi_0) Du\| + \alpha \sum_{F \in \mathscr{F}} ⨍_F |[I_{\mathrm{CR}} u]_F|^2\, dx.\end{aligned}$$

Proof. The discrete problem shows that

$$\begin{aligned} E_{\rm NC}(u_h) &= \min_{v_h\in\mathscr{A}_{\rm CR}} E_{\rm NC}(D_{\rm NC}v_h) \\ &\le E_{\rm NC}(I_{\rm CR}u) \\ &= \sum_{T\in\mathscr{T}}\int_T \Phi(D_{\rm NC}I_{\rm CR}u)\,dx + \alpha\sum_{F\in\mathscr{F}} \fint_F |[I_{\rm CR}u]_F|^2\,dx - \int_\Omega f\cdot I_{\rm CR}u\,dx. \end{aligned} \tag{1.18}$$

Lemma 1.3.3 for $A := Du$ and $B := I_{\rm CR}u$ and an integration over $T\in\mathscr{T}$ lead to

$$\begin{aligned} \lambda_1\|D\Phi(Du)-D\Phi(DI_{\rm CR}u)\|^2_{L^2(T)} &+ \int_T \Phi(DI_{\rm CR}u)\,dx \\ &\le \int_T \Phi(Du)\,dx + \int_T D\Phi(DI_{\rm CR}u) : D(u-I_{\rm CR}u)\,dx. \end{aligned}$$

Since $D\Phi(DI_{\rm CR}u)$ is constant on T, the projection property (1.13) shows that the last term on the right-hand side vanishes. This, (1.18), and Proposition 1.4.1 lead to

$$\begin{aligned} \lambda_1\|D\Phi(Du)&-D\Phi(D_{\rm NC}I_{\rm CR}u)\|^2 + E_{\rm NC}(u_h) \\ &\le \int_\Omega \Phi(D_{\rm NC}I_{\rm CR}u)\,dx + \alpha\sum_{F\in\mathscr{F}}\fint_F |[I_{\rm CR}u]_F|^2\,dx - \int_\Omega f\cdot I_{\rm CR}u\,dx \\ &= E(u) + \alpha\sum_{F\in\mathscr{F}}\fint_F |[I_{\rm CR}u]_F|^2\,dx + \int_\Omega f\cdot(u-I_{\rm CR}u)\,dx \\ &\le E(u) + \varkappa\|h_{\mathscr{T}}f\|\,\|(1-\Pi_0)Du\| + \alpha\sum_{F\in\mathscr{F}}\fint_F |[I_{\rm CR}u]_F|^2\,dx. \end{aligned}$$

□

Let $C_{\rm sr} := \max_{T\in\mathscr{T}} h_T^2/|T|$ be the shape-regularity constant.

Theorem 1.4.7 (guaranteed lower energy bound). *Any* $v\in\mathscr{A}$ *and*

$$C(f,v) := \left(8C_1E(v) + 8\gamma C_1\int_\Omega \det Dv\,dx + 8|\Omega|C_2 + 16C_1^2C_F^2\|f\|^2 + 5\|u_D\|^2\right)^{1/2}$$

satisfy

$$E_{\rm NC}(u_h) \le E(u) + C(f,v)(\varkappa\,\|h_{\mathscr{T}}f\| + 3\alpha C_{\rm sr}/\pi^2).$$

Proof. For any $v\in\mathscr{A}$, the term $\int_\Omega \det Dv\,dx$ does not depend on the particular choice of v but only depends on the boundary data u_D. Hence, the combination of Lemma 1.4.5–1.4.6 leads to

$$E_{\rm NC}(u_h) \le E(u) + \varkappa\|h_{\mathscr{T}}f\|\,C(f,v) + \alpha\sum_{F\in\mathscr{F}}\fint_F |[I_{\rm CR}u]_F|^2\,ds.$$

For any edge $F \in \mathscr{F}$, the Poincaré inequality along F and the the trace inequality [DE12, eqn (1.42)] reveal for the edge patch ω_F that

$$\fint_F |[I_{\mathrm{CR}} u]_F|^2 \, dx \le \pi^{-2} h_F \|\partial [I_{\mathrm{CR}} u]_F / \partial s\|^2_{L^2(F)} \le \pi^{-2} C_{\mathrm{sr}} \|D_{\mathrm{NC}} I_{\mathrm{CR}} u\|^2_{L^2(\omega_F)}.$$

The sum over all edges in $\mathscr{F}$ and Lemma 1.4.5 conclude the proof. □

Remark 1.4.8. The efficiency of the lower bound is topic of ongoing research. In particular, the requirement of a sufficiently large stabilization parameter α leads to an additive shift in the lower bound. The numerical tests below will investigate the dependence for a model situation.

1.4.3 Guaranteed Error Control for the Pseudo-stress

This section presents an application to guaranteed a posteriori error estimates for the pseudo-stress. Let $\tau := D\Phi(Du)$ and $\tau_h := D\Phi(D_{\mathrm{NC}} u_h)$ denote the exact and discrete pseudo-stress.

The computable a posteriori error estimator for $\|\tau - \tau_h\|$ utilizes the conforming companion operator $J_3 : \mathrm{CR}^1(\mathscr{T}) \to P_3(\mathscr{T}) \cap H^1(\Omega)$ from [CS15, Lemma 3.3] which satisfies, for any $v_h \in \mathrm{CR}^1(T)$ and any $T \in \mathscr{T}$, that

$$\int_T (v_h - J_3 v_h) \, dx = 0 \quad \text{and} \quad \int_T D_{\mathrm{NC}} (v_h - J_3 v_h) \, dx = 0.$$

Proposition 1.4.9 (guaranteed a posteriori error estimate). *The exact minimizer $u \in \mathscr{A}$ of E and its pseudo-stress $\tau := D\Phi(Du)$ and the discrete minimizer u_h of E_{NC} with discrete pseudo-stress $\tau_h := D\Phi(D_{\mathrm{NC}} u_h)$ satisfy*

$$\lambda_1 \|\tau - \tau_h\|^2 \le 2\left(E_{\mathrm{NC}}(u_h) - E(u) + \int_\Omega f \cdot (u_h - J_3 u_h) \, dx \right) + \frac{1}{\lambda_1} \|D_{\mathrm{NC}}(u_h - J_3 u_h)\|^2.$$

Proof. The convexity control from Lemma 1.3.3 for $A := D_{\mathrm{NC}} u_h$ and $B := Du$ and an integration over Ω lead to

$$\begin{aligned} \lambda_1 \|\tau - \tau_h\|^2 &\le \int_\Omega \Phi(D_{\mathrm{NC}} u_h) \, dx - \int_\Omega \Phi(Du) \, dx - \int_\Omega \tau : D_{\mathrm{NC}}(u_h - u) \, dx \\ &= E_{\mathrm{NC}}(u_h) - E(u) + \int_\Omega f \cdot (u_h - u) \, dx - \int_\Omega \tau : D_{\mathrm{NC}}(u_h - u) \, dx. \end{aligned}$$

The projection property of the companion operator J_3 and the discrete Euler-Lagrange equation reveal

$$\int_\Omega \tau : D_{\mathrm{NC}}(u_h - u) \, dx = \int_\Omega (1 - \Pi_0) \tau : D_{\mathrm{NC}}(u_h - J_3 u_h) \, dx + \int_\Omega f \cdot (J_3 u_h - u) \, dx.$$

The combination of the foregoing two displayed formulae with the Cauchy inequality results in

$$\lambda_1 \|\tau - \tau_h\|^2 \le E_{\mathrm{NC}}(u_h) - E(u) + \int_\Omega f \cdot (u_h - J_3 u_h)\, dx + \|(1-\Pi_0)\tau\| \, \|D_{\mathrm{NC}}(u_h - J_3 u_h)\|.$$

The Young inequality reads

$$\|(1-\Pi_0)\tau\| \, \|D_{\mathrm{NC}}(u_h - J_3 u_h)\| \le \frac{\lambda_1}{2} \|(1-\Pi_0)\tau\|^2 + \frac{1}{2\lambda_1} \|D_{\mathrm{NC}}(u_h - J_3 u_h)\|^2.$$

The first term on the right-hand side can be absorbed to obtain the claimed estimate. □

Remark 1.4.10. The term $E_{\mathrm{NC}}(u_h) - E(u)$ in the estimate of Proposition 1.4.9 can be estimated with the guaranteed lower energy bound of Theorem 1.4.7.

1.4.4 Numerical Experiments

This subsection numerically investigates the performance of the guaranteed lower energy bounds on the square domain and the Γ-shaped domain.

1.4.4.1 Numerical Realization

The material parameters under consideration read (cf. [CP00])

$$\kappa = 137, \quad \mu = 70, \quad A_1 = \begin{bmatrix} 0.3 & 0 \\ 0 & -0.1 \end{bmatrix}, \quad A_2 = \begin{bmatrix} -0.1 & 0 \\ 0 & -0.5 \end{bmatrix}, \quad w_1 = w_2 = 0.$$

The reference energy (referred to as "exact energy") stems from a computation with P_1-conforming finite elements on uniformly refined meshes and Aitken extrapolation. The numerical experiments have been carried with the Matlab routine `fmincon`.

In this simple model situation the problem appears to be well-posed for any choice of $\alpha > 0$, with constants that deteriorate as $\alpha \to 0$.

1.4.4.2 Square Domain

Let $\Omega = (0,1)^2$ be the square domain with $f \equiv 0$ and Dirichlet data

$$u_D|_{\partial\Omega}(x_1, x_2) := \begin{cases} (0, \quad \frac{1}{10}(1 - 2|x_1 - 0.5|)) & \text{if } x_2 = 1, \\ (0, \quad 0) & \text{else.} \end{cases}$$

The guaranteed lower energy bounds are displayed in Table 1.1. The exact energy reads 73.8253.

Table 1.1 Guaranteed lower energy bounds for the square domain. The exact energy reads 73.8253

$\sqrt{2}\times h=$	1	2^{-1}	2^{-2}	2^{-3}	2^{-4}	2^{-5}	2^{-6}
$\alpha=1$	5.2755	8.02748	13.3979	24.7867	36.6346	43.458	46.2624
$\alpha=0.1$	31.5893	31.8635	32.5132	35.0702	42.4706	54.2174	63.6917
$\alpha=0.05$	33.0521	33.1891	33.518	34.8862	39.5763	49.7449	60.9973

1.4.4.3 Γ-Shaped Domain

Let $\Omega = (-1,1)^2 \setminus ([0,1]\times[-1,0])$ be the Γ-shaped domain with $f\equiv 0$ and Dirichlet data

$$u_D|_{\partial\Omega}(x_1,x_2) := \begin{cases} (0, & \frac{1}{10}(1-|x_1|)) \quad \text{if } x_2 = 1,\\ (0, & 0) \qquad\qquad\quad \text{else.}\end{cases}$$

The guaranteed lower energy bounds are displayed in Table 1.2. The exact energy reads 103.9553.

Table 1.2 Guaranteed lower energy bounds for the Γ-shaped domain. The exact energy reads 103.9553.

$\sqrt{2}\times h=$	1	2^{-1}	2^{-2}	2^{-3}	2^{-4}	2^{-5}	2^{-6}
$\alpha=1$	21.1156	26.2638	37.2546	49.6945	58.2414	62.2028	63.6278
$\alpha=0.1$	56.2229	56.8688	59.367	66.5721	78.4894	89.545	96.0492
$\alpha=0.05$	58.1715	58.499	59.8371	64.3799	74.4118	86.5119	95.5195

1.4.4.4 Discussion of the Computational Results

In the numerical experiments, the problem seems to be stable for any α and the guaranteed lower bound becomes sharper for smaller values of α. On the other hand, the discrete Korn constant deteriorates for small α which implies that also the mesh-size h is required to be very small in order to achieve the lower energy bound. This effect can be observed in the numerical experiments, where the choice of $\alpha = 0.1$ outperforms the lower bound for $\alpha = 0.05$ on the triangulations under consideration.

1.5 Discontinuous Galerkin Method for Degenerate Convex Minimization Problems

This section discusses a discontinuous Galerkin FEM discretization for a degenerate convex model problem from topology optimization.

1.5.1 Optimal Design Benchmark

The *optimal design problem* analyzed here is to seek the distribution of two materials of prescribed amount for maximal torsion stiffness of an infinite bar of given cross section Ω. The volume fraction of the materials is given by the parameters $0 \leq \mu_1 \leq 1$ and $\mu_2 := 1 - \mu_1$.

The mathematical analysis of this problem [BC08] leads to the minimization problem

$$\min_{v \in H_0^1(\Omega)} E(v) \quad \text{with} \quad E(v) := \int_\Omega W(Dv)\,dx - F(v). \tag{1.19}$$

Since the problem is not convex, it is not clear from the beginning if there exists any solution u for the problem at all. The non-convex nature of the original problem leads to the occurrence of microstructures. In the prototypical numerical simulation of Figure 1.2 one observes the arrangement of the two materials in an interior region, a boundary layer and small transition layer between the two regions. These transition layers are the microstructures, in which we can only determine the amount of each material but not their position. The solution u describes the macroscopic, space-averaged state. The relaxation of problem (1.19) employs the convex envelope E^{**} and leads to the *degenerate convex minimization problem*

$$\sup_{\lambda \in \mathbb{R}} \min_{v \in H_0^1(\Omega)} E^{**}(\lambda, v), \tag{1.20}$$

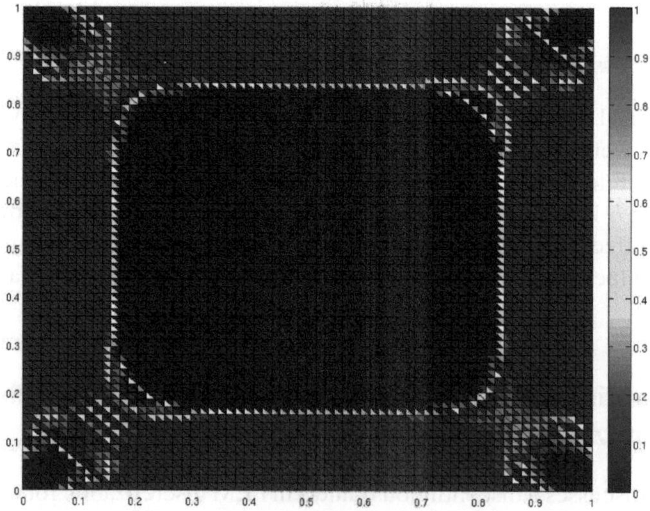

Fig. 1.2 Example of microstructures. The colour indicates the volume fraction between the two materials.

where λ is the Lagrangian parameter for the volume constraint of the two materials. Kohn and Strang [KS86a, KS86b, KS86c] proved that the original problem (1.19) and the relaxed form (1.20) are equivalent. The functional $E^{**}(v)$, defined for a fixed λ, has the form

$$E^{**}(v) := \int_\Omega W^{**}(Dv)dx - \int_\Omega fv\,dx, \tag{1.21}$$

with the convex envelope W^{**} of W. The energy density $W(F) := g_\lambda(|F|)$ is determined by the (scalar) stored energy function $g_\lambda(t)$ depicted in Figure 1.3 and defined by

$$g_\lambda(t) := \begin{cases} \mu_2(t^2/2-\lambda) & \text{for } t \le t_1 \\ \sqrt{t_1 t_2 \mu_1 \mu_2}t - \lambda(\mu_1+\mu_2) & \text{for } t_1 \le t \le t_2 \\ \mu_1(t^2/2-\lambda) & \text{for } t_2 \le t \end{cases}.$$

The minimization problem with (1.21) can be analyzed with the direct method of the calculus of variations [Dac08] leading to possible non-unique minimizers u of the problem (1.19). It can be shown [CP97, Fri94] that the stress-field $\sigma := DW(Du)$, which is also an interesting macroscopic quantity for the given problem itself, is unique in the sense that two solutions u_1 and u_2 have equal σ. Furthermore, local regularity of $\sigma \in H^1_{loc}(\Omega)$ holds on polygonal domains while there is global H^1 regularity of σ in case of a smooth boundary [CM02].

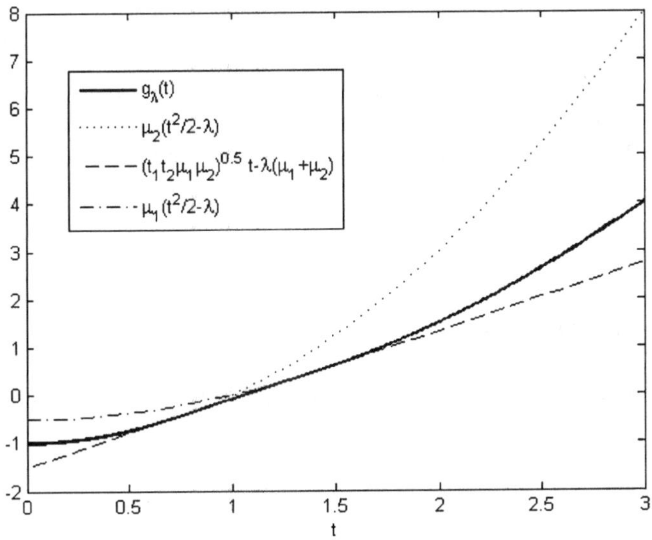

Fig. 1.3 The stored energy function $g_\lambda(t)$ for $\lambda = 0.2$, $\mu_1 = 1$ and $\mu_2 = 2$.

1.5.2 Discontinuous Galerkin Methods

The recent work [CL15] focuses on duality techniques based on mixed and nonconforming FEMs for the optimal design problem. In the DG-FEM the continuity condition over the edges is relaxed even more when compared with nonconforming FEMs, namely that there is no continuity condition at all. This leads to a situation where the values in the nodes of the triangulation can be *multi-valued*. This is schematically shown in Figure 1.4, where the black corners in the triangle symbolize that at each node multiple values are possible depending on the actual triangles $T \in \mathscr{T}(\Omega)$ to which the node belongs. Let $\mathscr{T}$ be a shape-regular triangulation of the bounded Lipschitz domain with polygonal boundary Ω. The DG-FEM space V_0 is then defined by

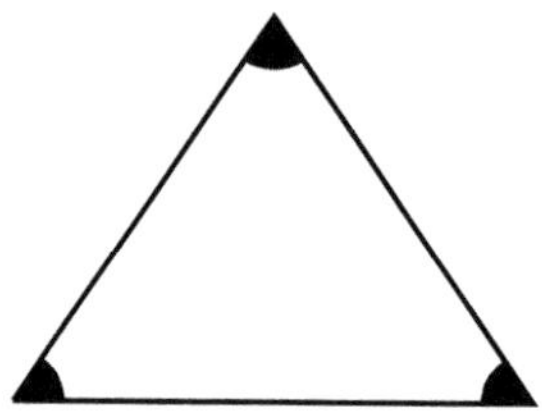

Fig. 1.4 Discontinuous Galerkin FEM

$$V(\mathscr{T}) := P_1(\mathscr{T}) \quad \text{and} \quad V_0(\mathscr{T}) := \{v \in V(\mathscr{T}) \ : \ v|_{\partial\Omega} = 0\}.$$

The functions in the trial space are not continuous and thus there exists jump terms in the *distributional gradient*, meaning that any continuous test functions $\phi \in [\mathscr{D}(\Omega)]^2$ and $v \in H^1(\mathscr{T})$ (the space of piecewise H^1 functions) satisfy

$$\int_\Omega v \operatorname{div}\phi \, dx = -\int_\Omega \phi \cdot \nabla_{NC} v \, dx + \sum_{F \in \mathscr{F}(\Omega)} \int_F \phi \cdot [v]_F ds.$$

1.5.3 Lifting Operator R

The lifting operator R maps $H^1(\mathscr{T})$ into $P_0(\mathscr{T};\mathbb{R}^2)$. It is defined for all $v \in H^1(\mathscr{T})$ and $\tau \in P_0(\mathscr{T};\mathbb{R}^2)$ by

$$\int_\Omega R(v) \cdot \tau dx = -\sum_{F \in \mathscr{F}(\Omega)} \int_E [v]_F \nu_F ds \cdot \langle \tau \rangle_F .$$

The discontinuous Galerkin (dG) method proposed here with the penalty parameter $\eta > 0$ employs the discrete energy functional

$$E_{\mathrm{dG}}(v) := \int_\Omega W(D_{NC}v + R(v))dx - F(v) + \eta \sum_{F\in\mathscr{F}(\Omega)} \left| \fint_F [v]_F ds \right|^2 . \qquad (1.22)$$

The term $\eta \sum_{F\in\mathscr{F}(\Omega)} | \fint_F [v]_F ds|^2$ penalizes big jumps over interior edges. The lifting operator R is needed to guarantee convergence of the numerical scheme, because of inconsistencies in the distributional gradient of functions that are not globally continuous. The discontinuous Galerkin FEM seeks $u_{\mathrm{dG}} \in V_0(\mathscr{T}_\ell)$ such that

$$u_h \in \arg\min\{E_{\mathrm{dG}}(v_h) : v_h \in V_0(\mathscr{T}_\ell)\}. \qquad (1.23)$$

The existence of discrete minimizers follows from growth conditions. In particular, any discrete minimizer u_h is characterized by the discrete Euler-Lagrange equation

$$\int_\Omega DW(D_{NC}u_h + R(u_h)) \cdot \nabla v_h dx + \eta \sum_{F\in\mathscr{F}(\Omega)} \fint_F [u_h]_F [v_h]_F ds = E(v_h). \qquad (1.24)$$

1.5.4 Connection with the Nonconforming Method

The following arguments employ the averaging operator $I^*_{\mathrm{CR}} : H^1(\mathscr{T}) \to \mathrm{CR}^1_0(\mathscr{T})$ defined, for any $v \in H^1(\mathscr{T})$, by

$$\fint_F I^*_{\mathrm{CR}} v\, ds = \fint_F \langle v \rangle_F \, ds \quad \text{for all } F \in \mathscr{F}(\Omega).$$

This operator gives rise to the following explicit representation of the lifting operator R.

Lemma 1.5.1. *Any $v_h \in H^1(\mathscr{T})$ satisfies*

$$D_{\mathrm{NC}} v_h + R(v_h) = I^*_{\mathrm{CR}} v_h.$$

Proof. The proof is an elementary consequence of the piecewise integration by parts. □

The following theorem establishes a connection between the dG method and the nonconforming P_1 method.

Theorem 1.5.2. *Let u_h be a minimizer of (1.23). Then $u_{\mathrm{CR}} := I^*_{\mathrm{CR}} u_h$ minimized E_{dG} over $\mathrm{CR}^1_0(\mathscr{T})$,*

$$u_{\mathrm{CR}} \in \arg\min\{E_{\mathrm{dG}}(v_{\mathrm{CR}}) \; : \; v_{\mathrm{CR}} \in \mathrm{CR}^1_0(\mathscr{T})\}.$$

Proof. Lemma 1.5.1 and the Euler-Lagrange equation (1.24) imply

$$\int_\Omega DW(D_{\mathrm{NC}} I^*_{\mathrm{CR}} u_h) \cdot \nabla v_{\mathrm{CR}} dx = F(v_{\mathrm{CR}}) \quad \text{for all } v_h \in V_0(\mathscr{T}). \tag{1.25}$$

This is the Euler-Lagrange equation for the Crouzeix-Raviart function $u_{\mathrm{CR}} := I^*_{\mathrm{CR}} u_h$. This caracterizes all the discrete minimizers and, hence, u_{CR} is a minimizer of E_{dG}. □

1.5.5 Adaptive Finite Element Method

The AFEM is based on the following refinement indicator. Given a discrete minimizer $u_\ell = u_h(\mathscr{T}_\ell) \in V_0(\mathscr{T}_\ell)$ define, for any $T \in \mathscr{T}_\ell$,

$$\eta_\ell^2(u_\ell, T) := \|h_T f\|^2_{L^2(T)} + \sum_{F \in \mathscr{F}(T)} h_F^{-1} \|[u_\ell]_F\|^2_{L^2(F)}.$$

This motivates the following adaptive mesh-refining.
Input: *Marking parameter* $0 < \theta \leq 1$, *initial mesh* $\mathscr{T}_0$
Set: For $\ell = 0, 1, 2, 3, \ldots$ **do**

1. *Compute a discrete minimizer* $u_\ell \in \operatorname{arg\,min} E_{\mathrm{dG}}(V_0(\mathscr{T}_\ell))$
2. *Compute refinement indicators* $(\eta_\ell^2(u_\ell, T) : T \in \mathscr{T}_\ell)$
3. *Generate a set of marked elements* $\mathscr{M}_\ell \subseteq \mathscr{T}_\ell$ *such that*

$$\sum_{T \in \mathscr{M}_\ell} \eta_\ell^2(u_\ell, T) \leq \theta \sum_{T \in \mathscr{T}_\ell} \eta_\ell^2(u_\ell, T)$$

4. *Generate a* $\mathscr{T}_{\ell+1}$ *with newest-vertex bisection based on the marked set* $\mathscr{M}$ *as in § 1.3.2.4.*

end do

1.5.6 Computational Experiments

A convergent *adaptive finite element method* (AFEM) in its primal form was introduced in [BC08] and a convergent adaptive mixed finite element method in [CGR12].

AFEMs refine the mesh locally to save degrees of freedom and thus reduce the computational cost when compared to a uniform overall mesh-refinement.

The numerical experiments consider two different domains, namely the Γ-shaped domain and the slit domain displayed in Figure 1.5.

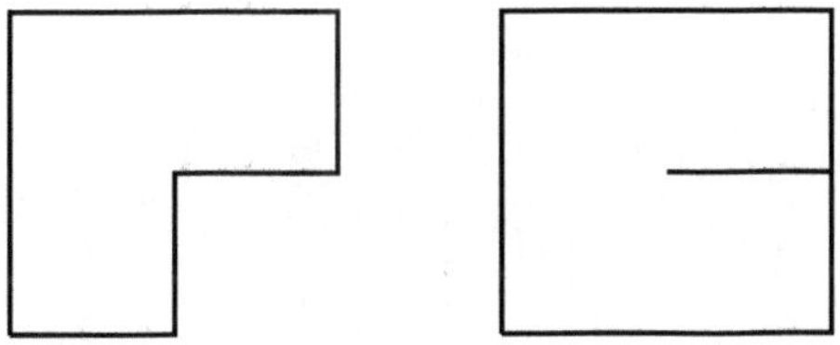

Fig. 1.5 Γ-shaped domain and slit domain

1.5.7 Γ-shaped Domain

The Aitken interpolation for the minmum energy of the problem leads to the reference energy $E = -0,244309$.

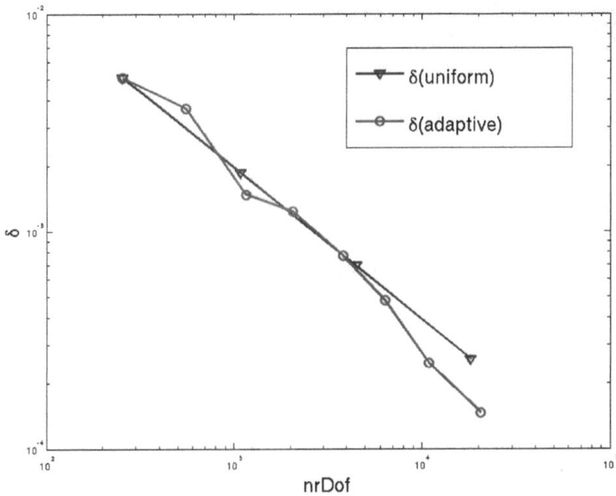

Fig. 1.6 Decay of the energy error δ_ℓ versus number degrees of freedom on the Γ-shaped domain

The domain Ω_2 has a reentrant corner at the origin which presumably limits the regularity of the exact solution. Therefore, a higher resolution of this area is necessary in order to obtain optimal approximations. The above error estimator in Figure 4.6 for the stress error behaves also like $N^{-1/2} = h$. Thus, linear convergence rates for the error bound is obtained. A local refinement towards the corners and the reentrant corner in the origin is observed, leading to an improvement convergence rate for the energy error δ_ℓ. In total, the numerical results show clearly that the AFEM-algorithm converges for the domain Ω. Furthermore, the energy error converges with a better rate for the adaptive mesh.

The microstructures show the expected structure, namely an interior region, a boundary layer and small transition layer between the two regions.

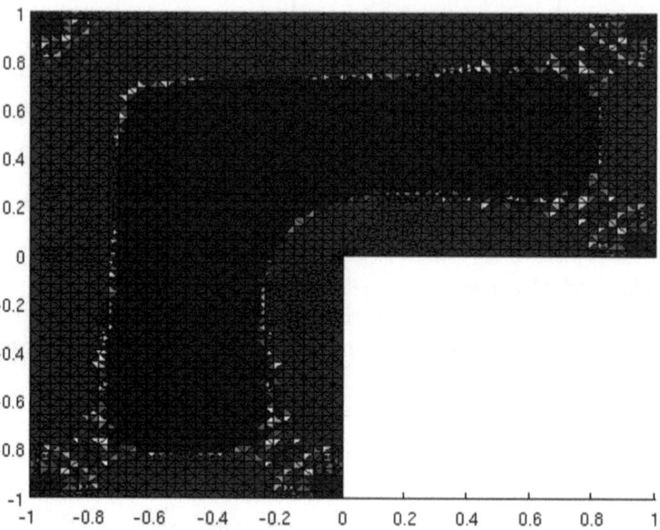

Fig. 1.7 Microstructures for $\mathscr{T}_7$ of Ω_2

1.5.8 Slit Domain

The Aitken interpolation for the minimum energy of the problem yields $E_A = -0,161548$.

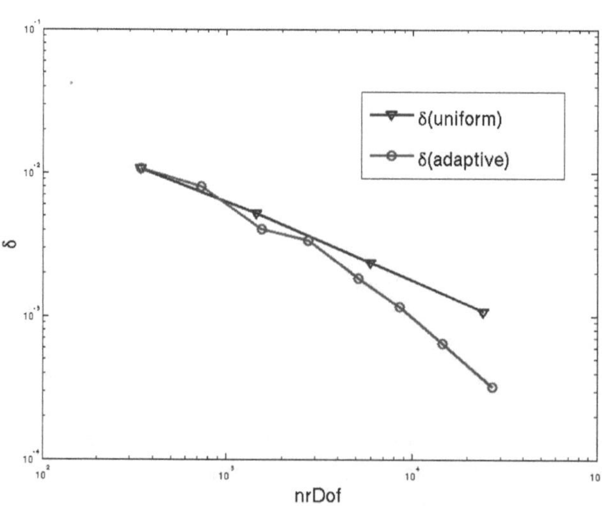

Fig. 1.8 Energy error δ_ℓ versus number degrees of freedom for the slit domain

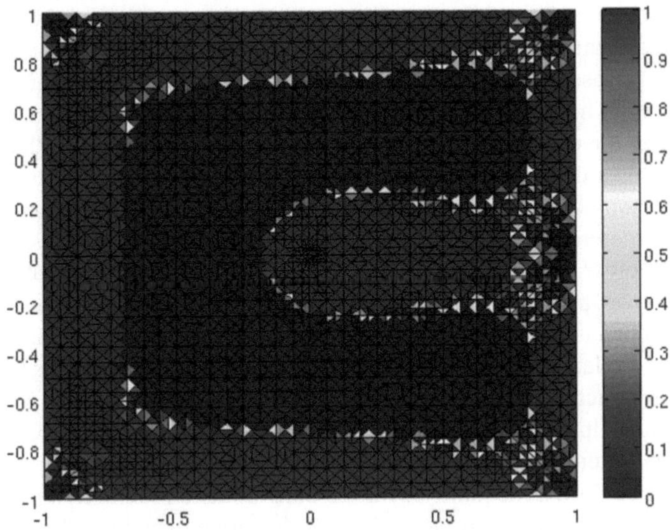

Fig. 1.9 Microstructures for $\mathscr{T}_7$ of Ω_3.

A strong local refinement towards the reentrant corner is observed. This leads to an improved convergence rate for the energy error because of the same reason mentioned for domain Ω_2. In total, the numerical results show clearly that the AFEM-algorithm converges for the domain Ω.

The microstructures show the expected structure, namely an interior region, a boundary layer and small transition layer between the two regions.

1.6 Conclusions and Outlook

The application of discontinuous Galerkin methods was motivated by stabilization techniques in [BC10, BC14] with penalty terms which imitate the surface energy in microstructures. Therein, at least the underlying data structures suggest to utilize the discontinuous Galerkin methods and their first applications in this work have been successful in the sense that those can be used for relaxed problems – at least when the relaxation hull is convex. However, the proposed scheme with its relation to nonconforming finite element methods lead to the alternative use of nonconforming or mixed finite element schemes and their discrete modification as in [CL15, CGR12]. This may finally lead to close the reliability-efficiency gap in the a posteriori error analysis, typically encountered when relaxation hulls are not strictly convex [CJ03]. There remains an open question for future research whether a refined efficiency analysis can provide sufficient conditions for this gap's closure.

The applications of post-modern calculus of variations and the power of the relaxation theory has been demonstrated throughout the research group and, for instance, led to relaxations of single-slip systems which can be employed in a rigorous computer simulation with a mathematical foundation that has to be developed. One amongst the many challenges is that the relaxed energy, thought even polyconvex in many cases, involves a non high-dimensional convex minimization on the discrete level where globally-convergent fast solvers are unknown and urgently needed. Despite the progress in [CD14], analyzed within this project, the first contribution ever towards the mathematical foundation of the numerical analysis with a relaxed and non-convex minimization problem, the treatment of pointwise constraints on the determinant is not justified. There are many heuristics, which are reasonable for other linearized problems and justifiable for very smooth solutions and strictly convex problems, that may be adopted. Their mathematical foundation remains as an open issue for future research that may enable the rigorous numerical treatment of the relaxation hulls in single- and multiple-slip systems arising in the enforced microstructures in plasticity. The nonstandard discretisation thereof may or may not enable upper and lower bound of the effective energy as in [CL15] in the future. The first attempts in this paper illustrate that this is feasible in principle but the interaction of Korn's inequality with the constraints on the determinant are not fully understood and suggest modifications of the strategy and the development of new discretisations and algorithms for future research.

Within the scope of finite plasticity, there remain various open questions when it comes to the time-evolution [HHM12, HM12, GKH15, Mie15]. For relaxation for time dependent plasticity, see [KK11]. One difficulty is the lack of strong convergence of the strain approximations and related internal variables even within one time-step. Even from the modeling point of view it is unclear whether a relaxation plus a stabilization via a surface energy or via higher-order model may be helpful. The generalization of [BC10, BC14] to non convex but polyconvex problems is one possibility to overcome those difficulties in the future.

The overall mathematical foundation of the various effective mathematical models developed within the research group for the microstructures in finite plasticity in total remains an open question. This project has provided partial answers towards this ultimative but open goal of the justification of a computer simulation in that important class of applications in computational calculus of variations.

Acknowledgements. This work was supported by the DFG research group 797 "Analysis and Computation of Microstructure in Finite Plasticity" through project CA 151/19. It is our pleasure to acknowledge fruitful discussions with Prof. G. Dolzmann within this DFG research project. The second author furthermore acknowledges the support by the DFG research center MATHEON and the Berlin Mathematical School. The third author acknowledges the support of the WCU program through KOSEF (R31-2008-000-10049-0) and the kind hospitality of the department for Computational Science and Engeneering of Yonsei University.

Appendix A. Computation of λ_1

This section is devoted to the explicit calculation of the constant λ_1 from Lemma 1.3.3. The quadratic form $T : \mathbb{M} \to \mathbb{R}$ reads

$$T(F) := \frac{1}{2}|\mathrm{sym}F|^2_{\mathbb{C}} - \frac{1}{2g}\big(\mathrm{sym}F : \mathbb{C}(A_1 - A_2)\big)^2 - \gamma \det F \quad \text{for any } F \in \mathbb{M}.$$

Throughout this appendix, identify $\mathbb{M}$ with $\mathbb{R}^4$ via

$$\begin{pmatrix} a & b \\ c & d \end{pmatrix} \mapsto (a,b,c,d).$$

Straightforward calculations prove for any $E \in \mathbb{S}$ that

$$\mathbb{C}E = \mathbf{A}E \quad \text{for } \mathbf{A} := \begin{bmatrix} \kappa+\mu & 0 & 0 & \kappa-\mu \\ 0 & 2\mu & 0 & 0 \\ 0 & 0 & 2\mu & 0 \\ \kappa-\mu & 0 & 0 & \kappa+\mu \end{bmatrix}.$$

Any $F \in \mathbb{M}$ satisfies

$$\det E = E^\top \mathbf{B} E \quad \text{for } \mathbf{B} := \begin{bmatrix} 0 & 0 & 0 & 1/2 \\ 0 & 0 & -1/2 & 0 \\ 0 & -1/2 & 0 & 0 \\ 1/2 & 0 & 0 & 0 \end{bmatrix}.$$

Furthermore,

$$\big(\mathrm{sym}F : \mathbb{C}(A_1 - A_2)\big)^2 = E^\top \mathbf{C} E$$

$$\text{for } \mathbf{C} := \begin{bmatrix} \eta_1^2(\kappa+\mu)+\eta_2^2(\kappa-\mu)+4\kappa\eta_1\eta_2 & 0 & 0 & (\eta_1+\eta_2)^2(\kappa^2+\mu^2) \\ 0 & 0 & 0 & 0 \\ 0 & 0 & 0 & 0 \\ (\eta_1+\eta_2)^2(\kappa^2+\mu^2) & 0 & 0 & \eta_1^2(\kappa-\mu)+\eta_2^2(\kappa+\mu)+4\kappa\eta_1\eta_2 \end{bmatrix}.$$

This implies

$$T(F) = E^\top \mathbf{D} E \quad \text{for } \mathbf{D} := \frac{1}{2}\mathbf{A}^{1/2} - \gamma \mathbf{B} - \frac{1}{2g}\mathbf{C}.$$

The derivative DT at $F \in \mathbb{M}$ acts on any $G \in \mathbb{M}$ as

$$DT(F;G) = 2F^\top \mathbf{D} G.$$

The matrix $\mathbf{D}$ is symmetric an, hence, has 4 real eigenvalues. The largest eigenvalue $\rho(\mathbf{D})$ of $\mathbf{D}$ satisfies

$$|DT(\cdot;A) - DT(\cdot;B)|^2 = |2\mathbf{D}(A-B)|^2 \le 4\rho(\mathbf{D})T(A-B).$$

Hence, the constant λ_3 of [CD14, display after (4.8)] equals $4\lambda_{\mathbf{D}}^{max}$. The constant λ_1 of [CD14] therefore reads $\lambda_1 = \max\{1/(4(\gamma_1^2+\gamma_2^2)), 4\lambda_{\mathbf{D}}^{max}\}$ as stated in Lemma 1.3.3. □

References

[BC08] Bartels, S., Carstensen, C.: A convergent adaptive finite element method for an optimal design problem. Numer. Math. 108(3), 359–385 (2008)

[BC10] Boiger, W., Carstensen, C.: On the strong convergence of gradients in stabilised degenerate convex minimisation problems. SIAM J. Numer. Anal. 47(6), 4569–4580 (2010)

[BC14] Boiger, W., Carstensen, C.: A posteriori error analysis of stabilised FEM for degenerate convex minimisation problems under weak regularity assumptions. Advanced Modeling and Simulation in Engineering Sciences 1(5) (2014)

[BJ87] Ball, J.M., James, R.D.: Fine phase mixtures as minimizers of energy. Arch. Rational Mech. Anal. 100(1), 13–52 (1987)

[BJ92] Ball, J.M., James, R.D.: Proposed experimental tests of a theory of fine microstructure and the two-well problem. Phil. Trans. R. Soc. Lond. A 338, 389–450 (1992)

[BKK00] Ball, J.M., Kirchheim, B., Kristensen, J.: Regularity of quasiconvex envelopes. Calc. Var. Partial Differential Equations 11(4), 333–359 (2000)

[Car01] Carstensen, C.: Numerical analysis of microstructure. In: Theory and Numerics of Differential Equations (Durham 2000). Universitext, pp. 59–126. Springer, Berlin (2001)

[Car08] Carstensen, C.: Convergence of an adaptive FEM for a class of degenerate convex minimization problems. IMA J. Numer. Anal. 28(3), 423–439 (2008)

[CD04] Carstensen, C., Dolzmann, G.: An a priori error estimate for finite element discretizations in nonlinear elasticity for polyconvex materials under small loads. Numer. Math. 97(1), 67–80 (2004)

[CD14] Carstensen, C., Dolzmann, G.: Convergence of adaptive finite element methods for a nonconvex double-well minimisation problem. Math. Comp. (2014)

[CDK11] Conti, S., Dolzmann, G., Kreisbeck, C.: Asymptotic behavior of crystal plasticity with one slip system in the limit of rigid elasticity. SIAM J. Math. Anal. 43, 2337–2353 (2011)

[CDK13] Conti, S., Dolzmann, G., Kreisbeck, C.: Relaxation of a model in finite plasticity with two slip systems. Math. Models Methods Appl. Sci. 23(11), 2111–2128 (2013)

[CDK15] Conti, S., Dolzmann, G., Kreisbeck, C.: Variational modeling of slip: From crystal plasticity to geological strata. In: Hackl, K., Conti, S. (eds.) Analysis and Computation of Microstructure in Finite Plasticity. LNACM, vol. 78, pp. 31–62. Springer, Heidelberg (2015)

[CG14] Carstensen, C., Gallistl, D.: Guaranteed lower eigenvalue bounds for the biharmonic equation. Numer. Math. 126(1), 33–51 (2014)

[CGR12] Carstensen, C., Günther, D., Rabus, H.: Mixed finite element method for a degenerate convex variational problem from topology optimization. SIAM J. Numer. Anal. 50(2), 522–543 (2012)

[Chi00] Chipot, M.: Elements of nonlinear analysis. Birkhäuser Advanced Texts: Basler Lehrbücher. [Birkhäuser Advanced Texts: Basel Textbooks]. Birkhäuser Verlag, Basel (2000)

[CHM02] Carstensen, C., Hackl, K., Mielke, A.: Non-convex potentials and microstructures in finite-strain plasticity. R. Soc. Lond. Proc. Ser. A Math. Phys. Eng. Sci. 458(2018), 299–317 (2002)

[CJ03] Carstensen, C., Jochimsen, K.: Adaptive finite element methods for microstructures? Numerical experiments for a 2-well benchmark. Computing 71(2), 175–204 (2003)

[CK88] Chipot, M., Kinderlehrer, D.: Equilibrium configurations of crystals. Arch. Rational Mech. Anal. 103(3), 237–277 (1988)

[CL15] Carstensen, C., Liu, D.J.: Nonconforming FEMs for an optimal design problem. SIAM J. Numer. Anal. (2015) (in press)

[CM02] Carstensen, C., Müller, S.: Local stress regularity in scalar nonconvex variational problems. SIAM J. Math. Anal. 34(2), 495–509 (2002)

[CP97] Carstensen, C., Plecháč, P.: Numerical solution of the scalar double-well problem allowing microstructure. Math. Comp. 66(219), 997–1026 (1997)

[CP00] Carstensen, C., Plecháč, P.: Numerical analysis of compatible phase transitions in elastic solids. SIAM J. Numer. Anal. 37(6), 2061–2081 (2000)

[CP01] Carstensen, C., Plecháč, P.: Numerical analysis of a relaxed variational model of hysteresis in two-phase solids. M2AN Math. Model. Numer. Anal. 35(5), 865–878 (2001)

[CS15] Carstensen, C., Schedensack, M.: Medius analysis and comparison results for first-order finite element methods in linear elasticity. IMA J. Numer. Anal. (published online, 2015), doi:10.1093/imanum/dru048

[Dac08] Dacorogna, B.: Direct methods in the calculus of variations, 2nd edn. Applied Mathematical Sciences, vol. 78. Springer, New York (2008)

[DE12] Di Pietro, D.A., Ern, A.: Mathematical aspects of discontinuous Galerkin methods. Mathématiques & Applications (Berlin), vol. 69. Springer, Heidelberg (2012)

[Dör96] Dörfler, W.: A convergent adaptive algorithm for Poisson's equation. SIAM J. Numer. Anal. 33(3), 1106–1124 (1996)

[Fri94] Friesecke, G.: A necessary and sufficient condition for nonattainment and formation of microstructure almost everywhere in scalar variational problems. Proc. Roy. Soc. Edinburgh Sect. A 124(3), 437–471 (1994)

[GKH15] Günther, C., Kochmann, D., Hackl, K.: Rate-independent versus viscous evolution of laminate microstructures in finite crystal plasticity. In: Hackl, K., Conti, S. (eds.) Analysis and Computation of Microstructure in Finite Plasticity. LNACM, vol. 78, pp. 63–88. Springer, Heidelberg (2015)

[HHM12] Hackl, K., Heinz, S., Mielke, A.: A model for the evolution of laminates in finite-strain elastoplasticity. ZAMM-Journal of Applied Mathematics and Mechanics/Zeitschrift für Angewandte Mathematik und Mechanik 92(11-12), 888–909 (2012)

[HL03] Hansbo, P., Larson, M.G.: Discontinuous Galerkin and the Crouzeix-Raviart element: application to elasticity. M2AN Math. Model. Numer. Anal. 37(1), 63–72 (2003)

[HM12] Hildebrand, F., Miehe, C.: Variational phase field modeling of laminate deformation microstructure in finite gradient crystal plasticity. Proc. Appl. Math. Mech. 12(1), 37–40 (2012)

[KK11] Kochmann, D., Hackl, K.: The evolution of laminates in finite crystal plasticity: a variational approach. Continuum Mechanics and Thermodynamics 23, 63–85 (2011)

[Koh91] Kohn, R.V.: The relaxation of a double-well energy. Contin. Mech. Thermodyn. 3(3), 193–236 (1991)

[KP91] Kinderlehrer, D., Pedregal, P.: Characterizations of Young measures generated by gradients. Arch. Rational Mech. Anal. 115(4), 329–365 (1991)

[KS86a] Kohn, R.V., Strang, G.: Optimal design and relaxation of variational problems. I. Comm. Pure Appl. Math. 39(1), 113–137 (1986)

[KS86b] Kohn, R.V., Strang, G.: Optimal design and relaxation of variational problems. II. Comm. Pure Appl. Math. 39(1), 139–182 (1986)

[KS86c] Kohn, R.V., Strang, G.: Optimal design and relaxation of variational problems. III. Comm. Pure Appl. Math. 39(3), 353–377 (1986)

[LC88] Lurie, K.A., Cherkaev, A.V.: On a certain variational problem of phase equilibrium. In: Material instabilities in Continuum Mechanics (Edinburgh, 1985–1986), pp. 257–268. Oxford Univ. Press, New York (1988)

[Lus96] Luskin, M.: On the computation of crystalline microstructure. Acta Numerica 5, 191–257 (1996)

[Mie15] Mielke, A.: Variational approaches and methods for dissipative material models with multiple scales. In: Hackl, K., Conti, S. (eds.) Analysis and Computation of Microstructure in Finite Plasticity. LNACM, vol. 78, pp. 125–156. Springer, Heidelberg (2015)

[MNS02] Morin, P., Nochetto, R.H., Siebert, K.G.: Convergence of adaptive finite element methods. SIAM Rev. 44(4), 631–658 (2002)

[MŠ99] Müller, S., Šverák, V.: Convex integration with constraints and applications to phase transitions and partial differential equations. J. Eur. Math. Soc. (JEMS) 1(4), 393–422 (1999)

[Pip91] Pipkin, A.C.: Elastic materials with two preferred states. Quart. J. Mech. Appl. Math. 44(1), 1–15 (1991)

[Ser96] Seregin, G.A.: The uniqueness of solutions of some variational problems of the theory of phase equilibrium in solid bodies. J. Math. Sci. 80(6), 2333–2348 (1996); Nonlinear boundary-value problems and some questions of function theory

[Ser98] Seregin, G.A.: A variational problem on the phase equilibrium of an elastic body. St. Petersbg. Math. J. 10, 477–506 (1998)

Chapter 2
Variational Modeling of Slip: From Crystal Plasticity to Geological Strata

Sergio Conti, Georg Dolzmann, and Carolin Kreisbeck

Abstract. Slip processes are soft modes of deformation, characteristic of a variety of layered materials. The layers can be at the atomic scale, as in the plastic deformation of crystalline lattices, or on a macroscopic scale, as in stacks of cards or sheets of paper and geological strata. The characteristic deformation processes involve sliding of the layers over one another, leading to a shear deformation with a specific orientation. If the forcing is not parallel to the layers, complex microstructures may form, which have a remarkable similarity over different systems and often consist of alternating shears of different sign. We review here recent results on the detailed analysis of slip processes in crystal plasticity based on the theory of relaxation, discuss the general variational framework for these microstructures, and compare with available experimental results in different systems. We then address the situation in which slip in several different directions may coexist in the same system, as frequently observed in plastically deformed crystals.

2.1 Introduction

Plastic deformation of crystals can be modeled as a combination of slip processes along specific directions, referred to as slip directions or slip systems, which depend on the crystallographic structure of the material under consideration. In an experimental set-up, typically only a few different slip systems are active at the same time and it is not uncommon to observe only one active slip system. This fact is related to the effect of latent (or cross) hardening, which penalizes deformations involving slip along more than one slip system at the same material point. Macroscopic plastic deformations are realized at a microscopic scale by fine mixtures of slip along

Sergio Conti
Institut für Angewandte Mathematik, Universität Bonn, 53115 Bonn, Germany

Georg Dolzmann · Carolin Kreisbeck
Fakultät für Mathematik, Universität Regensburg, 93040 Regensburg, Germany

S. Conti and K. Hackl (eds.), *Analysis and Computation of Microstructure in Finite Plasticity*,
Lecture Notes in Applied and Computational Mechanics 78, DOI: 10.1007/978-3-319-18242-1_2

different directions in different parts of the sample, see [OR99, CHM02, MSL02] for a modern energetic mathematical formulation of these mechanisms.

The orientation of these microstructures is to a large extent geometrically determined, the microstructures themselves can be understood as a variant of a buckling phenomenon, in which the neighboring layers impede each other. The geometric nature of the instability leading to the microstructure can be elucidated by establishing relations to and comparisons with other systems undergoing slip processes. Even though the physical processes behind pattern formation are quite different and involve widely different length and time scales, one observes striking geometric similarities.

In this report, we review recent progress in this area and draw a parallel between the subgrain microscale in materials science and macroscopic patterns in other systems briefly introduced in Section 2.2 below. In particular, we point out similarities between the formation of laminates in elasto-plastic materials due to external loading and the appearance of periodic kink band structures observed in rock strata under tectonic compression [PC90], see Sect. 2.2.1, and in compressed stacks of sheets of paper which are laterally confined [HPW00], see Sect. 2.2.2. We note that there are many other systems which exhibit similar structures, such as fiber reinforced elastomers, see for example [BF94, Fle97, PGPR09, SPC14].

Our approach to the mathematical description and analysis of these effects is energetic in nature. The investigation of variational integrals with nonconvex energy densities in the framework of the calculus of variations in general and nonlinear elasticity in particular offers an approach to the study of spontaneous appearance of microstructure in materials [Dac07, Mül99]. Elasto-plastic deformations involving a mixture of phases at different scales appear naturally as elements of minimizing sequences. More specifically, we restrict our attention to the fundamental aspect of the formation of single layers and to the response of layered materials to applied external forces. By a layered material we mean a material that consists of relatively stable layers, which are weakly coupled to one another. Typical examples include a deck of cards or a stack of sheets of paper. A crystal lattice can also be seen as a layered material with respect to several different orientations related to the crystallographic structure of the material. The assumption that only one orientation is relevant permits a substantial simplification of the analysis, focusing on a specific type of microstructure.

A general theoretical explanation for the development of fine structures from a variational viewpoint is that the system energy is not quasiconvex, and that structures at scales much smaller than the macroscopic length scale under consideration may reduce the energy of the system and are therefore energetically favorable [Mor52, Mor66]. Since these fine structures are known to have a crucial impact on the behavior of the system on the larger scale, it is important to determine the effective material response. This is the subject of the mathematical theory of relaxation [Dac07, Mül99], and we review some aspects of this theory in Section 2.4 as well. Recently interest has been also directed to an extension of this approach to time-dependent microstructures [KK11, HHM12], we refer to [Mie15, GKH15] for recent reviews. The concrete implementation of relaxation theory to problems in plasticity is still in its beginnings and will be a subject of important future research, we refer to [CGK15] for a review of the current state of the art.

Following the general discussion of the underlying concepts of relaxation theory, we review case studies in which the theoretical approach has been successfully applied. Our starting point are variational models in crystal plasticity with one active slip system, for which we refer to Section 2.5. The effective behavior of these models in the geometrically nonlinear setting was studied in [CT05, Con06, CDK13a, CDK13b, ACD09]. It turns out that the underlying relaxation mechanisms are based on simple laminates, which identifies them as the optimal structures also observable in experiments, see Sect. 2.2.3.

Based upon these results, we draw parallels between microstructures in plasticity and microstructures in the related systems described in Section 2.2. In fact, this energetic model transfers to the context of geological strata if the physical quantities are reinterpreted in a suitable way, see Sec.2.5.4 for details. In this sense, the simple laminated structures observed in elasto-plastic materials translate into chevron patterns for layered sedimentary rocks. Moreover, the results obtained by the foregoing analogy are in good agreement with experiments on the folding of a stack of paper under high compression parallel to the layers, see Section 2.2.2 for a description of the experimental set-up and results. The variational viewpoint outlined in Section 2.5.4 gives a new perspective on and complements the existing models in the field. A discrete model, which we describe in more detail in Section 2.3 and which serves us as a comparison, was introduced in [HPW00] and further developed and extended in [WHP04, WVHY12].

This report is concluded in Section 2.6 with an extension of the results to situations with multiple active slip systems, which correspond to different crystallographic orientations. This situation leads to a new type of microstructure, in which slip along different systems may interact. The richer set of possible microstructures renders the relaxation more demanding, indeed up to now a full relaxation has only been achieved in the case of two slip systems with hardening [CDK13b]. We present a partial relaxation result for the case of three slip systems at 120-degrees angles, which may be seen as a first step towards the analysis of the set of slip systems which appear in fcc or bcc crystals, details will be given elsewhere [CD]. A full relaxation result in a geometrically linear setting was obtained in [CO05].

2.2 Experimental Observation of Slip Microstructures in Nature

In this section we present three different material systems in which laminated structures can be observed. More specifically, we refer to chevron folds in rocks, kink band structure in stacks of sheets of papers under compression and layers at microscopic scales in single-slip crystal plasticity.

Fig. 2.1 Chevron folds, Cornwall, UK. From Wikipedia [Sma].

2.2.1 Chevron Folds in Rocks

Chevron folds are a characteristic structure which is often observed in stratified rocks, as for example turbidites, as illustrated in Fig. 2.1. These rocks have been formed by sedimentation and contain alternating layers with different mechanical properties. They can be visualized as a stack of layers, which are alternating strong and weak, as for example sandstone layers separated by shales. The observed zigzag patterns can be understood as originating from flat layers after longitudinal compression and a superimposed single, macroscopic rotation.

2.2.2 Kink Bands in Stacks of Paper under Compression

Fig. 2.2 shows an experiment on kink banding in layered structures [HPW00], which is often used as a simple reference for the effects happening in the geological context. The set-up is such that the specimen consisting of a pile of papers confined by flexible faces is compressed in the direction of the layers. The lateral confinement prevents the paper from macroscopically buckling away from the initial configuration.

Depending on the intensity of the load, different phases and regimes can be distinguished. The observation is as follows, referring for definiteness to the case of flexible confinement, (b) in Fig. 2.2. In the first stage of compression, slip in small bands appears (first picture). These bands have an orientation of about 20 degrees from the plane normal, in both directions (second picture). The bands where slip occurs are separated by large regions in which the material has not undergone slip and

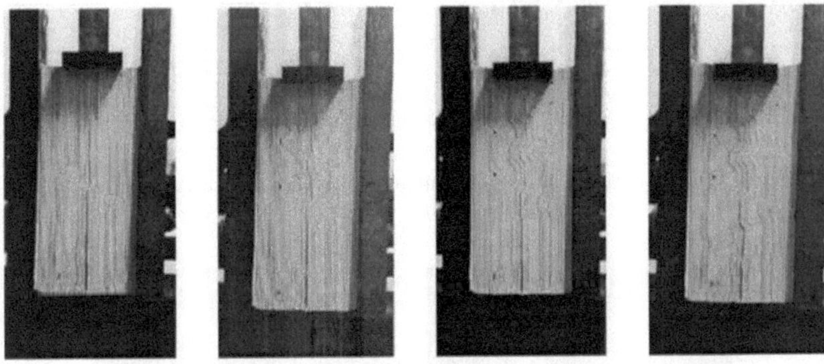

(a) Rigid Faces

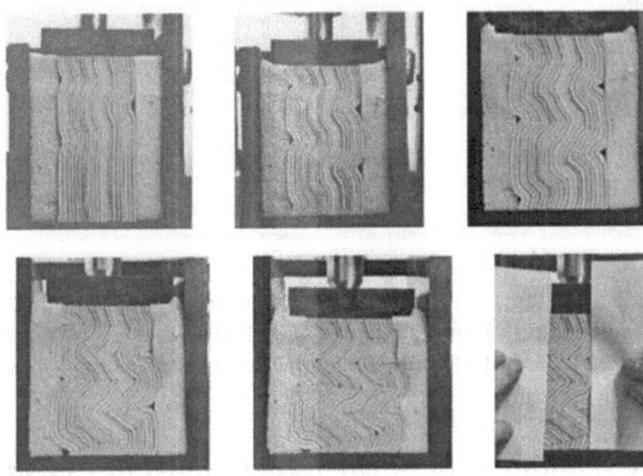

(b) Flexible Faces

Fig. 2.2 An experiment with compression of a stack of paper. Reproduced from [HPW00, Fig. 2] with permission by Elsevier.

the layers are simply translated to the left or to the right. With increasing compression the bands expand, reducing the size of the regions with no slip (left picture in the bottom line). Ultimately, the entire sample has undergone slip, in both possible directions (central picture in the bottom line). The overall structure is periodic and has sharp kinks between approximately affine regions.

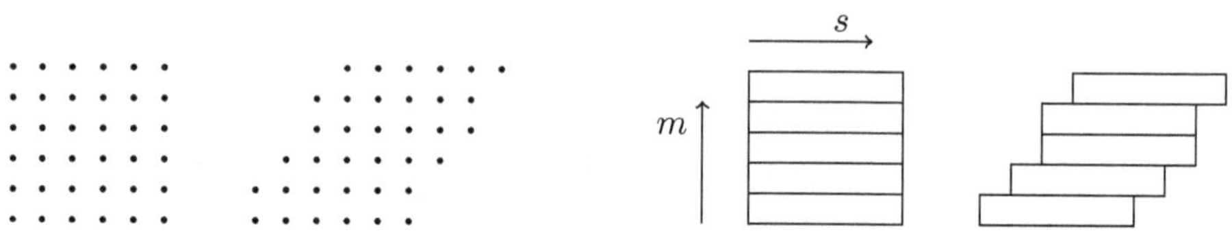

Fig. 2.3 Atomistic view of a slip system and analogy with deck of cards. In this figure, $s = e_1$ and $m = e_2$.

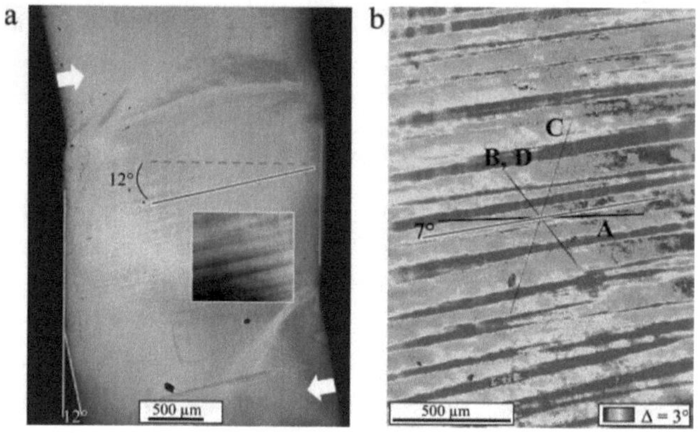

Fig. 2.4 Experimental pictures using a combination of digital image correlation and electron backscattering defraction. Reproduced from [DDMR09, Fig. 5] with permission from Elsevier. We also refer to [DRMD15] in this volume for an explanation of the experimental techniques and detailed findings.

2.2.3 Simple Laminates in Shear Experiments in Crystal Plasticity

Plastic deformation of crystals is due to the sliding of neighboring atomic planes with respect to each other, mediated by the motion of dislocations. It can be described by the slip plane normal m, the slip direction s (two mutually orthogonal unit vectors) and the amount of slip γ (a scalar), see Fig. 2.3. Macroscopically, the corresponding deformation gradient is a simple shear, $\mathrm{Id} + \gamma s \otimes m$.

The discrete symmetry of the crystal leads to the presence of few relevant slip systems, which normally correspond to the most densely packed planes and to the shortest interatomic distances inside these planes. In experiments, one typically observes that locally only one slip system is active, which is explained by a mechanism of cross hardening in the system. This may lead to complex microstructures, in which different simple shear deformations are mixed to generated macroscopic deformations which are not simple shear, see Fig. 2.4. Mathematically, these microstructures can be understood via a variational model within the deformation theory of plasticity [OR99, CHM02, MSL02], which is appropriate under monotonic loading conditions.

2.3 The Hunt-Peletier-Wadee Model for Kink Bands

In this paragraph we summarize the experimental observations and the mathematical models, which have been presented in a two-dimensional model originally introduced by Hunt, Peletier and Wadee in 2000 [HPW00]. This model focuses on the reproduction of the early stages of kink-band formation, in particular the development of the first single kink-band, see Fig. 2.2. For further developments we refer to [WVHY12, BFA98, BEH03, WHP04, HORL11, DPBH12] and the references therein.

The starting point is the idea that the observed strain concentration is not due to plastic yielding inside the layers, but results predominantly from the interplay of the following three effects: layer-parallel stiffness, lateral confinement, and (Coulomb) friction between the layers. In the geological situation described in Section 2.2.1, the overburden pressure plays the role of the lateral confinement. On the one hand, the layer-parallel stiffness provides energy storage that leads to a jump phenomenon observed at the beginning of the experiment, when the first kink-band forms. On the other hand, interlayer friction creates the lock-up states of a kink, which correspond to the experimental observation that a single kink once formed is stable and keeps on folding towards a maximal angle. The lateral confinement is responsible for chevron folds being favorable structures. Indeed, sharp kinks are optimal in the sense that they keep the voids between the layers as small as possible, and hence reduce the work against the lateral confinement, while at the same time they do not cost too much bending energy. For a single kink, this modeling assumption is motivated by a rigorous variational argument in [HPW00, Appendix]. This argument gives a reason for the appearance of kink-like microstructures. It is based on one-dimensional kinematics and does not, therefore, address the issue of the two-dimensional geometry of the folds.

Resting upon the three pillars outlined above, Hunt, Peletier and Wadee [HPW00] developed (under a few a priori assumptions like fixed kink-band width) a simple model that approximates bending structures with discrete linear segments connected by inline springs, forming a hinge with sharp corners and straight limbs, and takes external loading into account. One building block with the acting forces is depicted in Fig. 2.5.

The potential energy E of the system is the sum of the energy contributions of the single hinge segments. These again are composed of the (elastic) potential energy E_k of the spring with stiffness k in the hinge (with neutral state of parallel limbs), the (pseudo)energy contribution E_{fr} due to interlayer friction, as well as the work W_{lc} and W_{l} against the lateral confinement and the external loading, respectively. These assumptions lead to

$$E = E_k + E_{\mathrm{fr}} - W_{\mathrm{lc}} - W_{\mathrm{l}}. \tag{2.1}$$

An explicit computation minimizing the system energy yields the equilibrium states of this system. A stability analysis by means of the global Maxwell stability criterion casts light on the jump phenomenon that leads to the emergence of the first kink.

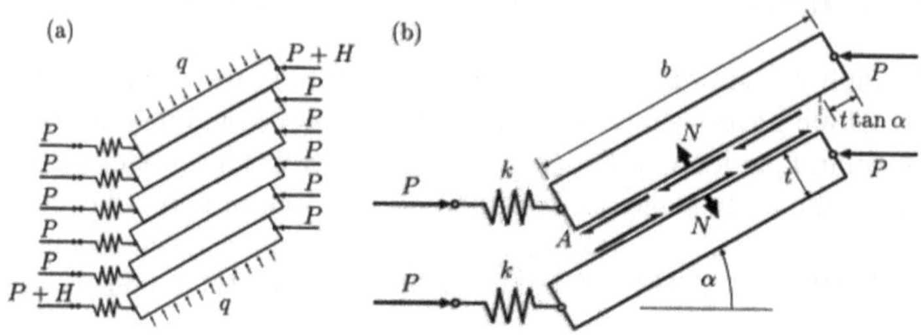

Fig. 2.5 A hinge segment with acting forces. Reproduced from [HPW00, Fig. 2] with permission from Elsevier.

In conclusion, this model provides good explanations for the effects observed in the experiment of Fig. 2.2, and opens the road for further investigations needed to clarify additional aspects. Firstly, the model focuses on the early stages up to the formation of the first kink. An extended discussion of the development of the first few kink-bands, among other modifications and improvements to the basic model, is given in [WVHY12]. We also refer to [BFA98], where the focus lies on kink-band evolution, especially on band broadening and transverse kink-band propagation in fiber composites. Secondly, the energy in (2.1) relies on the hypothesis that kinks form, as well as on a priori assumptions on the character of the kinks. The condition of fixed kink-band width is relaxed in both [WVHY12] and [BFA98], where it results from a minimizing procedure after adding a bending energy to the model, which leads to an intrinsic length scale.

2.4 Variational Modeling of Microstructure

In the following, we briefly review a general variational framework for the study of the spontaneous appearance of periodic microstructures in materials. The starting point is the observation that lack of convexity of energy densities in variational principles may cause oscillations of minimizing sequences; the fundamental tools are quasiconvexity and the theory of relaxation, which permits to understand the asymptotic behavior of minimizing sequences. We refer to [Mül99, Dac07, Rou97] for a more detailed presentation. We shall later discuss in Section 2.5 its application to layered materials, including crystal plasticity and the experiment on paper sheets presented in Fig. 2.2.

Spontaneously formed microstructures in materials can in many cases be understood as the physical realization of minimizing sequences for variational integrals which may not have minimizers, as for example in the large body of work on martensites, see [BJ87, CK88, BJ92, Mül99]. Indeed, the lack of a suitable convexity property of the integrand leads to a failure of lower semicontinuity of the variational functional with respect to the relevant topology. Consequently, the direct method in the calculus of variations, which is based on lower-semicontinuity, cannot be applied to prove the existence of minimizers subject to appropriate boundary conditions. In this situation, a study of the minimizing sequences gives valuable information about the behavior of the system under consideration. Simple problems, like the one proposed by Young [You69], which is frequently referred to as the sailor's problem, illustrate that oscillations can help to further reduce the energy.

A typical variational problem is given by

$$E(u) = \int_{\Omega} W(\nabla u(x))\,\mathrm{d}x \to \min, \qquad u \in \mathscr{D} = u_0 + W_0^{1,p}(\Omega;\mathbb{R}^n), \tag{2.2}$$

where $\Omega \subset \mathbb{R}^n$, $W : \mathbb{R}^{n\times n} \to \mathbb{R}$ is a continuous function with p-growth, i.e., there exist $p \in (1,\infty)$ and constants c, $C > 0$ such that

$$c|F|^p - C \leq W(F) \leq C(1+|F|^p) \quad \text{for all } F \in \mathbb{R}^{n\times n}.$$

Here $u_0 \in W^{1,p}(\Omega;\mathbb{R}^n)$ represents given Dirichlet boundary data. An important additional assumption on W in continuum mechanics is the assumption of material frame invariance, i.e., that $W(RF) = W(F)$ for all $R \in SO(n)$ and all $F \in \mathbb{R}^{n\times n}$.

The appropriate concept of convexity for variational functionals depending on gradient fields is quasiconvexity, as introduced by Morrey [Mor52, Mor66]. Under the p-growth conditions stated above, quasiconvexity of W is a necessary and sufficient condition for lower semicontinuity of E with respect to weak convergence in $W^{1,p}$. If W is quasiconvex, the direct method proves that (2.2) has solutions. Here we say that the function W is quasiconvex if among all possible local microstructures with affine boundary conditions Fx the homogeneous affine configuration $u(x) = Fx$ is optimal, in formulas

$$W(F) = \int_{(0.1)^n} W(\nabla u)\,\mathrm{d}x \leq \int_{(0,1)^n} W(F + \nabla\varphi)\,\mathrm{d}x \tag{2.3}$$

for all $\varphi \in W_0^{1,\infty}((0,1)^n;\mathbb{R}^n)$ and all $F \in \mathbb{R}^{n\times n}$. If W fails to be quasiconvex, relaxation theory [Dac07] suggests to replace (2.2) with the corresponding effective problem, in which the local structures are already averaged out so that it describes the macroscopic response of the system. Precisely, the relaxed problem is given by

$$E^{\mathrm{rlx}}(u) = \int_{\Omega} W^{\mathrm{qc}}(\nabla u(x))\,\mathrm{d}x \to \min, \qquad u \in \mathscr{D}. \tag{2.4}$$

Here W^{qc} denotes the quasiconvex envelope of W, meaning the largest quasiconvex function below W. Under the foregoing assumptions, it is known that

$$\inf_{u\in\mathscr{D}} E(u) = \min_{u\in\mathscr{D}} E^{\mathrm{rlx}}(u),$$

meaning that relaxation preserves the total energy of the system. Moreover, there is a one-to-one correspondence between minimizers of E^{rlx} and $W^{1,p}$-weak limits of minimizing sequences for E. An additional benefit of the knowledge of the relaxed

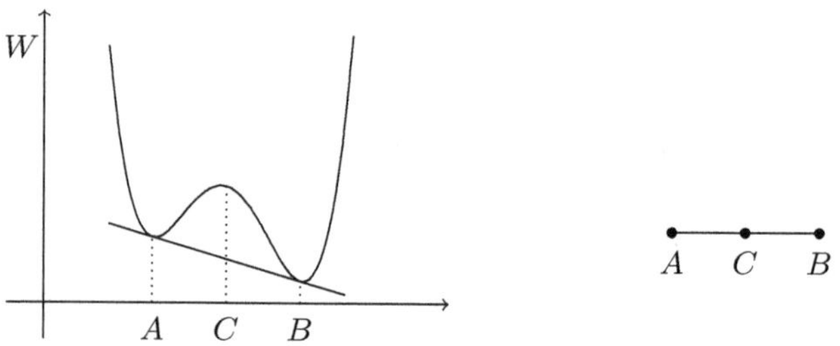

Fig. 2.6 Energy reduction by convexification through microstructure. Left panel: if the energy is not convex, the average C of two states A and B may have higher energy than the average of the energies of A and B. It is then energetically convenient to replace C by a fine-scale mixture of A and B, if kinematically possible. Right panel: this decomposition is often sketched as a "splitting" of C into A and B. See Fig. 2.7 for a sketch of one possible kinematic realization of the splitting, in the situation that $A-B$ is a rank-one matrix.

problem is that it allows for numerical simulations of elastic materials without a resolution of all the fine scale oscillation and may lead to a significant reduction of the computational costs [Car01, CDD02]. The main difficulty, however, is to find an explicit representation for W^{qc}. In fact, this amounts to solving an infinite-dimensional minimization problem and requires a good idea of the optimal microstructures governing the relaxation process. So far, a full analytical relaxation has been found only for very few special examples, see for instance [Koh91, DD02, CD14b].

In continuum mechanics, the condition of non-interpenetration of matter is often included by requiring that $W(F)$ is infinite whenever $\det F \leq 0$. In order to avoid infinite compression of materials one also adds the assumption that $W(F)$ diverges to infinity if $\det F \searrow 0$. Obviously, the p-growth conditions formulated above cannot hold for energy densities with these physically relevant properties and the general theory of relaxation cannot be applied. An extension of this theory with a suitable definition of W^{qc} and appropriate growth conditions has been recently obtained, see [KRW13, CD14a].

We conclude this section by pointing out the fundamental mechanism which relates nonconvex energies to oscillations in minimizing sequences. A typical shape of the energy density W as it appears frequently in nonlinear elasticity theory can be seen in Fig. 2.6, depicted along a specific direction in the space of matrices, i.e. $t \mapsto F_t \in \mathbb{R}^{n\times n}$ for a scalar parameter $t \in \mathbb{R}$. Assume that this curve is in fact a rank-one line, in the sense that

$$F_t = A + tR$$

with $F_0 = A$, $F_1 = B$ and $F_1 - F_0 = R = a \otimes b$ for some $a, b \in \mathbb{R}^n$ with $|b| = 1$. Then it is possible to construct for each convex combination $F = \lambda F_0 + (1-\lambda)F_1$ with $\lambda \in (0,1)$ and every $k \in \mathbb{N}$ a Lipschitz-function u_k on $(0,1)^n$ with $\nabla u_k \in \{F_0, F_1\}$ for almost every $x \in (0,1)^n$. More precisely, we set

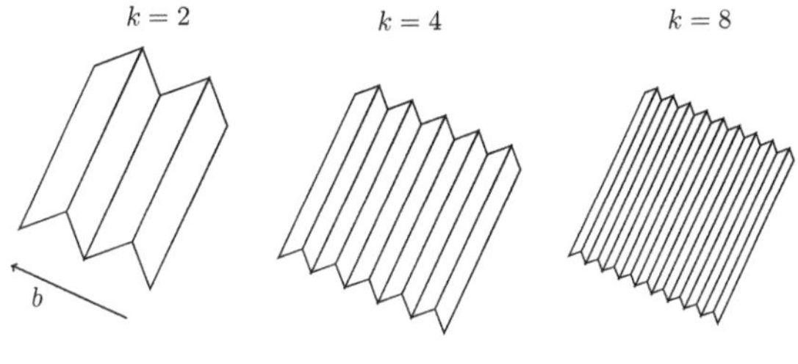

Fig. 2.7 Laminates with different periods as elements of a minimizing sequence of the form given in (2.5). The orientation of the layers is determined by the rank-one direction and given by the vector b. The deformation gradient is constant in each layer and lies in $\{F_0, F_1\}$.

$$u_k(x) = Fx + \frac{1}{k} h\big(k(x \cdot b)\big)a, \qquad x \in (0,1)^n, \tag{2.5}$$

where $h : \mathbb{R} \to \mathbb{R}$ is the continuous one-periodic function with $h' = \lambda - 1$ in $(0, \lambda)$ and $h' = \lambda$ in $(\lambda, 1)$. As Fig. 2.7 shows, b describes the orientation of the oscillations by being the normal on the planar jump discontinuities of ∇u_k, while a and k are the amplitude and the scale of the oscillation, respectively. As F is the average of the oscillations between F_0 and F_1, one has $u_k \overset{*}{\rightharpoonup} Fx$ in $W^{1,\infty}((0,1)^n;\mathbb{R}^n)$. In the following, we refer to the construction in (2.5) as a simple laminate. Notice that the way the functions u_k are constructed in (2.5) does not make $\varphi_k = u_k - Fx$ admissible trial functions for (2.3). A modification of u_k by a cut-off using affine interpolation to achieve the boundary conditions Fx, though, produces only an error (measured in the $W^{1,2}$-norm) of order $\mathscr{O}(k^{-1})$. We denote this new function by v_k.

The foregoing computations show that

$$W^{\mathrm{qc}}(F) \leq \int_{(0,1)^n} W(\nabla v_k)\,\mathrm{d}x \sim \lambda W(F_+) + (1-\lambda)W(F_-) + \mathscr{O}(k^{-1}).$$

Hence, the simple laminate construction gives an upper bound on W^{qc}. An improved upper bound can be obtained by iterating the construction, leading to pictures with layers within layers and to the concept of lamination-convex and rank-one-convex envelopes, see [Dac07, Mül99]. Constraints on the determinant can also be included by appropriately modifying the cutoff procedure, see [MŠ99, Con08]. A matching lower bound can be found by the concept of polyconvexity [Mor52, Bal77, Dac07, Mül99], which is based on the idea that the integral of the determinant of a gradient field only depends on the boudary values. To the best of our knowledge, all explicit computations of W^{qc} for physically relevant problems are based on this strategy.

2.5 Models in Crystal Plasticity with One Active Slip System

This section applies the strategy for the reduction of energy described in Section 2.4 to the model energies which have been proposed for single crystals and their plastic deformation. We begin in Section 2.5.1 with the variational formulation of crystal plasticity in a broader framework and focus then on the situation with rigid elasticity and one active slip system. The mathematical relaxation result is presented and discussed in Section 2.5.2, the physical significance of the microstructure is explained in Section 2.5.3. In Section 2.5.4. we then compare the mathematical results with the formation of kink bands in rocks described in Section 2.2.1. We conclude the treatment of single-slip processes with the discussion of more comprehensive models which extend and complement the ones presented here in various directions, including elastic energy (Section 2.5.5) and higher order regularizations related to geometrically necessary dislocations (Section 2.5.6). Multiple slip is addressed in Section 2.6.

2.5.1 Variational Formulation of Crystal Plasticity

Let the function $u : \Omega \to \mathbb{R}^n$ be the deformation of an elasto-plastic body with reference configuration $\Omega \subset \mathbb{R}^n$. As a fundamental assumption of crystal plasticity we use the multiplicative decomposition of the deformation gradient in the sense of Kröner [Krö60] and Lee [Lee69], i.e.,

$$F = \nabla u = F^e F^p, \tag{2.6}$$

where the elastic part F^e captures local rotation and stretching, and the plastic part F^p describes the irreversible deformations encoding the history of plastic flow. For a discussion of the significance of the multiplicative decomposition (2.6) we refer also to [LL67, SO85, MS92, CG01, Mie03, MM06, RC14] and references therein. In crystals, plastic deformation happens along well-characterized slip systems (s,m) consisting of a slip plane with normal $m \in \mathbb{R}^n$ and a slip direction described by $s \in \mathbb{R}^n$, with s and m orthogonal unit vectors as discussed in Section 2.2.3. The orientation and number of the slip systems is specified by the crystallographic structure of the lattice. For instance, for fcc crystals in three dimensions one has twelve slip systems.

Since models with only one slip system are more easily accessible to analytical treatment, they have been subject of a number of works [CT05, Con06, CDK09, CDK11, CDK13a]. Besides their reduced complexity, single-slip models are interesting as they describe the early stages of plastic deformation under the assumption of an ideal single crystal as an initial configuration, i.e., when the first dislocations emerge and start to move. In the single-slip framework the plastic part F^p has the form

$$F^p = \mathrm{Id} + \gamma s \otimes m, \qquad \gamma \in \mathbb{R}, \tag{2.7}$$

which corresponds to a simple shear deformation with γ the amount of shear. A related micromechanical model commonly visualized as a deck of cards (or just as well a pile of paper) is depicted in Fig. 2.3.

Since elastic strains in metals are much smaller than the plastic ones, one often restricts to the elastically rigid case [OR99], in which F^e is locally assumed to be a rotation, i.e. $F^e \in SO(n)$ pointwise. Then, the total deformation gradient F lies (pointwise) in the set

$$\mathcal{M} = \{F \in \mathbb{R}^{n\times n} : F = R(\mathrm{Id} + \gamma s \otimes m),\ R \in SO(n),\ \gamma \in \mathbb{R}\}. \tag{2.8}$$

We adopt here the time-discrete variational approach to elasto-plasticity proposed by [OR99, CHM02, MSL02], restricting ourselves to the first time step. This has to be seen as a simplification that rules out preexisting microstructures and is appropriate for (locally) monotonic loading conditions. The system energy in integral form of the first incremental problem is related to the energies that appeared in the theory of finite elasticity (see Section 2.4). Precisely,

$$E(u) = \int_\Omega W(\nabla u)\,\mathrm{d}x + l(u) \tag{2.9}$$

with l the external loading and W the condensed energy density. The latter involves three different contributions optimized over all possible multiplicative decompositions, i.e.,

$$W(F) = \min_{F=F^eF^p} \{W_e(F^e) + W_p(F^p) + \mathrm{Diss}(F^p)\}, \qquad F \in \mathbb{R}^{n\times n}. \tag{2.10}$$

Here W_e is the elastic energy density, which in the case of rigid elasticity is given by $W_e(F^e) = 0$ if $F^e \in SO(n)$ and $W_e = \infty$ otherwise, W_p is the hardening function and Diss is the dissipative energy contribution. In the following, we choose linear hardening, meaning

$$W_p(F^p) = \begin{cases} \kappa|\gamma|^2 & \text{if } F^p = \mathrm{Id} + \gamma s \otimes m \text{ for some } \gamma \in \mathbb{R}, \\ \infty & \text{otherwise,} \end{cases}$$

and specify

$$\mathrm{Diss}(F^p) = \begin{cases} \tau|\gamma| & \text{if } F^p = \mathrm{Id} + \gamma s \otimes m \text{ for some } \gamma \in \mathbb{R}, \\ \infty & \text{otherwise,} \end{cases}$$

with $\tau, \kappa \geq 0$ given material constants, where τ stands for the critical shear stress and κ is the hardening modulus. Here s and m are regarded as fixed vectors. The side condition can, of course, be enforced by only one of W_p and Diss; we include it in both functions to express the fact that for the present purposes they need only be defined on the set $\mathcal{M}$.

Thus, for W one obtains after performing the local minimization that

$$W(F) = W_{\tau,\kappa}(F) = \begin{cases} \tau|\gamma| + \kappa|\gamma|^2 & \text{if } F \in \mathcal{M}, \\ \infty & \text{otherwise.} \end{cases} \tag{2.11}$$

The quantity γ is computed from F using (2.8). Since $F = R(\mathrm{Id} + \gamma s \otimes m)$ implies $|F|^2 = 2 + \gamma^2$, it is easy to see that $|\gamma|$ is uniquely determined from F. This energy density is non-convex and even non-quasiconvex due to geometrical softening implied by the local rotations of F^e.

2.5.2 Relaxation Results in Crystal Plasticity with One Slip System

We discuss here the known results on the relaxation of the condensed $W_{\tau,\kappa}$ defined in (2.11). The underlying relaxation mechanism, and the laminates which are constructed to prove the upper bounds, give information on the physically optimal microstructures. At this point, it is an open problem to find an explicit formula for $W^{\mathrm{qc}}_{\tau,\kappa}$ with $W_{\tau,\kappa}$ the full energy density including dissipation and hardening. However, several results have been obtained if only one of the two contributions is present, namely, for $W_{\tau,0}$ and $W_{0,\kappa}$, which we present below. For notational simplicity we only discuss the two-dimensional case here, for generalizations to three dimensions we refer to [CT05, CDK13a]. The quasiconvex hull of a set $K \subset \mathbb{R}^{n\times n}$ can be understood as the set of possible averages of gradient fields taking values in K, see [Mül99, Dac07] for a precise definition.

Theorem 2.5.1 (From [CT05, Con06]). *The quasiconvex hull of the set defined in (2.8) is*

$$\mathscr{N} = \{F \in \mathbb{R}^{2\times 2} : \det F = 1 \text{ and } |Fs| \le 1\}, \tag{2.12}$$

in formulas $\mathscr{N} = \mathscr{M}^{\mathrm{qc}}$. For $F \in \mathscr{N}$,

$$W^{\mathrm{qc}}_{0,\kappa}(F) = \kappa(|Fm|^2 - 1), \tag{2.13}$$

and

$$W^{\mathrm{qc}}_{\tau,0}(F) = \tau\sqrt{|F|^2 - 2}, \tag{2.14}$$

while $W^{\mathrm{qc}}_{0,\kappa}(F) = W^{\mathrm{qc}}_{\tau,0}(F) = \infty$ for $F \in \mathbb{R}^{2\times 2} \setminus \mathscr{N}$.

A comparison between the results in the regimes $\tau = 0$ and $\kappa = 0$ shows that the sets of macroscopic strains that can be achieved with finite energy are identical, and given by $\mathscr{N}$ defined in (2.12). On $\mathscr{N}$ the formulas $W^{\mathrm{qc}}_{\tau,0}$ and $W^{\mathrm{qc}}_{0,\kappa}$, however, differ qualitatively, which suggests that distinct microstructure govern the respective relaxation process.

We sketch the basic ideas of the proofs of (2.13) and (2.14), since they are instructive for a good understanding of the fine microscopical patterns that can be observed in experiments, as those discussed in Section 2.2. In both cases simple laminates as in Section 2.4, are optimal, they will have different characteristics, though.

The proofs rely on matching upper and lower bounds for $W^{\mathrm{qc}}_{\tau,0}(F)$ and $W^{\mathrm{qc}}_{0,\kappa}(F)$ with $F \in \mathbb{R}^{2\times 2}$, respectively. We first deal with the case $F \notin \mathscr{N}$. Assume that a function $u \in W^{1,\infty}((0,1)^2;\mathbb{R}^2)$ is given, such that $u(x) = Fx$ on the boundary and $W_{\tau,\kappa}(\nabla u) \in L^1((0,1)^2)$. Then $\det \nabla u = 1$ and $|\nabla us| \le 1$ pointwise. This implies $\det F = 1$ and $|Fs| \le 1$, against the assumption $F \notin \mathscr{N}$. Therefore $W^{\mathrm{qc}}_{\tau,\kappa}(F) = \infty$.

The case $F \in \mathscr{N}$ requires more work. The upper bound is based on the explicit construction of an appropriate rank-one line through F, so that F can be written as a convex combination of rank-one matrices in $\mathscr{M}$. These two matrices represent the set of admissible gradients for the laminate construction leading to trial functions admissible for the minimization problem in the definition of quasiconvexity. We point out that one particular difficulty is the fact that the energy densities are extended valued, in contrast to the setting introduced in the previous section. This is a technically demanding point which can be treated using the constrained convex integration approach developed in [MŠ99], we refer to [CT05] for details. The lower bound, on the other hand, guarantees the optimality of the construction and follows easily by convexity.

We give here a sketch of the mathematical derivation of the laminates entering the upper bound, and discuss later in Section 2.5.3 the heuristic motivation and the physical significance of the constructions.

We start with the quadratic-growth case of $W_{0,\kappa}$. The optimal rank-one line through $F \in \mathscr{N} \setminus \mathscr{M}$ is given by

$$t \mapsto F_t = F + tFm \otimes s, \qquad t \in \mathbb{R}.$$

Notice that the right-hand side in (2.13) is constant along F_t and coincides with $W_{0,\kappa}$ on $\mathscr{M}$. Moreover, one can find $F_- = F_{t_-}$, $F_+ = F_{t_+}$ with $t_- < 0 < t_+$ such that $F_-, F_+ \in \mathscr{M}$ with the corresponding plastic slips γ_- and $\gamma_+ =: \gamma_*$. Then, $F_\pm m = Fm$ implies $|\gamma_\pm|^2 = |Fm|^2 - 1$, so that $\gamma_- = -\gamma_+$, meaning that the amount of slip in the two phases of the resulting simple laminate are the same, but oriented in opposite directions. The geometry in matrix space is illustrated in Fig. 2.8 b). In this figure, we use the representation

$$(F^T F)^{1/2} = \begin{pmatrix} a+b & c \\ c & a-b \end{pmatrix} = U, \qquad a,b,c \in \mathbb{R}, \tag{2.15}$$

where a is given by $a = \sqrt{1+b^2+c^2}$, since $\det F = 1$ and U denotes the square root of the matrix $F^T F$. We fix for definiteness $s = e_1$ and $m = e_2$. If $F \in \mathscr{M}$, then $|Fe_1| = 1$ and the length of the first row of U is equal to one. Hence,

$$U = \begin{pmatrix} \cos\phi & \sin\phi \\ \sin\phi & z \end{pmatrix}, \quad \phi \in [0, 2\pi], z \in \mathbb{R},$$

and the constraint $\det U = 1$ gives $z = (1 + \sin^2\phi)/\cos\phi$ and $b < 0$ in the representation of (2.15). In these coordinates, the set $\mathscr{M}$ corresponds to the blue curve open to $-\infty$, as seen in the figure.

We now turn to the linear-growth case of $W^{\mathrm{qc}}_{\tau,0}$. The relaxation mechanism also in this situation based on simple laminates, but it is a different laminate. Consider the rank-one line

$$t \mapsto F_t = F + tFy^\perp \otimes y, \qquad t \in \mathbb{R},$$

where the unit vector $y \in \mathbb{R}^2$ is chosen such that $(Fy \cdot Fy^\perp)^2 = |F|^2 - 2$ and $|Fy^\perp| = 1$ (there are two solutions for y up to the choice of a sign), see Fig. 2.8 a).

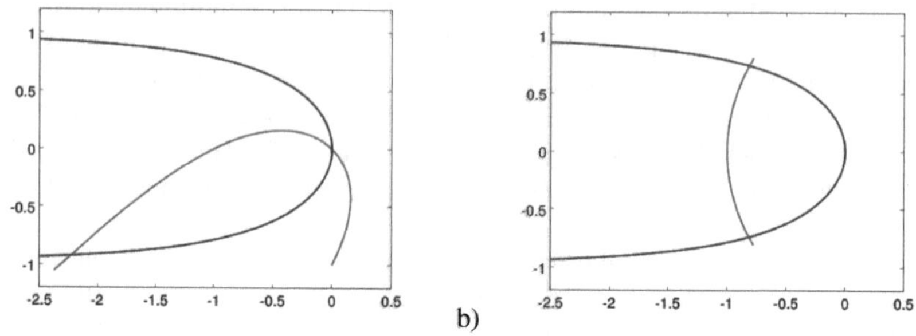

Fig. 2.8 Construction of rank-one lines for a) $W_{\tau,0}$ and b) $W_{0,\kappa}$ with $s = e_1$ and $m = e_2$, in the coordinates (b,c) introduced in (2.15). The blue curve denotes the matrices in $\mathscr{M}$, the red curve the rank-one line $t \mapsto F_t$. Note that the rank-one line through the matrix F with $b = -1$, $c = 0$ intersects the set on which $W_{\tau,0}$ and $W_{0,\kappa}$ are finite in two points. The relaxed energy is finite on the rank-one segments generated by all these pairs of points and infinite elsewhere.

This guarantees that the right-hand side in (2.14) is affine on F_t for $t \in [t_-, t_+]$. Then, by solving a quadratic equation, one finds for $F \in \mathscr{N} \setminus \mathscr{M}$ two matrices $F_- = F_{t_-}$, $F_+ = F_{t_+}$ with $t_- < 0 < t_+$ such that $F_-, F_+ \in \mathscr{M}$ with $\gamma_* := \gamma_+$. If $Fy \cdot Fy^\perp \geq 0$ (similarly for $Fy \cdot Fy^\perp \leq 0$), it can be shown that t_- equals $-(Fy \cdot Fy^\perp)$, which entails $F_- \in SO(2)$ or equivalently $\gamma_- = 0$. Hence, one phase of the optimal laminate does not experience any plastic deformation but results from pure elastic deformation in form of local rotation.

2.5.3 Heuristic Origin of the Laminates

To understand the mechanical origin of the laminates appearing in Theorem 2.5.1 it is useful to consider the following thought experiment, illustrated in Fig. 2.9. A material with a single slip system, with $s = e_1$ and $m = e_2$, is subject to tension in direction m, that is, normal to the slip plane. If the material only deforms plastically, any deformation gradient $F^p = \mathrm{Id} + \gamma e_1 \otimes e_2$ does not change the distance between two planes orthogonal to e_2, see the second sketch in Fig. 2.9 b). However, many segments joining the two planes are made longer. For example, the segment joining $(0,0)$ with $(0,1)$, which has unit length, is mapped to the segment joining $(0,0)$ with $(\gamma,1)$, which has length $\sqrt{1+\gamma^2} > 1$. Combining this deformation with a rotation, one can have the two points $(0,0)$ and $(0,1)$ move to the points $(0,0)$ and $(0,\sqrt{1+\gamma^2})$, thereby increasing their distance to order γ^2. This deformation is, however, not appropriate for other parts of the sample, since it leads to a macroscopic rotation, see Fig. 2.9 b).

A more complex pattern permits to use the horizontal slip to make the sample longer in the vertical direction, see Fig. 2.10. The plastic deformation is in this case not uniform. The material divides into two parts; one of them slips to the left, the other one to the right. The two images of the central segment are both longer than the original segment, therefore one can rotate them so that they coincide (third

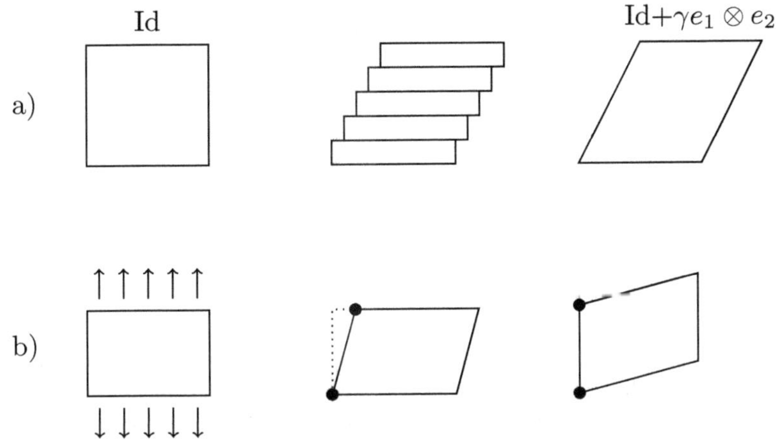

Fig. 2.9 Geometry for the orthogonal forcing problem, leading to the laminates of Fig. 2.10. a) Material description: One has a single slip system, with horizontal slip direction and vertical normal. b) Forcing: The material is in tension in the direction normal to the slip planes.

figure in Fig. 2.10 a)). If the operation is not performed on two but on many subsets of the sample, all separated by parallel lines, a fine structure arises which permits to elongate the probe in the vertical direction, and shorten it in the horizontal one, see the last sketch in Fig. 2.10 a). The resulting pattern is a typical laminate, as illustrated in Fig. 2.7. This construction turns out to be the optimal microstructure for $W_{0,\kappa}$.

In the case of $W_{\tau,0}$ a different microstructure arises, see in Fig. 2.10 b). In this situation, only one part of the domain transforms plastically, the other one is only rotated. Again, if the domain subdivision is fine enough, the macroscopic deformation is approximately a compression in the horizontal direction and a stretching in the vertical one. A related discussion, focused on the phenomenon of geometrical softening in plasticity, is given in [OR99].

In order to understand qualitatively the difference between the two laminates, it is important to notice that the effect used here is purely nonlinear and would not be present in a geometrically linear theory. Indeed, the elongation is in both cases second-order in the slip γ, as the Taylor series

$$\sqrt{1+\gamma^2} = 1 + \frac{1}{2}\gamma^2 + O(\gamma^4)$$

shows. Normally, nonlinear phenomena are larger if the intensity is large, which can be achieved by concentration. To make this precise, let us first consider the energy $W_{\tau,0}$, which has linear growth. For a given amount of energy $e > 0$ invested in the deformation, the material can produce any distribution of γ with average e/τ (in the sense of the L^1-norm). Since the elongation is quadratic in γ, it is larger if the plastic slip is concentrated on a small part of the sample: a large slip on a small

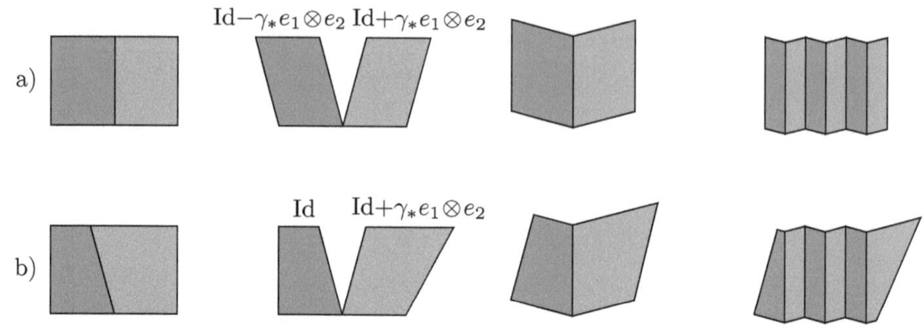

Fig. 2.10 Illustration of laminate formation in the two regimes. a) $W_{0,\kappa}$ with quadratic growth: In this case there is only one possible direction for the lamination, namely, along $m = e_2$. b) $W_{\tau,0}$ with linear growth: In this case there are two possible directions, symmetric with respect to $m = e_2$.

volume generates a larger nonlinear effect than a small slip over a large volume. In other words, the average of the square is larger than the square of the average. This is the reason why it is natural to expect that, in the case of the energy $W_{\tau,0}$, slip and energy are localized in a part of the sample. The localization cannot be complete, as would be the case in a shear band, due to elastic compatibility requirements. However, two symmetric patterns are possible, see Fig. 2.11.

The case of $W_{0,\kappa}$ is different: here the energy is second-order in γ, therefore it is not the average of γ which should be considered fixed, but the average of its square. In this case we consider the Taylor series in γ^2, as above expanding up to the first term which is nonlinear. Since our variable is γ^2, the first nonlinear term is γ^4. In particular, the expression

$$\sqrt{1+\gamma^2} = 1 + \frac{1}{2}\gamma^2 - \frac{1}{8}\gamma^4 + O(\gamma^6)$$

is sublinear in γ^2. This means, that for a given amount of energy invested in the deformation, the maximal stretch is obtained if the slip $|\gamma|$ is uniformly distributed, hence the material chooses not to localize and $|\gamma|$ is uniform. Only its sign, which is not seen by the energy, oscillates. The same effect is also present for $p > 2$, as was shown in [CDK11]. In particular, for all $p \geq 2$ the same microstructure is optimal.

The two constructions presented above are quite different, and it is not clear which microstructure is the correct one if both terms are present in the energy. In fact, the formula for $W^{\mathrm{qc}}_{\tau,\kappa}$ is unknown in the general case. However, for models with small plastic slip $\tau|\gamma|$ will be the dominating term, so that $W_{\tau,0}$ serves as a good approximation to the actual energy density, while for sufficiently large γ the linear term may be neglected and we may work with $W_{0,\kappa}$. This expectation is backed up by numerical simulations. Indeed, a transition between one regime and the other was observed in numerical simulations in [BCHH04, CCO08], which were performed with an energy where both τ and κ are positive. Precisely, it was observed that for small

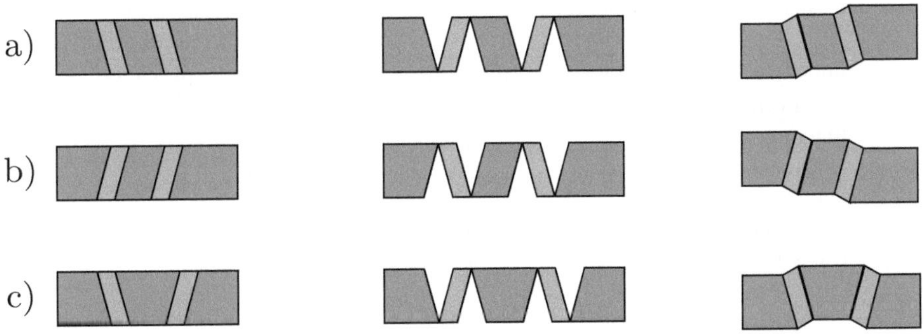

Fig. 2.11 In the linear-growth regime of $W_{\tau,0}$, if the sample is thin with respect to the number of slip bands, two different orientations of the bands may coexist. a) is the periodic pattern of Fig. 2.10, which is possible for any sample geometry and any volume fraction; b) is the symmetric variant, which is completely equivalent; c) shows a pattern in which the two directions are mixed.

deformations the sign of γ oscillates between zero and a positive value, whereas for large deformations, the oscillation is between two opposite values. Additional structures involving also elastic components were seen in [BCHH04, CCO08], since the elastic coefficients were finite in the numerical relaxation.

2.5.4 Relation to Kink Bands in Rocks

We come back to the kink bands presented in Section 2.2.1 and 2.2.2. We propose a variational, two-dimensional model for compression of layered materials, which bears a close relationship to the approach of crystal plasticity discussed in Section 2.5.1. Our model does not include any kinematic assumption on the geometry of the deformation. It is stationary, and based on an energy-dissipation functional, much like in the case of plasticity. The microstructure is then understood as a mathematical consequence of the non-convexity of the governing energy.

Let $u : \Omega \to \mathbb{R}^2$ be the deformation of a cross section $\Omega \subset \mathbb{R}^2$ of the stack of paper or a configuration of geological strata, within a plane-strain approximation. The orientation of the layers is described by the pair of vectors (e_1, e_2) with e_2 being the normal on layers and e_1 being layer-parallel. We assume that the deformation gradient $F = \nabla u$ decomposes into

$$F = \nabla u = F^r F^s,$$

where $F^s = \mathrm{Id} + \gamma e_1 \otimes e_2$ with $\gamma \in \mathbb{R}$ describes interlaminae shearing and $F^r \in SO(2)$ accounts for local changes in orientation of the material layers. In analogy to Section 2.5.1, we consider the energy E given by

Table 2.1 Analogies between models in geology and models in crystal plasticity

model in structural geology of strata	model in single-slip crystal plasticity
layer orientation (e_1, e_2)	slip system (s, m)
interlayer shearing F^s	plastic slip F^p
change of layer-orientation F^r	local lattice rotations F^e
amount of shearing γ	amount of plastic slip γ
lock-up parameter κ	hardening modulus κ
static friction coefficient τ	critical shear stress τ

$$E(u) = \int_\Omega W(\nabla u)\,dx + l(u),$$

with l the external load due to tectonic compression or external loading and confining forces, and

$$\begin{aligned} W(F) &= \min_{F=F^rF^s} \{W_r(F^r) + H(F^s) + \mathrm{Diss}(F^s)\} \\ &= \begin{cases} \tau|\gamma| + \kappa|\gamma|^2 & \text{if } F \in \mathcal{M}, \\ \infty & \text{otherwise,} \end{cases} \end{aligned}$$

where $\tau, \kappa \geq 0$ are material constants. The dissipation is physically understood as originating from the static interlayer friction, and τ is the friction coefficient. Hardening, modeled by the quadratic hardening function H, is a consequence of the increased stiffness of prefolded structures, so we call κ the lock-up parameter.

This model is from a mathematical point of view identical to the one introduced in the previous section, see Tab. 2.1 for an overview of the relevant identifications. An application of the mathematical results of Section 2.5.2 to this new context allows one to predict the appearance of an arrangement of chevron folds, which constitutes the analog to compatible laminate structures in crystal plasticity. The relevant laminates are the same as discussed in Section 2.5.3. This fully two-dimensional variational energetic approach extends the ideas of the one-dimensional analysis of [HPW00, WHP04].

Fig. 2.12. shows a comparison of the laminates of Fig. 2.10 with the experiment on the stack of paper discussed in Fig. 2.2 In particular, one observes also experimentally the asserted transition between the two regimes, where either friction or lock-up prevails. This shows that in the beginning, interlayer friction is the dominant effect, while for stronger compression the hardening contribution due to locking of the layers takes over.

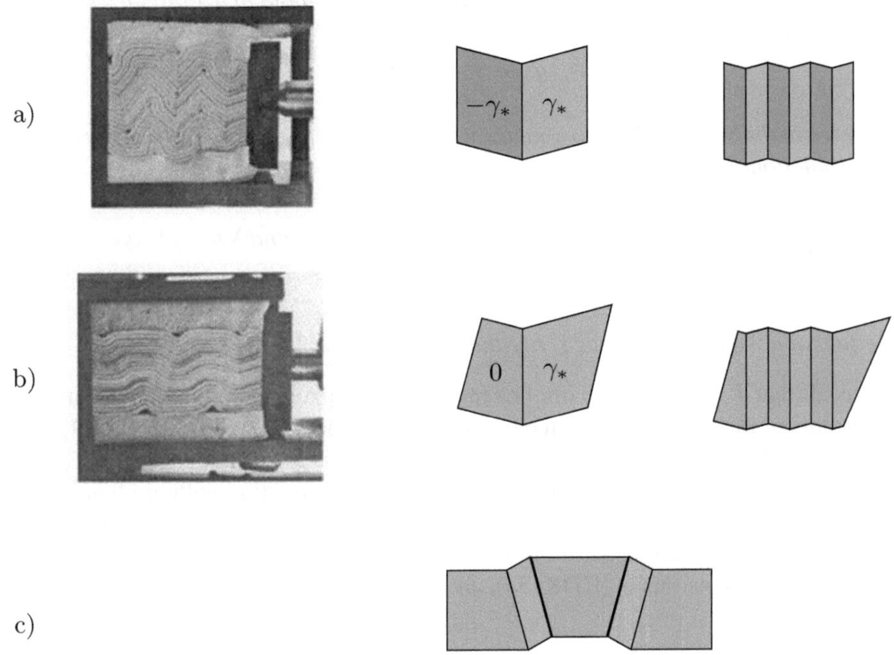

Fig. 2.12 Comparison of the laminates of Fig. 2.10 with the patterns in compressed paper from Fig. 2.2 (reproduced from [HPW00, Fig. 2], with permission from Elsevier). a) Large-compression deformation, compared with the laminate for $W_{0,\kappa}$, with slip $\pm\gamma_*$. b) Small compression, compared with the laminate for $W_{\tau,0}$, which has $\gamma = 0$ and $\gamma = \gamma_*$. It is apparent that in the experiment, just like in c), the slip has concentrated to a part of the domain. However, the interfaces are not all parallel, but instead use the two possible orientations, leaving macroscopic regions in between, as discussed in Fig. 2.11 and illustrated in c).

2.5.5 *Elastic Approximation*

The results discussed in Section 2.5.2 are based on the assumption that the elasticity is hard, so that the elastic strain F^e is a rotation at each material point. This assumption can be justified considering the limit of a sequence of functionals with larger and larger elastic coefficients. Precisely, we set

$$W^{\varepsilon}_{0,\kappa}(F) = \inf_{\gamma\in\mathbb{R}}\{\frac{1}{\varepsilon}\,\mathrm{dist}^2(F(\mathrm{Id}-\gamma s\otimes m), SO(2)) + \kappa|\gamma|^2\},$$

where $\varepsilon > 0$ is a small parameter, representing the ratio between the typical elastic coefficient and the hardening coefficient. Although for lower growth energies of this type may relax to zero [CDK09], the variational problems generated by $W^{\varepsilon}_{0,\kappa}$ converge, in the appropriate sense, to the rigid-elastic problem discussed above.

Theorem 2.5.2 (From [CDK11]). *Let $\Omega \subset \mathbb{R}^2$ be a bounded Lipschitz set. The functionals*

$$E_\varepsilon(u) = \int_\Omega W^\varepsilon_{0,\kappa}(\nabla u)dx$$

converge in the sense of Γ-convergence with respect to the strong convergence in L^1 to the functional

$$E_0(u) = \begin{cases} \int_\Omega W^{\mathrm{qc}}_{0,\kappa}(\nabla u)dx & \text{if } u \in W^{1,2}(\Omega;\mathbb{R}^2) \text{ and } \nabla u \in \mathcal{N} \text{ a.e.} \\ \infty & \text{otherwise.} \end{cases}$$

The function $W_{0,\kappa}$ and the set $\mathcal{N}$ are those of Theorem 2.5.1.

We recall that Γ-convergence implies that the minimizers of E_0 are the accumulation points of the minimizing sequences of the sequence E_ε, see [Bra02, DM93] for details. One key ingredient in the proof of this result is the statement that if $E_\varepsilon(u_\varepsilon)$ is bounded, then u_ε has a subsequence converging to a function with gradient in $\mathcal{N}$ almost everywhere, in particular, with unit determinant. This assertion is based on a generalization of the div-curl Lemma, see [CDM11]. Theorem 2.5.2 was generalized to three dimensions in [CDK13a], and to two slip systems in [CDK13b].

2.5.6 Higher-Order Regularizations

Even though the model introduced above predicts, depending on the application in mind, zigzag patterns of kink-bands or microscopic laminate structures, it does not have an intrinsic length scale, so that arbitrarily fine oscillations can occur, or even have to be expected. To overcome this problem one can incorporate a suitable regularization term that penalizes fast oscillations. In the case of crystal plasticity, this is usually done via the theory of strain-gradient plasticity. In a geometrically linear framework, this corresponds to adding the expression

$$\int_\Omega |\operatorname{curl} F^p| \, \mathrm{d}x$$

to the system energy in (2.9), which corresponds to the line-tension energy of geometrically necessary dislocations [CG01, MM06]. For a mathematical derivation of this expression from microscopic models, see for example [GM06, GLP10, CGM11] in a geometrically linear setting and [SZ12, MSZ14, MSZ15] in a geometrically nonlinear context.

The macroscopic effect of this regularization term was addressed for example in [CO05, AD14, AD15] in various simplified models. In the current geometrically nonlinear single-slip context, a first result was obtained in [Sch14]. These works build upon a large body of mathematical literature on singularly perturbed nonconvex variational problems, which often leads to the identification of self-similar optimal microstructures via the study of the scaling of the energy, and which started with the works of Kohn and Müller on microstructure in shape-memory alloys [KM92, KM94].

2.6 Beyond One Slip-System

As discussed in the introduction, latent hardening leads to the observation that locally only one slip system may be active. However, in real materials, there are quite a few slip systems available and the assumption that laminates can be formed only within one slip system is too restrictive. In this section, we go beyond this assumption and present relaxation results for models in which several different slip systems are available. In such a situation, the relaxed energy has typically different regimes, some in which only one slip system is used and some in which different slip systems interact.

2.6.1 *Two Slip Systems in a Plane*

We first consider a model with two orthogonal slip systems in a plane which are given by $e_1 \otimes e_2$ and $e_2 \otimes e_1$, where e_1, e_2 denotes the standard basis in $\mathbb{R}^2$. If only the dissipation term is present, then the energy is given by

$$\widehat{W}_{2s}(F) = \begin{cases} |\gamma| & \text{if } F = R(\mathrm{Id} + \gamma e_i \otimes e_i^{\perp}), \ \ \gamma \in \mathbb{R}, \ \ i = 1,2, \ \ R \in SO(2), \\ \infty & \text{otherwise.} \end{cases} \tag{2.16}$$

In this case, the rank-one convex and the polyconvex envelope can be determined explicitly (for the definitions we refer to [Mül99, Dac07]). The quasiconvex envelope, however, is not yet known. Additionally, the rank-one convex envelope cannot be computed using simple laminates, and not even with a finite iteration of the lamination construction, but instead requires a construction with laminates of infinite order.

Theorem 2.6.1 (From [ACD09]). *The rank-one convex envelope $\widehat{W}_{2s}^{\mathrm{rc}}$ of $\widehat{W}_{2s}$ defined in (2.16) is given by*

$$\widehat{W}_{2s}^{\mathrm{rc}}(F) = \begin{cases} (\lambda_2 - \lambda_1)(F) & \textit{if } \det F = 1, \min\{|Fe_1|, |Fe_2|\} \leq 1, \\ \psi(|Fe_1|, |Fe_2|) & \textit{if } \det F = 1, 1 \leq |Fe_1| \leq |Fe_2|, \\ \psi(|Fe_2|, |Fe_1|) & \textit{if } \det F = 1, 1 \leq |Fe_2| \leq |Fe_1|, \\ \infty & \textit{if } \det F \neq 1, \end{cases}$$

where

$$\psi(\alpha, \beta) = \int_1^{\alpha} \frac{2s^2}{\sqrt{s^4 - 1}}\, ds + \frac{1}{\alpha}\left(\sqrt{\alpha^2\beta^2 - 1} - \sqrt{\alpha^4 - 1}\right).$$

The polyconvex envelope $\widehat{W}_{2s}^{pc}$ *of* $\widehat{W}_{2s}$ *defined in (2.16) is given by*

$$\widehat{W}_{2s}^{pc}(F) = \max_{\theta \in [0,\pi/2]} \sqrt{|F|^2 + 2|Fe_1 \cdot Fe_2| \sin(2\theta) + 2\cos(2\theta)} - 2\cos\theta \,.$$

Further, $\widehat{W}_{2s}^{pc} \leq \widehat{W}_{2s}^{qc} \leq \widehat{W}_{2s}^{rc}$*, but for some matrices F one has* $\widehat{W}_{2s}^{pc} \neq \widehat{W}_{2s}^{rc}$*.*

Here $\lambda_1(F)$ and $\lambda_2(F)$ denote the signed singular values of F, i.e., the ordered eigenvalues of U in the polar decomposition $F = QU$, $Q \in SO(2)$, $U = U^T$. They are identified uniquely by the conditions

$$\lambda_1^2(F) + \lambda_2^2(F) = |F|^2 \,, \qquad \lambda_1(F)\lambda_2(F) = \det F \,, \qquad \lambda_2 \geq |\lambda_1| \,.$$

For the proof and a more specific discussion of the difference between $\widehat{W}_{2s}^{rc}$ and $\widehat{W}_{2s}^{pc}$ we refer to [ACD09].

We now turn to the case of quadratic growth and define

$$W_{2s}(F) = \begin{cases} |\gamma|^2 & \text{if } F = R(\mathrm{Id} + \gamma e_i \otimes e_i^\perp), \ \gamma \in \mathbb{R}, \ i = 1,2, \ R \in SO(2), \\ \infty & \text{otherwise.} \end{cases} \tag{2.17}$$

In this situation, a full relaxation result is available.

Theorem 2.6.2 (From [CDK13b]). *The quasiconvex envelope of the function* W_{2s} *defined in (2.17) is given for* $F \in \mathbb{R}^{2\times 2}$ *by*

$$W_{2s}^{qc}(F) = \begin{cases} (|Fe_1|^2 - 1) & \text{if } \det F = 1, |Fe_2| \leq 1\,, \\ (|Fe_2|^2 - 1) & \text{if } \det F = 1, |Fe_1| \leq 1\,, \\ \psi(\max\{|Fe_1 + Fe_2|, |Fe_1 - Fe_2|\}) & \text{if } \det F = 1, |Fe_1|, |Fe_2| > 1\,, \\ \infty & \text{if } \det F \neq 1\,, \end{cases}$$

where $\psi(t) = (\sqrt{(t^2-1)_+} - 1)_+^2$, $t \in \mathbb{R}$*, and where we use for* $a \in \mathbb{R}$ *the notation* $(a)_+^2 = \max\{a, 0\}^2$*. The same holds for the rank-one convex and polyconvex envelopes,* $W_{2s}^{pc} = W_{2s}^{rc} = W_{2s}^{qc}$*.*

The proof, including extensions to the case of p-growth with $p \geq 2$ and three dimensions, is given in [CDK13b]. The relevant laminates are sketched in Figure 2.13.

2.6.2 Three Slip Systems in a Plane

Our last example concerns the situation in which three slip systems are active. We assume that

$$W_{3s}(F) = \begin{cases} |\gamma|^2 & \text{if } F = R(\mathrm{Id} + \gamma v_i \otimes v_i^\perp), \ \gamma \in \mathbb{R}, \ i = 1,2,3, \ R \in SO(2), \\ \infty & \text{otherwise,} \end{cases} \tag{2.18}$$

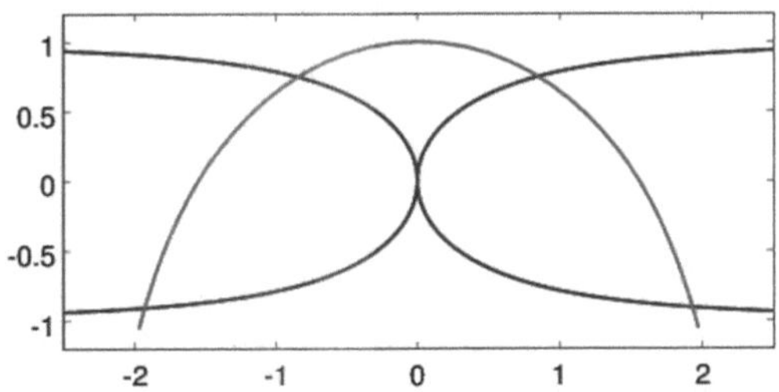

Fig. 2.13 Construction of the relaxation for W_{2s} which uses first-order laminates only. The slip system $e_1 \otimes e_2$ corresponds to the curve which is open to $-\infty$, the slip system $e_2 \otimes e_1$ to the curve which is open to ∞. In addition to the rank-one lines shown in Fig. 2.8 which use only one slip system, the relaxation uses a laminate which is supported on both slip systems. The rank-one line through the matrix $F = (0,1)$ in a, b, c coordinates intersects the set on which W_{2s} is finite in four points. The relaxed energy is finite on the surface $\det F = 1$.

where

$$v_1 = \begin{pmatrix} 1 \\ 0 \end{pmatrix}, \quad v_2 = \begin{pmatrix} -1/2 \\ \sqrt{3}/2 \end{pmatrix}, \quad v_3 = \begin{pmatrix} -1/2 \\ -\sqrt{3}/2 \end{pmatrix}. \tag{2.19}$$

We define $\mathscr{N}^{(i)}$ as the sets corresponding to the set defined in (2.12) for the three slip systems,

$$\mathscr{N}^{(i)} = \{F \in \mathbb{R}^{2\times 2} : \det F = 1, |Fv_i| \leq 1\},$$

and $\mathscr{A}$ as the complement of their union,

$$\mathscr{A} = \{F \in \mathbb{R}^{2\times 2} : \det F = 1, |Fv_i| > 1 \text{ for } i = 1,2,3\}.$$

Further, we let $\mathscr{A}_*$ be the set of points which is contained in none or at least two of the $\mathscr{N}^{(i)}$,

$$\mathscr{A}_* = \mathscr{A} \cup \bigcup_{i=1,2,3} (\mathscr{N}^{(i+1)} \cap \mathscr{N}^{(i+2)})$$

and $\mathscr{N}_*$ the rest, i.e., the set of points contained in exactly one of the $\mathscr{N}^{(i)}$.

$$\mathscr{N}_* = \bigcup_{i=1,2,3} \mathscr{N}^{(i)} \setminus (\mathscr{N}^{(i+1)} \cup \mathscr{N}^{(i+2)}).$$

See Fig. 2.14 for an illustration. We use the same representation as in 2.15. Here and in the following, the index i is understood cyclically, in the sense that $\mathscr{N}^{(i)} = \mathscr{N}^{(i+3)}$ and so on.

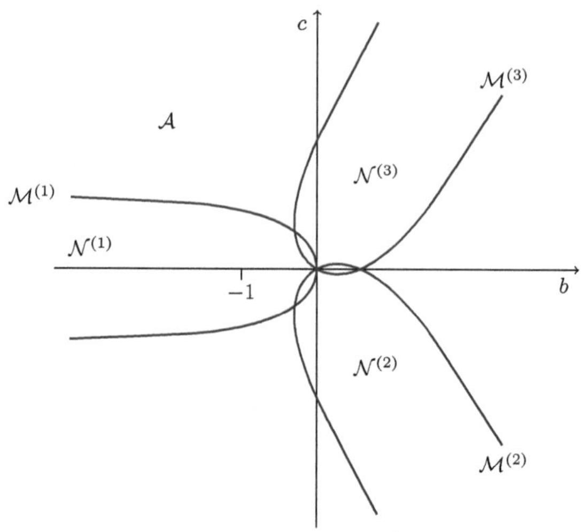

Fig. 2.14 Representation of the phase diagram for the system with three slip-systems with 120° degree angles in the (b,c) plane, with the coordinates defined in (2.15). The three red curves represent the matrices of single-slip type, i.e., the sets $\mathscr{M}^{(i)} = \{F = R(\mathrm{Id} + \gamma v_i \otimes v_i^{\perp})\}$. The regions $\mathscr{N}^{(i)}$ are "inside" the curves, the region $\mathscr{A}$ is the one "outside".

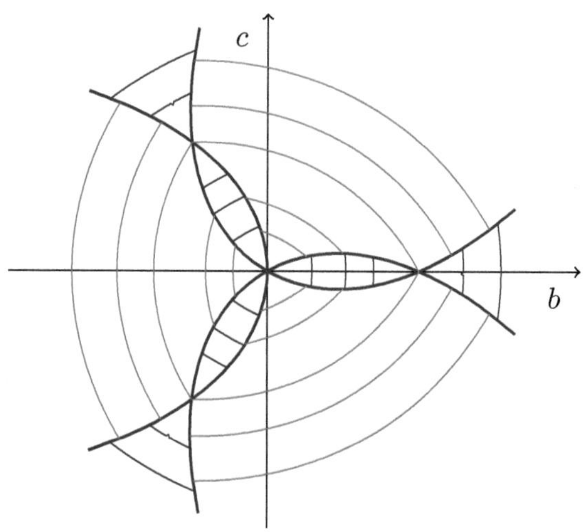

Fig. 2.15 Blow-up of the central region of the phase diagram of Fig. 2.14 with some of the relevant rank-one lines drawn.

Our main result is the following.

Theorem 2.6.3. *Let W_{3s} be as in (2.18). Then*

$$g^{**}(\max_i |Fv_i^\perp|) \le W_{3s}^{qc}(F) \le g(\max_i |Fv_i^\perp|) \qquad \text{if } F \in \mathscr{N}_*,$$

$$W_{3s}^{qc}(F) = h(\max_i |Fv_i|) \qquad \text{if } F \in \mathscr{A}_*,$$

and

$$W_{3s}^{qc}(F) = \infty \qquad \text{if } \det F \neq 1.$$

Here, h and g are defined by

$$h(z) = \frac{4}{3}z^2 - \frac{2}{3} - \frac{4}{3}\sqrt{z^2 - 3/4},$$

and

$$g(z) = \min\{z^2 - 1, f(z)\},$$

where

$$f(z) = 2 + 4z^2 - 2\sqrt{12z^2 - 3},$$

and g^{**} is the convex envelope of g.

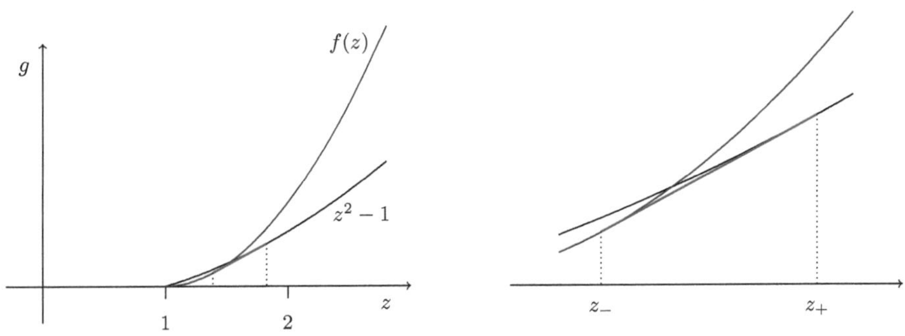

Fig. 2.16 The function g and its convex envelope. The red segment is the double-tangent construction, which differs from g in the interval (z_-, z_+). The right-hand-side panel shows a blowup of this region. The difference $g - g_{**}$ is maximal at the cusp, which occurs at $z = \sqrt{7/3}$, and is of about 7%. The two functions coincide for $z \le z_- \simeq 1.385$, which corresponds to a strain of about 40%.

The strategy of the proof is to construct upper bounds by simple laminates and lower bounds by polyconvexity. The lower bounds follow easily from the definitions of f and g^{**}. The upper bounds are all obtained from first-order laminates, which are different in different regions, see Fig. 2.15 for an illustration. This is the reason for the definition of $\mathscr{A}_*$ and $\mathscr{N}_*$. The upper and lower bound s given do not coincide in a region of relatively large strains, see Fig. 2.16. We remark that for moderate strains, as for example for all F with $\det F = 1$ and $|F - \text{Id}| \le 0.469$, the two bounds coincide and the relaxation is explicitly known.

Acknowledgements. This work was partially supported by the Deutsche Forschungsgemeinschaft through the Research Unit FOR 797 *"Analysis and computation of microstructure in finite plasticity"*, projects CO 304/4-2 (first author) and DO 633/2-2 (second and third author).

References

[ACD09] Albin, N., Conti, S., Dolzmann, G.: Infinite-order laminates in a model in crystal plasticity. Proc. Roy. Soc. Edinburgh A 139, 685–708 (2009)

[AD14] Anguige, K., Dondl, P.W.: Relaxation of the single-slip condition in strain-gradient plasticity. Preprint, arXiv:1402.0114 (2014)

[AD15] Anguige, K., Dondl, P.W.: Energy estimates, relaxation, and existence for strain-gradient plasticity with cross-hardening. In: Hackl, K., Conti, S. (eds.) Analysis and Computation of Microstructure in Finite Plasticity. LNACM, vol. 78, pp. 157–174. Springer, Heidelberg (2015)

[Bal77] Ball, J.M.: Convexity conditions and existence theorems in nonlinear elasticity. Arch. Rational Mech. Anal. 63, 337–403 (1976/1977)

[BCHH04] Bartels, S., Carstensen, C., Hackl, K., Hoppe, U.: Effective relaxation for microstructure simulations: algorithms and applications. Comput. Methods Appl. Mech. Engrg. 193(48-51), 5143–5175 (2004)

[BEH03] Budd, C.J., Edmunds, R., Hunt, G.W.: A nonlinear model for parallel folding with friction. R. Soc. Lond. Proc. Ser. A Math. Phys. Eng. Sci. 459(2036), 2097–2119 (2003)

[BF94] Budiansky, B., Fleck, N.A.: Compressive kinking of fiber composites: A topical review. Appl. Mech. Rev. 47(6S), S246–S250 (1994)

[BFA98] Budiansky, B., Fleck, N., Amazigo, J.: On kink-band propagation in fiber composites. J. Mech. Phys. Solids 46(9), 1637–1653 (1998)

[BJ87] Ball, J.M., James, R.D.: Fine phase mixtures as minimizers of the energy. Arch. Ration. Mech. Analysis 100, 13–52 (1987)

[BJ92] Ball, J.M., James, R.D.: Proposed experimental tests of a theory of fine microstructure and the two-well problem. Phil. Trans. R. Soc. Lond. A 338, 389–450 (1992)

[Bra02] Braides, A.: Γ-convergence for beginners. Oxford Lecture Series in Mathematics and its Applications, vol. 22. Oxford University Press, Oxford (2002)

[Car01] Carstensen, C.: Numerical analysis of microstructure. In: Theory and Numerics of Differential Equations (Durham 2000). Universitext, pp. 59–126. Springer, Berlin (2001)

[CCO08] Carstensen, C., Conti, S., Orlando, A.: Mixed analytical-numerical relaxation in finite single-slip crystal plasticity. Cont. Mech. Thermod. 20, 275–301 (2008)

[CD] Conti, S., Dolzmann, G.: Relaxation in crystal plasticity with three active slip systems (in preparation)

[CD14a] Conti, S., Dolzmann, G.: On the theory of relaxation in nonlinear elasticity with constraints on the determinant. Arch. Rat. Mech. Anal. (2014) doi: 10.1007/s00205-014-0835-9

[CD14b] Conti, S., Dolzmann, G.: Relaxation of a model energy for the cubic to tetragonal phase transformation in two dimensions. Math. Models. Metods App. Sci. 24, 2929–2942 (2014)

[CDD02] Conti, S., DeSimone, A., Dolzmann, G.: Soft elastic response of stretched sheets of nematic elastomers: a numerical study. J. Mech. Phys. Solids 50, 1431–1451 (2002)

[CDK09] Conti, S., Dolzmann, G., Klust, C.: Relaxation of a class of variational models in crystal plasticity. Proc. Roy. Soc. London A 465, 1735–1742 (2009)

[CDK11] Conti, S., Dolzmann, G., Kreisbeck, C.: Asymptotic behavior of crystal plasticity with one slip system in the limit of rigid elasticity. SIAM J. Math. Anal. 43, 2337–2353 (2011)

[CDK13a] Conti, S., Dolzmann, G., Kreisbeck, C.: Relaxation and microstructure in a model for finite crystal plasticity with one slip system in three dimensions. Disc. Cont. Dyn. Systems S 6, 1–16 (2013)

[CDK13b] Conti, S., Dolzmann, G., Kreisbeck, C.: Relaxation of a model in finite plasticity with two slip systems. Math. Models Methods Appl. Sci. 23, 2111–2128 (2013)

[CDM11] Conti, S., Dolzmann, G., Müller, S.: The div-curl lemma for sequences whose divergence and curl are compact in $W^{-1,1}$. Comptes Rendus Math. 349, 175–178 (2011)

[CG01] Cermelli, P., Gurtin, M.E.: On the characterization of geometrically necessary dislocations in finite plasticity. J. Mech. Phys. Solids 49(7), 1539–1568 (2001)

[CGK15] Carstensen, C., Gallistl, D., Krämer, B.: Numerical algorithms for the simulation of finite plasticity with microstructures. In: Hackl, K., Conti, S. (eds.) Analysis and Computation of Microstructure in Finite Plasticity. LNACM, vol. 78, pp. 1–30. Springer, Heidelberg (2015)

[CGM11] Conti, S., Garroni, A., Müller, S.: Singular kernels, multiscale decomposition of microstructure, and dislocation models. Arch. Rat. Mech. Anal. 199, 779–819 (2011)

[CHM02] Carstensen, C., Hackl, K., Mielke, A.: Non-convex potentials and microstructures in finite-strain plasticity. R. Soc. Lond. Proc. Ser. A Math. Phys. Eng. Sci. 458(2018), 299–317 (2002)

[CK88] Chipot, M., Kinderlehrer, D.: Equilibrium configurations of crystals. Arch. Rational Mech. Anal. 103, 237–277 (1988)

[CO05] Conti, S., Ortiz, M.: Dislocation microstructures and the effective behavior of single crystals. Arch. Rat. Mech. Anal. 176, 103–147 (2005)

[Con06] Conti, S.: Relaxation of single-slip single-crystal plasticity with linear hardening. In: Gumbsch, P. (ed.) Multiscale Materials Modeling, pp. 30–35. Fraunhofer IRB, Freiburg (2006)

[Con08] Conti, S.: Quasiconvex functions incorporating volumetric constraints are rank-one convex. J. Math. Pures Appliquees 90, 15–30 (2008)

[CT05] Conti, S., Theil, F.: Single-slip elastoplastic microstructures. Arch. Ration. Mech. Anal. 178(1), 125–148 (2005)

[Dac07] Dacorogna, B.: Direct methods in the calculus of variations, vol. 78. Springer (2007)

[DD02] DeSimone, A., Dolzmann, G.: Macroscopic response of nematic elastomers via relaxation of a class of SO(3)-invariant energies. Arch. Ration. Mech. Anal. 161(3), 181–204 (2002)

[DDMR09] Dmitrieva, O., Dondl, P.W., Müller, S., Raabe, D.: Lamination microstructure in shear deformed copper single crystals. Acta Materialia 57(12), 3439–3449 (2009)

[DM93] Dal Maso, G.: An introduction to Γ-convergence. In: Progress in Nonlinear Differential Equations and their Applications, vol. 8. Birkhäuser Boston Inc., Boston (1993)

[DPBH12] Dodwell, T.J., Peletier, M.A., Budd, C.J., Hunt, G.W.: Self-similar voiding solutions of a single layered model of folding rocks. SIAM J. Appl. Math. 72(1), 444–463 (2012)

[DRMD15] Dmitrieva, O., Raabe, D., Müller, S., Dondl, P.W.: Microstructure in plasticity, a comparison between theory and experiment. In: Hackl, K., Conti, S. (eds.) Analysis and Computation of Microstructure in Finite Plasticity. LNACM, vol. 78, pp. 205–218. Springer, Heidelberg (2015)

[Fle97] Fleck, N.A.: Compressive failure of fiber composites. Adv. Appl. Mech. 33, 43–117 (1997)

[GKH15] Günther, C., Kochmann, D.M., Hackl, K.: Rate-independent versus viscous evolution of laminate microstructures in finite crystal plasticity. In: Hackl, K., Conti, S. (eds.) Analysis and Computation of Microstructure in Finite Plasticity. LNACM, vol. 78, pp. 63–88. Springer, Heidelberg (2015)

[GLP10] Garroni, A., Leoni, G., Ponsiglione, M.: Gradient theory for plasticity via homogenization of discrete dislocations. J. Eur. Math. Soc. (JEMS) 12(5), 1231–1266 (2010)

[GM06] Garroni, A., Müller, S.: A variational model for dislocations in the line tension limit. Arch. Ration. Mech. Anal. 181, 535–578 (2006)

[HHM12] Hackl, K., Heinz, S., Mielke, A.: A model for the evolution of laminates in finite-strain elastoplasticity. ZAMM-Journal of Applied Mathematics and Mechanics/Zeitschrift für Angewandte Mathematik und Mechanik 92(11-12), 888–909 (2012)

[HORL11] Hobbs, B.E., Ord, A., Regenauer-Lieb, K.: The thermodynamics of deformed metamorphic rocks: a review. Journal of Structural Geology 33(5), 758–818 (2011)

[HPW00] Hunt, G.W., Peletier, M.A., Wadee, M.A.: The Maxwell stability criterion in pseudo-energy models of kink banding. Journal of Structural Geology 22(5), 669–681 (2000)

[KK11] Kochmann, D., Hackl, K.: The evolution of laminates in finite crystal plasticity: a variational approach. Continuum Mechanics and Thermodynamics 23, 63–85 (2011)

[KM92] Kohn, R.V., Müller, S.: Branching of twins near an austenite-twinned-martensite interface. Phil. Mag. A 66, 697–715 (1992)

[KM94] Kohn, R.V., Müller, S.: Surface energy and microstructure in coherent phase transitions. Comm. Pure Appl. Math. 47, 405–435 (1994)

[Koh91] Kohn, R.V.: The relaxation of a double-well energy. Contin. Mech. Thermodyn. 3(3), 193–236 (1991)

[Krö60] Kröner, E.: Allgemeine Kontinuumstheorie der Versetzungen und Eigenspannungen. Arch. Rational Mech. Anal. 4, 273–334 (1960)

[KRW13] Koumatos, K., Rindler, F., Wiedemann, E.: Orientation-preserving Young measures. Preprint arXiv:1307.1007 (2013)

[Lee69] Lee, E.H.: Elastic-plastic deformation at finite strains. Journal of Applied Mechanics 36, 1–5 (1969)

[LL67] Lee, E.H., Liu, D.T.: Finite strain elastic-plastic theory with application to plane wave analysis. Journal of Applied Physics 38, 19–27 (1967)

[Mie03] Mielke, A.: Energetic formulation of multiplicative elasto-plasticity using dissipation distances. Contin. Mech. Thermodyn. 15(4), 351–382 (2003)

[Mie15] Mielke, A.: Variational approaches and methods for dissipative material models with multiple scales. In: Hackl, K., Conti, S. (eds.) Analysis and Computation of Microstructure in Finite Plasticity. LNACM, vol. 78, pp. 125–156. Springer, Heidelberg (2015)

[MM06] Mielke, A., Müller, S.: Lower semicontinuity and existence of minimizers in incremental finite-strain elastoplasticity. ZAMM Z. Angew. Math. Mech. 86(3), 233–250 (2006)

[Mor52] Morrey, C.B.: Quasi-convexity and the lower semicontinuity of multiple integrals. Pacific Journal of Mathematics 2(1), 25–53 (1952)

[Mor66] Morrey, C.B.: Multiple integrals in the calculus of variations. In: Die Grundlehren der Mathematischen Wissenschaften, vol. 130. Springer-Verlag New York, Inc., New York (1966)

[MS92] Miehe, C., Stein, E.: A canonical model of multiplicative elasto-plasticity. formulation and aspects of the numerical implementation. European Journal of Mechanics A/Solids 11, 25–43 (1992)

[MŠ99] Müller, S., Šverák, V.: Convex integration with constraints and applications to phase transitions and partial differential equations. J. Eur. Math. Soc (JEMS) 1, 393–442 (1999)

[MSL02] Miehe, C., Schotte, J., Lambrecht, M.: Homogeneization of inelastic solid materials at finite strains based on incremental minimization principles. application to the texture analysis of polycrystals. J. Mech. Phys. Solids 50, 2123–2167 (2002)

[MSZ14] Müller, S., Scardia, L., Zeppieri, C.I.: Geometric rigidity for incompatible fields and an application to strain-gradient plasticity. Indiana Univ. Math. J. 63, 1365–1396 (2014)

[MSZ15] Müller, S., Scardia, L., Zeppieri, C.I.: Gradient theory for geometrically nonlinear plasticity via the homogenization of dislocations. In: Hackl, K., Conti, S. (eds.) Analysis and Computation of Microstructure in Finite Plasticity. LNACM, vol. 78, pp. 175–204. Springer, Heidelberg (2015)

[Mül99] Müller, S.: Variational models for microstructure and phase transitions. In: Bethuel, F., et al. (eds.) Calculus of Variations and Geometric Evolution Problems. Springer Lecture Notes in Math., vol. 1713, pp. 85–210. Springer (1999)

[OR99] Ortiz, M., Repetto, E.A.: Nonconvex energy minimization and dislocation structures in ductile single crystals. J. Mech. Phys. Solids 47(2), 397–462 (1999)

[PC90] Price, N.J., Cosgrove, J.W.: Analysis of geological structures. Cambridge University Press (1990)

[PGPR09] Pimenta, S., Gutkin, R., Pinho, S., Robinson, P.: A micromechanical model for kink-band formation: Part I - experimental study and numerical modelling. Comp. Sci. Tech. 69(7-8), 948–955 (2009)

[RC14] Reina, C., Conti, S.: Kinematic description of crystal plasticity in the finite kinematic framework: a micromechanical understanding of $F = F^e F^p$. J. Mech. Phys. Solids 67, 40–61 (2014)

[Rou97] Roubíček, T.: Relaxation in optimization theory and variational calculus. de Gruyter Series in Nonlinear Analysis and Applications, vol. 4. Walter de Gruyter & Co., Berlin (1997)

[Sch14] Schubert, T.: Scaling relation for low energy states in a single-slip model in finite crystal plasticity. ZAMM Z. Angew. Math. Mech. (2014)

[Sma] Smalljm. Wikipedia, `http://commons.wikimedia.org/wiki/File:Millook_cliffs_enh.jpg` (downloaded on November 28, 2014). Copyright CC BY 3.0

[SO85] Simo, J., Ortiz, M.: A unified approach to finite deformation elastoplastic analysis based on the use of hyperelastic constitutive equations. Comput. Methods Appl. Mech. Engrg. 49(2), 221–245 (1985)

[SPC14] Siboni, M.H., Ponte Castañeda, P.: Fiber-constrained, dielectric-elastomer composites: finite-strain response and stability analysis. J. Mech. Phys. Solids 68, 211–238 (2014)

[SZ12] Scardia, L., Zeppieri, C.: Line-tension model for plasticity as the Γ-limit of a nonlinear dislocation energy. SIAM J. Math. Anal. 44, 2372–2400 (2012)

[WHP04] Wadee, M.A., Hunt, G.W., Peletier, M.A.: Kink band instability in layered structures. J. Mech. Phys. Solids 52(5), 1071–1091 (2004)

[WVHY12] Wadee, M.A., Völlmecke, C., Haley, J.F., Yiatros, S.: Geometric modelling of kink banding in laminated structures. Philosophical Transactions of the Royal Society A: Mathematical, Physical and Engineering Sciences 370(1965), 1827–1849 (2012)

[You69] Young, L.C.: Lectures on the calculus of variations and optimal control theory. W. B. Saunders Co. (1969)

Chapter 3
Rate-Independent versus Viscous Evolution of Laminate Microstructures in Finite Crystal Plasticity

Christina Günther, Dennis M. Kochmann, and Klaus Hackl

Abstract. In this chapter we investigate the variational modeling of the evolution of inelastic microstructures by the example of finite crystal plasticity with one active slip system. For this purpose we describe the microstructures by laminates of first order. We propose an analytical partial relaxation of an incompressible neo-Hookean energy formulation, keeping the internal variables and geometric microstructure parameters fixed, thus approximating the relaxed energy by an upper bound of the rank-one-convex hull. Based on the minimization of a Lagrange functional, consisting of the sum of rate of energy and dissipation potential, we derive an incremental strategy to model the time-continuous evolution of the laminate microstructure. Special attention is given to the three distinct cases of microstructure evolution, initiation, rotation, and continuous change. We compare a rate-independent approach with another one that employs viscous regularization which has certain advantages concerning the numerical implementation. Simple shear and tension/compression tests will be shown to demonstrate the differences between both approaches and to show the physical implications of the models introduced.

3.1 Introduction

Laminate microstructures represent a frequent phenomenon in inelastic materials. Essentially they may be understood as material instabilities caused by a lack

Christina Günther · Klaus Hackl
Lehrstuhl für Mechanik - Materialtheorie,
Ruhr-Universität Bochum, 44801 Bochum, Germany
e-mail: {christina.guenther,klaus.hackl}@rub.de

Dennis M. Kochmann
California Institute of Technology Graduate Aerospace Laboratories,
Pasadena, CA 91125, USA
e-mail: kochmann@caltech.edu

S. Conti and K. Hackl (eds.), *Analysis and Computation of Microstructure in Finite Plasticity*,
Lecture Notes in Applied and Computational Mechanics 78, DOI: 10.1007/978-3-319-18242-1_3

of convexity of the corresponding free energy. They were initially observed in the case of shape memory alloys, and this discovery led to the development of an intricate and very successful theory of energy relaxation, described for example in [Dac82, BJ87, Bha03]. For a long time, similar laminate patterns have been observed to be produced by the motion of dislocations in single crystals [CM95, CSMC05, DDMD09]. But it was possible to model them only after significant progress had been made in the variational formulation of evolution laws for inelastic materials in [OR99, CKM02, MCJ03, ML03, MLG04]. These works made it possible to apply the variational concept of energy relaxation, especially via lamination, to dislocation microstructures as well.

After the underlying framework has been established progress was made in various directions. A rigorous mathematical foundation of the variational methods used was established by Mielke and co-authors in a series of papers [Mie02, Mie03, MS08, Mie04, MM09]. Closed form relaxed envelopes were found by Conti, Dolzmann and co-authors in [CT05, CO05, CDK09, CDK11, CDK13], and by Anguige and Dondl in [AD14]. Finally, numerically based treatments of dislocation microstructures may be found in the works of Carstensen, Hackl, Miehe and co-authors [BCHH04, BCC$^+$06, CHHO06, CSO08, BC10, MGB07, MRF10, Mie14].

In this work, we focus on the time-continuous evolution of laminate microstructures governed by variational principles. In contrast to the papers cited above we will employ a partially relaxed energy maintaining the inelastic microstructure parameters as variables and only eliminating elastically fast variables via minimization. This energy will then be supplemented by a corresponding relaxed dissipation potential which will allow to derive evolution equations for the microstructure parameters. We will undertake a comparison of rate-independent and viscous evolution laws. In this context we will stress three essential stages of the material process: initiation, rotation, and continuous evolution.

The present paper has the aim to review recent research of the authors on the subject. It reports results published in [HK08, HK09, HK10, HK11, KH10, KK11, HHM12, HHK14] along with novel results in a common context. This means that some passages of this report appear in similar form within those works.

3.2 Variational Modeling of Microstructures

The energetic state of an inelastic material be described by a total energy of the form

$$\mathscr{E}(t,\mathbf{u},\mathbf{z}) = \int_\Omega \Psi(\nabla\mathbf{u},\mathbf{z})\,\mathrm{d}V - \ell(t,\mathbf{u}), \tag{3.1}$$

where Ψ is the specific Helmholtz free energy, $\mathbf{u}$ the displacement vector, $\mathbf{z}$ a collection of internal variables, $\ell(t,\mathbf{u})$ represents the potential of external forces and Ω is the body's volume.

According to the principle of minimum potential energy, the actual displacement field follows from

$$\mathbf{u} = \operatorname{argmin}\left\{ \mathscr{E}(t,\mathbf{u},\mathbf{z}) \,\middle|\, \mathbf{u} = \mathbf{u}_0 \text{ on } \Gamma_{\mathbf{u}} \right\}, \tag{3.2}$$

where $\Gamma_{\mathbf{u}}$ denotes a subset of the body's boundary $\partial\Omega$. The internal variables can be computed from the principle of the minimum of dissipation potential [CKM02, HF08] according to

$$\dot{\mathbf{z}} = \operatorname{argmin}\left\{ \mathscr{L}(\mathbf{u},\mathbf{z},\dot{\mathbf{z}}) \,\middle|\, \dot{\mathbf{z}} \right\}, \tag{3.3}$$

where we introduced the Lagrange functional

$$\mathscr{L}(\mathbf{u},\mathbf{z},\dot{\mathbf{z}}) = \frac{\mathrm{d}}{\mathrm{d}t}\Psi(\nabla\mathbf{u},\mathbf{z}) + \Delta(\mathbf{z},\dot{\mathbf{z}}). \tag{3.4}$$

Here $\Delta(\mathbf{z},\dot{\mathbf{z}})$ is the so-called dissipation potential and the dot denotes differentiation with respect to time. This Lagrange functional consists of the sum of elastic power and dissipation due to changes of the internal state of the material [CKM02, OR99]. Minimization of the Lagrange functional hence determines the changes of the internal variables. Note, that in contrast to (3.1) the functional in (3.4) is purely local, i.e., the minimization principle (3.3) holds point-wise in the space of internal variables. Of course, the integrated form can be used as well if necessary. This fact will be employed later on when we derive relaxed dissipation potentials.

If the free energy density is not quasiconvex it is unfavorable for the material to accommodate an imposed macroscopic deformation gradient by a homogeneous deformation field but rather by forming microstructural patterns that mix different homogeneous states of minimal energy. The material body, aiming to reduce its energy, does not respond by means of a homogeneous deformation state but breaks up into multiple domains at local energy minima in such a way that it is compatible with the overall imposed deformation or any given boundary conditions.

Here we will focus on so-called laminate microstructures, which in the case of dislocation patterns is supported by experimental evidence [DDMD09]. For conciseness we will restrict ourselves to first-order laminates. Everything stated in subsequent sections can be extended to general laminates in an essentially straightforward manner.

A laminate of first order is characterized by N volume fractions λ_i separated by parallel planes with normal vector $\mathbf{b}$, as sketched in Fig. 3.1. To every volume fraction i there corresponds a value $\mathbf{z}_i$ of the internal variables. Moreover, in every volume fraction we have a deformation gradient $\mathbf{F}_i$ which we write as

$$\mathbf{F}_i = \mathbf{F}(\mathbf{I} + \mathbf{a}_i \otimes \mathbf{b}). \tag{3.5}$$

This formulation ensures that deformation gradients differ only by tensors of rank one, enforcing compatibility at laminate interfaces and hence ensuring the existence of a corresponding deformation field $\mathbf{u}$. We need to impose the volume average of the deformation gradient

$$\sum_{i=1}^{N} \lambda_i \mathbf{F}_i = \mathbf{F}, \tag{3.6}$$

which is equivalent to

$$\sum_{i=1}^{N} \lambda_i \mathbf{a}_i = \mathbf{0}. \tag{3.7}$$

Let us consider the normal vector **b** as ingrained into the material since any change of **b** would require a change of the internal variables and thus lead to dissipation. The amplitudes $\mathbf{a}_i$ on the other hand can be changed purely elastically. This suggests to define a semi-relaxed energy by

$$\Psi^{\mathrm{rel}}(\mathbf{F},\boldsymbol{\lambda},\mathbf{z},\mathbf{b}) = \inf\Big\{ \sum_{i=1}^{N} \lambda_i \Psi(\mathbf{F}_i,\mathbf{z}_i) \,\Big|\, \mathbf{a}_i \;:\; \sum_{i=1}^{N} \lambda_i \mathbf{a}_i = \mathbf{0} \Big\}, \tag{3.8}$$

where we introduced the abbreviations $\boldsymbol{\lambda} = \{\lambda_1,\dots,\lambda_N\}$ and $\mathbf{z} = \{\mathbf{z}_1,\dots,\mathbf{z}_N\}$. Note that the energy is only partially relaxed since full relaxation would require further minimization with respect to the internal variables (and in particular with respect to **b**). If we assume that the lamination respects the ordering $\{1,\dots,N\}$ and that the normal vector **b** remains fixed, the relaxation of the dissipation is given by

$$\Delta^*(\boldsymbol{\lambda},\mathbf{z},\dot{\boldsymbol{\lambda}},\dot{\mathbf{z}}) = \sum_{i=1}^{N} \lambda_i \Delta(\mathbf{z}_i,\dot{\mathbf{z}}_i) + \inf\Big\{ \sum_{i,j=1}^{N} \Delta\lambda_{ij} D(\mathbf{z}_i,\mathbf{z}_j) \,\Big|\, \Delta\lambda_{ij} \;:$$
$$\sum_{i=1}^{N} \Delta\lambda_{ij} = \dot{\lambda}_j, \sum_{j=1}^{N} \Delta\lambda_{ij} = \dot{\lambda}_i, \Delta\lambda_{ij} = 0 \text{ for } |(i-j)\,\mathrm{mod}\,N| \neq 1 \Big\}. \tag{3.9}$$

Here we have employed the so-called dissipation distance [Mie03] defined by

$$D(\mathbf{z}_0,\mathbf{z}_1) = \inf\Big\{ \int_0^1 \Delta(\mathbf{z}(s),\dot{\mathbf{z}}(s))\,\mathrm{d}s \,\Big|\, \mathbf{z}(0) = \mathbf{z}_0, \mathbf{z}(1) = \mathbf{z}_1 \Big\}. \tag{3.10}$$

Now we may apply the principle of the minimum of the dissipation potential to the functionals in (3.8) and (3.9) in order to obtain evolution equations for λ and **z** for fixed **b** via

$$\{\dot{\boldsymbol{\lambda}},\dot{\mathbf{z}}\} = \operatorname{argmin}\big\{ \mathscr{L}(\mathbf{F},\boldsymbol{\lambda},\dot{\boldsymbol{\lambda}},\mathbf{z},\dot{\mathbf{z}},\mathbf{b}) \,\big|\, \dot{\boldsymbol{\lambda}},\dot{\mathbf{z}} \big\}, \tag{3.11}$$

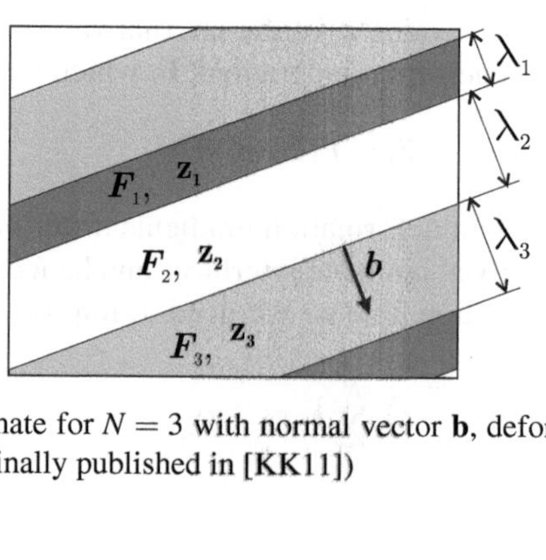

Fig. 3.1 First-order laminate for $N = 3$ with normal vector **b**, deformation gradients $\mathbf{F}_i$ and internal variables $\mathbf{z}_i$ (originally published in [KK11])

where we introduced the Lagrange functional

$$\mathscr{L}(\mathbf{F},\boldsymbol{\lambda},\dot{\boldsymbol{\lambda}},\mathbf{z},\dot{\mathbf{z}},\mathbf{b}) = \frac{\mathrm{d}}{\mathrm{d}t}\Psi^{\mathrm{rel}}(\mathbf{F},\boldsymbol{\lambda},\mathbf{z},\mathbf{b}) + \Delta^*(\boldsymbol{\lambda},\mathbf{z},\dot{\boldsymbol{\lambda}},\dot{\mathbf{z}}). \tag{3.12}$$

3.3 Single Slip Crystal Plasticity

For conciseness we will restrict ourselves here to a model of crystal plasticity involving one single slip system. For a full description of the general case, see [KK11].

The split of the deformation gradient into an elastic part $\mathbf{F}_\mathrm{e}$ and an irreversible, plastic part $\mathbf{F}_\mathrm{p}$ yields the standard multiplicative decomposition $\mathbf{F} = \mathbf{F}_\mathrm{e}\mathbf{F}_\mathrm{p}$. We will assume an incompressible neo-Hookean material possessing an energy of the form

$$\Psi(\mathbf{F}_\mathrm{e},\mathbf{p}) = \frac{\mu}{2}\left(\mathrm{tr}\,\mathbf{F}_\mathrm{e}^T\mathbf{F}_\mathrm{e} - 3\right) + \kappa|p|^\alpha, \qquad \det\mathbf{F} = 1, \tag{3.13}$$

where p denotes a hardening variable, and $\mu > 0$ is the shear modulus and $\kappa > 0$ the hardening modulus.

Plastic deformation is accommodated by dislocation gliding along specific active slip systems. Each slip system is characterized by its unit vectors $\mathbf{s}$ and $\mathbf{m}$ ($|\mathbf{s}| = |\mathbf{m}| = 1$, $\mathbf{s}\cdot\mathbf{m} = 0$), where $\mathbf{s}$ characterizes the slip direction and $\mathbf{m}$ denotes the unit vector normal to the slip plane. For a single active slip system the plastic deformation gradient can then be calculated as

$$\mathbf{F}_\mathrm{p}^{-1} = \mathbf{I} - \gamma\mathbf{s}\otimes\mathbf{m}, \tag{3.14}$$

where γ denotes the plastic slip. For the hardening variable we consider the flow rule [CKM02]

$$\dot{p} = |\dot{\gamma}| \tag{3.15}$$

with the initial condition $p(0) = 0$ (virgin initial state).

Dissipation occurs as a result of dislocation motion and is hence linked to changes of the plastic slips. For the dissipation potential with only one active slip system we simply assume [CKM02, HK08]

$$\Delta(\dot{\gamma}) = r|\dot{\gamma}| + \frac{s}{2}\dot{\gamma}^2, \tag{3.16}$$

where r is the critical resolved shear stress and s represents a viscosity parameter.

With Ψ and Δ given, the local material behavior is now completely determined by the variational principle of the minimum of the dissipation potential (3.3).

3.4 Partial Analytical Relaxation via Lamination

Let us assume a first-order laminate microstructure with N domains having interfaces with unit normal $\mathbf{b}$. We define the deformation gradient in domain i according to (3.5). To every volume fraction i there correspond values of the internal variables

γ_i and p_i. To ensure incompressibility of each laminate domain, we must enforce that for every domain i we have $\det \mathbf{F}_i = 1$, which is equivalent to

$$\mathbf{a}_i \cdot \mathbf{b} = 0. \tag{3.17}$$

Taking into account the constraints (3.7) and (3.17) by introducing Lagrange multipliers $\boldsymbol{\Lambda}$ and ρ_i, the semi-relaxed energy can be written as

$$\begin{aligned}&\Psi^{\text{rel}}(\mathbf{F}, \lambda_i, \gamma_{ij}, p_{ij}, \mathbf{b}) \\ &= \inf\Big\{ \frac{\mu}{2} \sum_i^N \lambda_i \left[\operatorname{tr} \mathbf{C}_{\mathrm{e},i} - 3 - 2\boldsymbol{\Lambda} \cdot \mathbf{a}_i - 2\rho_i \mathbf{a}_i \cdot \mathbf{b}\right] + \kappa \sum_i^N \lambda_i |p_i|^{\alpha}) \,\Big|\, \mathbf{a}_i \Big\}.\end{aligned} \tag{3.18}$$

We denote by $\mathbf{C}_{\mathrm{e},i} = \mathbf{F}_{\mathrm{e},i}^{\mathrm{T}} \mathbf{F}_{\mathrm{e},i}$ the elastic right Cauchy-Green tensor in domain i with, following (3.14) and (3.5),

$$\mathbf{F}_{\mathrm{e},i} = \mathbf{F}_i \mathbf{F}_{\mathrm{p},i}^{-1} = \mathbf{F}(\mathbf{I} + \mathbf{a}_i \otimes \mathbf{b})(\mathbf{I} - \sum_j^n \gamma_{ij} \mathbf{s}_j \otimes \mathbf{m}). \tag{3.19}$$

Minimization in (3.18) with respect to the unknown quantities $\mathbf{a}_i$ gives the relaxed energy

$$\begin{aligned}&\Psi^{\text{rel}}(\mathbf{F}, \lambda_i, \gamma_i, p_i, \mathbf{b}) = \\ &\kappa \sum_i^N \lambda_i |p_i|^{\alpha} + \frac{\mu}{2} \left[\frac{1}{\sum_i^N \frac{\lambda_i}{\mathbf{b}_i \cdot \mathbf{b}}} \left(\sum_j^N \sum_k^N \frac{\lambda_j \lambda_k \mathbf{b}_j \cdot \mathbf{C} \mathbf{b}_k}{\mathbf{b}_j \cdot \mathbf{b} \, \mathbf{b}_k \cdot \mathbf{b}} - \frac{1}{\mathbf{b} \cdot \mathbf{C}^{-1} \mathbf{b}} \right) \right. \\ &\left. + \sum_i^N \lambda_i \left(\frac{\mathbf{b}_i \cdot \mathbf{b}}{\mathbf{b} \cdot \mathbf{C}^{-1} \mathbf{b}} - \frac{\mathbf{b}_i \cdot \mathbf{C} \mathbf{b}_i}{\mathbf{b}_i \cdot \mathbf{b}} \right) + \sum_i^N \lambda_i \operatorname{tr} \left(\mathbf{F}_{\mathrm{p},i}^{-T} \mathbf{C} \mathbf{F}_{\mathrm{p},i}^{-1} \right) - 3 \right],\end{aligned} \tag{3.20}$$

where

$$\mathbf{b}_i = \mathbf{b} - \gamma_i (\mathbf{b} \cdot \mathbf{m}\, \mathbf{s} + \mathbf{b} \cdot \mathbf{s}\, \mathbf{m}) + \gamma_i^2 \mathbf{b} \cdot \mathbf{s}\, \mathbf{s}, \tag{3.21}$$

see [KK11] for details.

For instructiveness of the following examples, let us finally reduce the present model to a two-domain laminate ($N = 2$) and define the volume fraction of domain 2 by λ.

The relaxed dissipation potential, based on (3.16) still has to be determined. For a first order laminate consisting of two domains, the relaxed dissipation potential due to the change of plastic slip can be easily found by weighting the dissipation potential for each domain, given in (3.16), with the respective volume fraction. The summation yields the total relaxed dissipation potential due to plastic slip as

$$\Delta_1^*(\lambda, \dot{\gamma}_i) = (1 - \lambda) \left[r |\dot{\gamma}_1| + \frac{s}{2} \dot{\gamma}_1^2 \right] + \lambda \left[r |\dot{\gamma}_2| + \frac{s}{2} \dot{\gamma}_2^2 \right]. \tag{3.22}$$

Any change of volume fraction in the domains of the laminate is a dissipative process, since the corresponding plastic slips in the altered areas have to change as well.

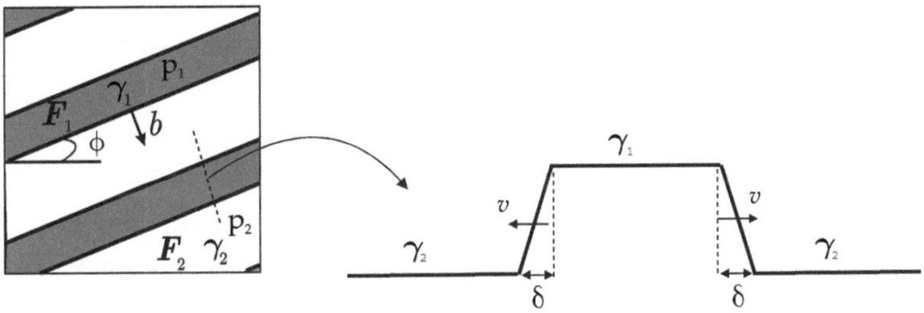

Fig. 3.2 Viscous transition zone at the moving interface between regions with different internal variables

Hence, it contributes to the entire dissipation potential. We assume here that the viscous part enters the dissipation by a viscous transition zone between the domains instead of a sharp interface between the domains (Figure 3.2). This zone moves with velocity v in the normal direction and has the width δ (volume ratio). The velocity can then be expressed in terms of the rate of λ by

$$v = \frac{\dot{\lambda}}{2}. \tag{3.23}$$

Exploiting the intercept theorem, we find

$$\dot{\gamma} = \frac{|\gamma_2 - \gamma_1|}{2\delta}\dot{\lambda} \tag{3.24}$$

which allows us to calculate the relaxed dissipation potential due to the change of volume fraction as

$$\Delta_2^*\left(\gamma_i, \dot{\lambda}\right) = r|\gamma_2 - \gamma_1|\left|\dot{\lambda}\right| + \frac{s}{2}\frac{(\gamma_2 - \gamma_1)^2}{2\delta}\dot{\lambda}^2. \tag{3.25}$$

The combined dissipation potential is then given as

$$\Delta^*(\lambda, \gamma_i, \dot{\lambda}, \dot{\gamma}_i) = \Delta_1^*(\lambda, \dot{\gamma}_i) + \Delta_2^*\left(\gamma_i, \dot{\lambda}\right), \tag{3.26}$$

and the Lagrange functional reads

$$\mathscr{L}(\mathbf{F}, \lambda, \gamma_i, p_i, \dot{\lambda}, \dot{\gamma}_i, \mathbf{b}) = \frac{\mathrm{d}}{\mathrm{d}t}\Psi^{\mathrm{rel}}(\mathbf{F}, \lambda, \gamma_i, p_i, \mathbf{b}) + \Delta^*(\lambda, \gamma_i, \dot{\lambda}, \dot{\gamma}_i). \tag{3.27}$$

Here, one of the major differences of the present model to previous approaches becomes apparent from the last term in (3.26): a change of the volume fractions (here, of λ) causes dissipation. However, we do not consider the dissipation required to transform some region with originally no plastic history into a part of the increasing domain (an increase of domain 2 would then mean $\dot{\lambda} > 0$ and $\Delta^* = r|\dot{\lambda}\,\gamma_2|$). Instead, we correctly account for the transformation of some part originally belonging to domain 1 into a part of the increasing domain 2. Therefore, the amount of dissipation depends on the microstructure at the beginning of each time step.

3.5 Rate-Independent Evolution

We have now all ingredients in hand to model the evolution of laminate microstructures. We will start with the rate-independent case, setting $s = 0$ in the expressions for the dissipation potentials given in (3.22) and (3.25). We discuss separately the three stages of initiation, rotation, and continuous evolution if laminate microstructures.

3.5.1 Evolution Equations

Via the principle of the minimum of the dissipation potential given in (3.3) we arrive at evolution equations for λ and γ_i, i.e., the stationarity conditions from minimizing the above Lagrange functional, which read

$$0 \in \frac{\partial \Psi^{\mathrm{rel}}}{\partial \lambda} + \frac{\partial \Delta^*}{\partial \dot{\lambda}}, \tag{3.28}$$

$$0 \in \frac{\partial \Psi^{\mathrm{rel}}}{\partial \gamma_i} + \frac{\partial \Psi^{\mathrm{rel}}}{\partial p_i} \operatorname{sign} \dot{\gamma}_i + \frac{\partial \Delta^*}{\partial \dot{\gamma}_i}, \quad \text{for all } 1 \le i \le N. \tag{3.29}$$

There is an aspect concerning the evolution of the hardening parameters p_i we still have to discuss. Any change of λ results in mixing the formerly pure domains in a small part of the body; e.g. an increase of λ in the two-domain laminate will raise the volume fraction of domain 2 (with history variables p_{2j}) by adding certain regions which were previously associated with domain 1 and hence exhibited history variable values p_{1j}. As a consequence, the hardening histories p_{2j} should be updated to account for a mixture of the two domains (see Fig. 3.3). We propose to obtain the updated p_i-values by computing the energetic average of the original values weighted by the volume fractions. For example for a single active slip system and for $\lambda_{n+1} = \lambda_n + \Delta\lambda$ and $\Delta\lambda > 0$ we have

$$(\lambda_n + \Delta\lambda) p^{\alpha}_{2,n+1} = \lambda_n \, p^{\alpha}_{2,n} + \Delta\lambda \, p^{\alpha}_{1,n}, \qquad p_{1,n+1} = p_{1,n}, \tag{3.30}$$

and analogously for $\Delta\lambda < 0$

$$(1 - \lambda_n - \Delta\lambda) p^{\alpha}_{1,n+1} = (1 - \lambda_n) p^{\alpha}_{1,n} - \Delta\lambda \, p^{\alpha}_{2,n}, \qquad p_{2,n+1} = p_{2,n}. \tag{3.31}$$

The corresponding formulations for multiple active slip systems are analogous and assume no interaction between the different slip systems at this step (cross-hardening is accounted for by the choice of Ψ_{p}). So, we omit these lengthy equations here for conciseness; see [HK09, KK11] for details.

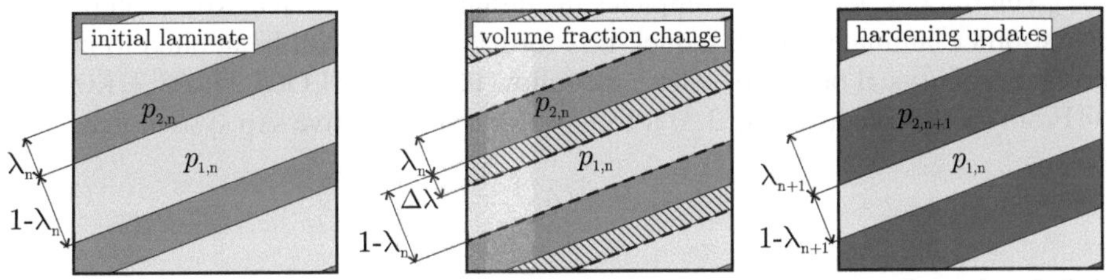

Fig. 3.3 Schematic view of the update procedure: as the volume fractions change in the initial laminate (here, the volume fraction of domain 2, λ, increases), the updated, internal hardening parameters are obtained from energetic averaging of the initial values (originally published in [KK11]).

3.5.2 Laminate Rotation

As can be seen from Fig. 3.4, once a laminate microstructure exists a change of the laminate orientation (i.e., a change of **b**) results in changes of the plastic slip in certain regions of the deformed body and is hence associated with a specific amount of dissipation given by

$$D_{\mathbf{b}}(\boldsymbol{\lambda},\mathbf{z}) = \sum_{i,j=1}^{N} \lambda_i \lambda_j D(\mathbf{z}_i,\mathbf{z}_j). \tag{3.32}$$

The dissipation $D_{\mathbf{b}}$ is proportional to the area of those regions which change their domain membership upon rotation and to the dissipation distance required to turn a region of domain i into a part of domain j. We assume that a jump in orientation will take place as soon as it becomes energetically favorable. This gives

$$\inf\left\{\Psi^{\text{rel}}(\mathbf{F},\boldsymbol{\lambda},\mathbf{z},\mathbf{b}_{n+1}) - \Psi^{\text{rel}}(\mathbf{F},\boldsymbol{\lambda},\mathbf{z},\mathbf{b}_n) \,\middle|\, \mathbf{b}_{n+1} \,:\, |\mathbf{b}_{n+1}| = 1\right\} + D_{\mathbf{b}}(\boldsymbol{\lambda},\mathbf{z}) \leq 0 \tag{3.33}$$

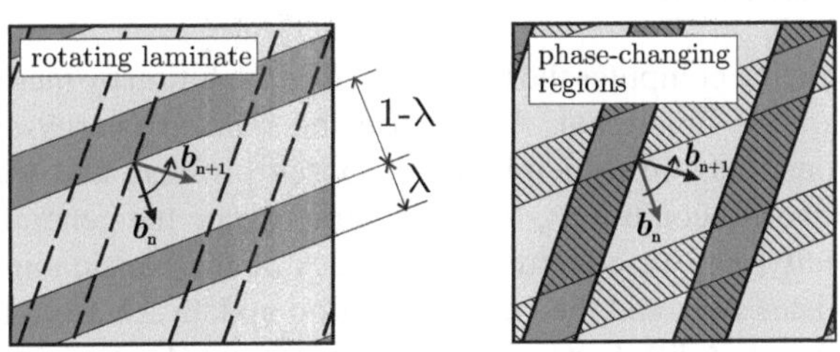

Fig. 3.4 Rotation of the original laminate (for simplicity with only two domains with volume fractions λ and $1-\lambda$) from the old normal vector $\mathbf{b}_n$ to the new normal vector $\mathbf{b}_{n+1}$. The right graphic highlights the hatched regions which have changed their domain membership upon rotation and hence caused dissipation (originally published in [KK11]).

for given $\mathbf{b}_n, \boldsymbol{\lambda}, \mathbf{z}$. Eq. (3.33) completes the description of the inelastic evolution of a first-order laminate. In the sequel, this formal concept will be applied to model problems of crystal plasticity, where we follow ideas of [MLG04, HK08, HK09].

From (3.32) (see also Fig. 3.3) it follows for a single active slip system that

$$D_{\mathbf{b}}(\lambda, \gamma_1, \gamma_2) = 2r\lambda(1-\lambda)|\gamma_1 - \gamma_2|. \tag{3.34}$$

3.5.3 Laminate Initiation

A crucial issue is the initiation of the laminate microstructure from the originally uniform crystal. We can treat this laminate initiation as follows. At the beginning of each time increment, one computes the driving force $q_\lambda = -\partial\Psi/\partial\lambda$ with respect to the volume fractions in the limit of a marginal amount of domain 2, i.e. for single-slip

$$q_0(\mathbf{F}, \gamma_1, \gamma_2, p_1, p_2, \mathbf{b}) = \lim_{\lambda \to 0} q_\lambda(\mathbf{F}, \lambda, \gamma_1, \gamma_2, p_1, p_2, \mathbf{b}). \tag{3.35}$$

Maximizing this driving force with respect to $\mathbf{b}_{n+1}$ and $\gamma_{2,n+1}$, one can determine the energetically favored values of these quantities in the arising domain 2, i.e.

$$(\mathbf{b}_{n+1}, \gamma_{2,n+1}) = \\ \operatorname{argmax}\left\{ q_0(\mathbf{F}_{n+1}, \gamma_{1,n}, \gamma_{2,n+1}, p_{1,n}, p_{2,n+1}, \mathbf{b}_{n+1}) \;\middle|\; p_{2,n+1} = |\gamma_{2,n+1}|,\ |\mathbf{b}_{n+1}| = 1 \right\}. \tag{3.36}$$

One then determines the actual value of λ_{n+1} by solving

$$r|\gamma_{1,n} - \gamma_{2,n+1}| = q_\lambda(\mathbf{F}_{n+1}, \lambda_{n+1}, \gamma_{1,n}, \gamma_{2,n+1}, p_{1,n}, p_{2,n+1} = |\gamma_{2,n+1}|, \mathbf{b}_{n+1}). \tag{3.37}$$

If there exists a solution λ_{n+1}, a laminate forms with domain 2 having the determined values of λ_{n+1}, $\gamma_{2,n+1}$ and $\mathbf{b}_{n+1}$.

3.5.4 Numerical Scheme

Our numerical scheme computes the microstructure evolution by incrementally minimizing the Lagrange functional. As we use the relaxed energy and dissipation potential, this constitutes in principle a well-posed problem and we can resort to solving the stationarity conditions. We here demonstrate the general procedure for single-slip plasticity, i.e. we compute the updates of the plastic slips $\Delta\gamma_i$, the history variable updates Δp_i, and the volume fraction update $\Delta\lambda$ from the stationarity conditions (3.28) and (3.29). For a given load increment $[\mathbf{F}_n, \mathbf{F}_{n+1}]$, each step starts with the current state as initial values $\lambda_n, \gamma_{i,n}, p_{i,n}$, and solves the stationarity conditions in order to update all internal variables at time t_{n+1} with known load $\mathbf{F}_{n+1}$. For the initially homogeneous material the interface normal $\mathbf{b}_{n+1}$ as well as the internal variables of the originating second laminate domain, λ_{n+1}, $\gamma_{2,n+1}$ and $p_{2,n+1}$, are determined from (3.36). Once a laminate has formed, the evolution of the internal variables λ, γ_i and p_i is computed using a staggered scheme. In a first step a time-discretized version of (3.28) is solved for the increment $\Delta\lambda$ for fixed γ_1 and γ_2. Afterwards, p_1 and p_2 are updated via (3.30) or (3.31). Then, in a second step, (3.29) are solved for the increments $\Delta\gamma_1$ and $\Delta\gamma_2$ for fixed λ. Finally, the updated values of $\lambda, \gamma_1, \gamma_2, p_1, p_2$ are transferred to the next time-step.

Note that the order of solving the stationarity conditions is only of minor importance as long as the the load increment is kept small, which we tacitly assume. With increasing load increments the order of solution gains influence; in particular, the initial laminate formation requires very small increments to capture the actual onset of lamination and thus the correct variables in the newly forming laminate domain.

3.6 Simulation of Rotating Laminates

Due to the special nature of the dissipation term (3.32) and the criterion (3.33) laminate rotation tends to occur in a discontinuous manner, remnant of the stick-slip behavior of dry friction. In order to study this phenomenon a simplified model adapted to this task has been introduced in [HHM12]. We repeat these results here, adapted to the notation of the present work.

Let us consider a particular case of the model introduced in Section 3.3 characterized by the fact that the plastic slip is only allowed to assume two distinct values given by $\gamma = \pm\gamma_c$. Moreover, for simplicity we set $\kappa = 0$ in (3.13), i.e. allow no hardening, and $s = 0$ in (3.16), i.e. restrict ourselves to the rate-independent case. This may constitute a model for a (fictitious) shape-memory-alloy possessing only two martensitic variants with transformation or Bain strains given by $I \pm \gamma_c\, \mathbf{s}\otimes\mathbf{m}$.

Under these assumptions the relaxed dissipation potential reduces to

$$\Delta^*(\dot{\lambda}) = 2r\gamma_c|\dot{\lambda}|. \tag{3.38}$$

The dissipation distance for rotation (3.34) becomes

$$D_{\mathbf{b}}(\lambda) = 4r\gamma_c\lambda(1-\lambda). \tag{3.39}$$

The laminate energy, i.e. the argument of the minimization in (3.18) can be specified as

$$\begin{aligned}\Psi^{\mathrm{lam}}(\mathbf{F},\lambda,\mathbf{a},\mathbf{b}) = {} & \lambda\,\Psi(\mathbf{F}(I+(1-\lambda)\mathbf{a}\otimes\mathbf{b})(I+\gamma_c\mathbf{s}\otimes\mathbf{m})) \\ & +(1-\lambda)\,\Psi(\mathbf{F}(I-\lambda\,\mathbf{a}\otimes\mathbf{b})(I-\gamma_c\mathbf{s}\otimes\mathbf{m})),\end{aligned} \tag{3.40}$$

where $\mathbf{a}\cdot\mathbf{b} = 0$. In the plane-strain case, we can specify the quantities above as $\mathbf{b} = (\cos\phi, \sin\phi)$ and $\mathbf{a} = a_0\,(-\sin\phi, \cos\phi)$, $\mathbf{m} = (\cos\bar{\phi}, \sin\bar{\phi})$, $\mathbf{s} = (-\sin\bar{\phi}, \cos\bar{\phi})$. The laminate microstructure is then fully described by the parameters a_0, λ and ϕ.

In the absence of dissipation the energy will be minimized with respect to all possible laminates resulting in a relaxed energy of the form

$$\Psi^{\mathrm{rel,full}}(\mathbf{F},\lambda,\phi) = \inf\{\,\Psi^{\mathrm{lam}}(\mathbf{F},\lambda,\mathbf{a})\,|\,\lambda\in[0,1], a_0\in\mathbb{R}, \phi\in[0,\pi)\,\}. \tag{3.41}$$

However, with dissipation present in the model we calculate the partially relaxed energy as

$$\Psi^{\mathrm{rel}}(\mathbf{F},\lambda,\phi)=\inf\{\,\Psi^{\mathrm{lam}}(\mathbf{F},\lambda,\mathbf{a})\,|\,a_0\in\mathbb{R}\,\}. \tag{3.42}$$

The principle of the minimum of the dissipation potential (3.3) now gives

$$0=\frac{\partial\Psi^{\mathrm{lam}}}{\partial a_0},\quad 0\in\frac{\partial\Psi^{\mathrm{lam}}}{\partial\lambda}+r\,\mathrm{sign}\,\dot{\lambda}. \tag{3.43}$$

In a time-incremental setting this can be written as

$$0=\frac{\partial\Psi^{\mathrm{lam}}}{\partial a_{0n}},\quad 0\in\frac{\partial\Psi^{\mathrm{lam}}}{\partial\lambda_n}+r\,\mathrm{sign}\,(\lambda_n-\lambda_{n-1}), \tag{3.44}$$

where the subscripts $n-1$ and n denote the values of the specific quantities at the beginning and at the end of the time-increment considered.

The condition for rotation of the laminate (3.33) becomes

$$f(\mathbf{F}_n,\lambda_n,\phi_{n-1},\phi_n)$$
$$=\Psi^{\mathrm{rel}}(\mathbf{F}_n,\lambda_n,\phi_n)-\Psi^{\mathrm{rel}}(\mathbf{F}_n,\lambda_n,\phi_{n-1})+4r\gamma_c\lambda_n(1-\lambda_n)<0. \tag{3.45}$$

Note that in [HHM12] a more general expression is given where a change in the parameter λ during rotation is taken into account. The evolution of ϕ is now given as

$$\phi_n=$$
$$\begin{cases}\operatorname{argmin}\{\,f(\mathbf{F}_n,\lambda_n,\phi_{n-1},\phi)\,|\,\phi\,\} & \text{for } \inf\{\,f(\mathbf{F}_n,\lambda_n,\phi_{n-1},\phi)\,|\,\phi\,\}<0\\ \phi_{n-1} & \text{else}\end{cases}. \tag{3.46}$$

Given $\mathbf{F}_n$, $a_{0,n-1}$, λ_{n-1}, ϕ_{n-1}, the equations (3.43) and (3.46) can be solved for λ_n, $a_{0,n}$, ϕ_n. This allows one to compute the evolution of λ and ϕ in a time-incremental manner.

As an example we present a simple shear test of the form $\mathbf{F}=\begin{pmatrix}1 & 0\\ \xi(t) & 1\end{pmatrix}$, where $\xi(t)=t$ for $t\in[0,2]$. The model parameters chosen are: $\mu=75\mathrm{MPa}$, $\gamma_c=0.2$, $2r\gamma_c=1\mathrm{MPa}$, $\bar{\phi}=\pi/10$. Hence, the inelastic shearing deformation is misaligned with respect to the applied shear.

In Figure 3.5 the laminate rotation angle ϕ is displayed as a function of ξ, once as result of the minimization in (3.41), and once as result of the time-incremental procedure in (3.46). The same is done for the volume ratio λ as function of ξ in Figure 3.6. In Figure 3.7 the difference in λ of the results from (3.43) and (3.46) is shown.

It can be seen that ϕ starts to deviate from the solution found by minimization, until finally the inequality (3.45) is satisfied. Then the minimization result is retrieved in a sudden way. This process repeats itself in a stick-slip-type behavior. After every jump in ϕ, the variable λ remains constant within a certain interval, until the differential inclusion in (3.43) becomes nontrivial again.

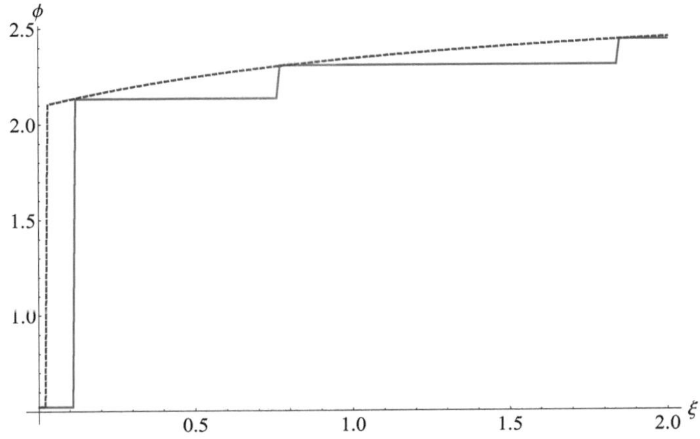

Fig. 3.5 Evolution of ϕ as function of ξ. Dashed line: minimizer of $\Psi^{\mathrm{rel,full}}$ in (3.41), solid line: solution via time-incremental evolution.

3.7 Viscous Evolution

The rate-independent model captures the microstructural characteristics of a first order laminate as demonstrated in [HK08, KK11]. However, a drawback of this model is that the numerical calculations are cumbersome and often unstable. These problems occur because calculating the evolution of the microstructure involves a global minimization in every time step. However, the functional to be minimized typically has extremely many local minima. Therefore, this approach leads to good

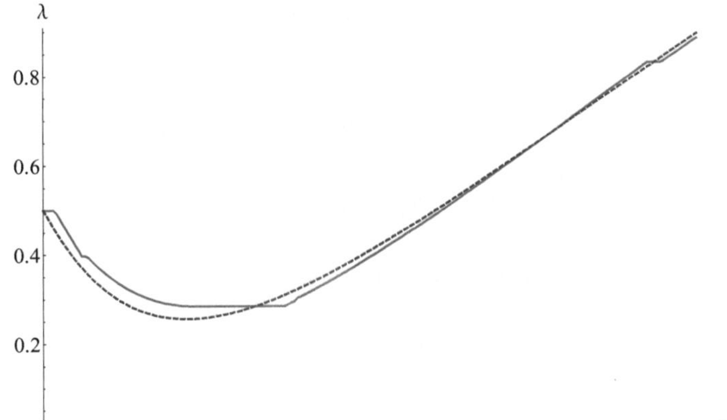

Fig. 3.6 Evolution of λ as function of ξ. Dashed line: minimizer of $\Psi^{\mathrm{rel,full}}$ in (3.41), solid line: solution via time-incremental evolution.

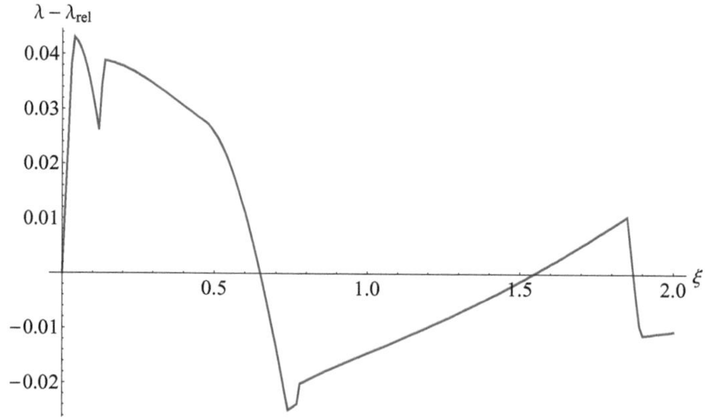

Fig. 3.7 Difference of λ as function of ξ between the minimizer of $\Psi^{\mathrm{rel,full}}$ in (3.41) and the solution via time-incremental evolution.

results on the material point level but the application on macroscopic specimens is questionable.

The model can be rendered purely local in all aspects by extending the dissipation potential by quadratic terms in the rates of the internal variables, i.e. by setting $s \neq 0$ in in (3.22) and (3.25). Note that no change is necessary concerning the relaxed energy (3.20).

3.7.1 Evolution Equations

The evolution equations for the history variables are found using the principle of the minimum of the dissipation potential (3.3). For the respective dissipative quantities λ and γ_i, these read

$$0 \in \frac{\partial \Psi^{\mathrm{rel}}}{\partial \lambda} + r\,|\gamma_2 - \gamma_1|\,\mathrm{sign}\,\dot{\lambda} + \frac{s}{2\delta}\,(\gamma_2 - \gamma_1)^2\,\dot{\lambda}, \tag{3.47}$$

$$0 \in \frac{\partial \Psi^{\mathrm{rel}}}{\partial \gamma_i} + r\lambda_i\,\mathrm{sign}\,\dot{\gamma}_i + s\lambda_i\,\dot{\gamma}_i. \tag{3.48}$$

Equations (3.47) and (3.48) can be solved analytically to give

$$\dot{\lambda} = -\frac{2\delta}{s\,(\gamma_1 - \gamma_2)^2}\left(\left|\frac{\partial \Psi^{\mathrm{rel}}}{\partial \lambda}\right| - r\,|\gamma_1 - \gamma_2|\right)_+ \mathrm{sign}\,\frac{\partial \Psi^{\mathrm{rel}}}{\partial \lambda}, \tag{3.49}$$

and

$$\dot{\gamma}_i = -\frac{1}{s\lambda_i}\left(\left|\frac{\partial \Psi^{\mathrm{rel}}}{\partial \gamma_i}\right| - r\lambda_i\right)_+ \mathrm{sign}\,\frac{\partial \Psi^{\mathrm{rel}}}{\partial \gamma_i}, \tag{3.50}$$

where $(\cdot)_+$ denotes the positive part of the respective expression.

The update of the hardening parameters p_i is then performed in the same manner as in the rate-independent case according to (3.30) and (3.31).

3.7.2 *Laminate Rotation*

Our aim is to replace the global criterion (3.33) by a local formulation. For this purpose, let the normal vector **b** be parametrized by its angle ϕ with respect to a reference direction. Then for fixed $\mathbf{F}, \lambda_i, \gamma_i, p_i$ the relaxed energy can be written as

$$\psi^{\mathrm{rel}}(\phi) = \Psi^{\mathrm{rel}}(\mathbf{F}, \lambda, \gamma_i, p_i, \mathbf{b}). \tag{3.51}$$

Condition (3.33) can be expressed as

$$\inf\{\, \psi(\phi_{n+1}) - \psi(\phi_n) \mid \phi_{n+1} \} + 2r\lambda(1-\lambda)|\gamma_1 - \gamma_2| \leq 0. \tag{3.52}$$

In order to be consistent with the viscous approach we would like the evolution of ϕ to be governed by a dissipation potential of the form

$$\Delta_\phi^{\mathrm{rel}} = a(\lambda, \gamma_1, \gamma_2)\, |\dot{\phi}| + b(\lambda, \gamma_1, \gamma_2)\, \dot{\phi}^2. \tag{3.53}$$

Unfortunately, there is up to date no canonical way of determining the functions $a(\lambda, \gamma_1, \gamma_2)$ and $b(\lambda, \gamma_1, \gamma_2)$. We therefore have a this point to resort to heuristic arguments. Inspired by the expressions in (3.34) and (3.25) we set

$$\begin{aligned} a(\lambda, \gamma_1, \gamma_2) &= \alpha D_{\mathbf{b}}(\lambda, \gamma_1, \gamma_2) = 2\alpha r\lambda\,(1-\lambda)|\gamma_1 - \gamma_2|\,|\dot{\phi}|, \\ b(\lambda, \gamma_1, \gamma_2) &= \qquad\qquad\qquad\quad \beta\lambda\,(1-\lambda)(\gamma_1 - \gamma_2)^2\,\dot{\phi}^2, \end{aligned} \tag{3.54}$$

introducing positive parameters α, β. These terms have the advantage of possessing the correct limit behavior for $\lambda \to 0$, $\lambda \to 1$ and $\gamma_1 - \gamma_2 \to 0$.

Using the proposed form of the dissipation potential, an evolution equation for ϕ, analogous in form to (3.49) and (3.50), is obtained as

$$\dot{\phi} = -\frac{1}{2\beta\,(1-\lambda)(\gamma_1 - \gamma_2)^2} \left(\left| \frac{\partial \psi^{\mathrm{rel}}}{\partial \phi} \right| - 2\alpha r\lambda\,(1-\lambda)|\gamma_1 - \gamma_2| \right)_+ \operatorname{sign} \frac{\partial \psi^{\mathrm{rel}}}{\partial \phi}. \tag{3.55}$$

Similar to the rate-independent case, the hardening parameters p_i have to be updated according to the approach in [KK11], as given in (3.30) and (3.31).

3.7.3 *Laminate Initiation*

The Equations (3.49, 3.50, 3.55) in combination with (3.30) or (3.31), and (3.15) form the final system of model equations that have to be solved. However, a closer inspection of these equations reveals that (3.49) and (3.50) are not well-defined in the case $\lambda = 0$ or $\lambda = 1$, or $\gamma_1 = \gamma_2$. Hence, in correspondence to the rate-independent model, an initiation scheme for the microstructure in terms of an initial volume fraction and plastic slip is necessary.

We will do this in a similar fashion as in subsection 3.5.3, with the substantial difference, that the initiated value $\gamma_{2,n+1}$ in the newly created laminate might be arbitrarily far from $\gamma_{1,n}$, while now it will only be allowed to deviate from it by a small amount, resulting in a continuous evolution of all internal variables.

To be precise, let us compute the driving force for laminate initiation as

$$q_1(\mathbf{F}, \gamma_1, p_1, \mathbf{b}) = \lim_{\substack{\lambda \to 0 \\ \gamma^* \to 0}} \frac{1}{\gamma^*} q_\lambda(\mathbf{F}, \lambda, \gamma_1, \gamma_2 = \gamma_1 + \gamma^*, p_1, p_2 = p_1 + |\gamma^*|, \mathbf{b}), \quad (3.56)$$

where once again $q_\lambda = -\partial \Psi / \partial \lambda$. Note the substantial difference to (3.35), where γ_2 could assume arbitrary values.

Maximizing this driving force with respect to $\mathbf{b}_{n+1}$ gives the direction in which the laminate is formed, i.e.

$$\mathbf{b}_{n+1} = \operatorname{argmax} \left\{ q_1(\mathbf{F}_{n+1}, \gamma_{1,n}, p_{1,n}, \mathbf{b}_{n+1}) \,\middle|\, |\mathbf{b}_{n+1}| = 1 \right\}. \quad (3.57)$$

In a subsequent step we set $\gamma^* = \gamma_{\text{ini}} \operatorname{sign} q_1(\mathbf{F}_{n+1}, \gamma_{1,n}, p_{1,n}, \mathbf{b}_{n+1})$, where $\gamma_{\text{ini}} \ll 1$ is a small fixed value, and determines λ_{n+1} by solving

$$r|\gamma_{\text{ini}}| = q_\lambda(\mathbf{F}_{n+1}, \lambda_{n+1}, \gamma_{1,n}, \gamma_{2,n+1} = \gamma_{1,n} + \gamma_{\text{ini}}, p_{1,n}, p_{2,n+1} = |\gamma_{2,n+1}|, \mathbf{b}_{n+1}). \quad (3.58)$$

If there exists a solution λ_{n+1}, a laminate forms with domain 2 having the determined values of λ_{n+1}, $\gamma_{2,n+1}$ and $\mathbf{b}_{n+1}$.

3.8 Comparison of the Laminate Evolution for the Rate-Independent Case and the Viscosity Limit

For the computations shown in the sequel we will use the so-called viscosity limit for comparison, i.e. the limit of the viscous evolution for vanishing loading velocities. Numerically this will be realized by calculating a fixed number of time-increments with constant external load w, before w is increased again. We control the loading velocity by introducing a factor ϑ. This quantity is an integer that gives the number of updates that we perform for a fixed external load w. The loading velocity is then given as $\Delta w/(\vartheta \Delta t)$. In the examples presented here we choose $\vartheta = 10$.

The numerical schemes outlined above (rate-independent and viscous approach) can be applied to arbitrary materials, for which the relaxed energy density is known. As the energy density employed in the preceding sections requires incompressible material behavior, we restrict our following examples to volume-preserving deformation paths only. We present results from applying the developed procedures to different exemplary problems.

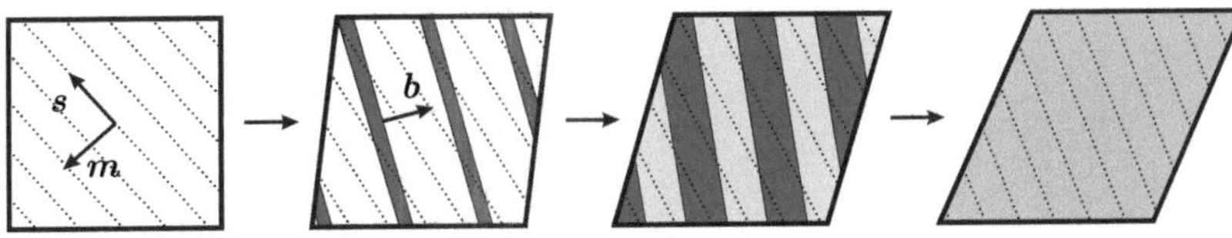

Fig. 3.8 Microstructure development of a first-order laminate for a simple shear test with a single active slip system. Two bifurcated laminate domains arise with common surface normal **b**. The newly nucleated domain 2 exhibits finite plastic slip already at the onset of lamination whereas the original domain 1 remains elastic first and eventually yields plastically, too. Volume fractions develop to finally recover the crystal in a stable homogeneous state with uniform plastic slip.

For all calculations we use $\mu = 2\,\text{GPa}$, $r = 1\,\text{MPa}$ and $\alpha = 4$ in (3.13). The active slip systems are characterized for plane problems by the orientation angle $\bar{\phi}$ denoting the angle of the slip direction **s** with the x-axis.

Figure 3.8 illustrates the general nature of solutions obtained for those problems considered here. First, the crystal behaves in a homogeneous elastic manner. At the onset of lamination, a second domain arises out of the originally uniform single crystal. This newly nucleated domain exhibits a finite amount of plastic slip already, whereas the original domain may still evolve elastically, and it occupies only a small volume fraction of the crystal. (Depending on the non-aligned slip system, plastic flow may also occur before the onset of lamination.) Upon further loading both domains eventually exhibit plastic flow and all internal variables evolve. Finally, only one domain remains as soon as the external strain reaches values at the relaxed and the unrelaxed energy coincide, leaving the crystal in a homogeneous stable state with uniform plastic slip and uniform hardening variables.

The first example shown investigates the microstructure evolution during a plane-strain simple shear test parametrized by the macroscopic deformation gradient

$$\mathbf{F} = \begin{pmatrix} 1 & w & 0 \\ 0 & 1 & 0 \\ 0 & 0 & 1 \end{pmatrix}. \tag{3.59}$$

This deformation is schematically sketched in Figure 3.9. The results for both presented approaches are determined with $\kappa = 0.1\,\text{GPa}$, the first approach is computed with constant load increments of $\Delta w = 5 \cdot 10^{-4}$ up to a maximum of $w_{\max} = 2.8$. The exact step size of the load increment is here of minor importance as long as the increment is kept small. (This is of particular importance for finding the initial laminate.) For the viscous approach, the step size is considered as $\Delta w = 1 \cdot 10^{-3}$. The slip system for both approaches is oriented under an angle of $\bar{\phi} = 135°$. For the viscous approach the parameters are considered as $s = rt^\star$ with $t^\star = 0.01\,\text{s}$, and $\delta = rj$ with $j = 0.1\,\text{mm}^3/\text{N}$.

Because of the non-aligned slip system the material stability of the homogeneous deformation is lost for both approaches and microstructures arise. The evolution of the volume fraction of the second domain λ and the plastic slips γ_1 and γ_2 for both

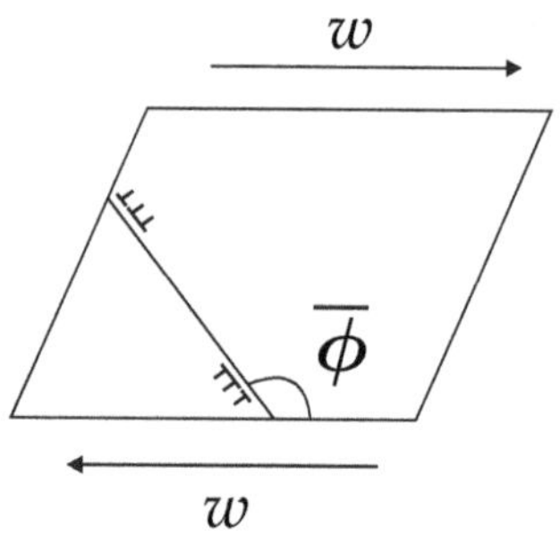

Fig. 3.9 Plane-strain simple shear test

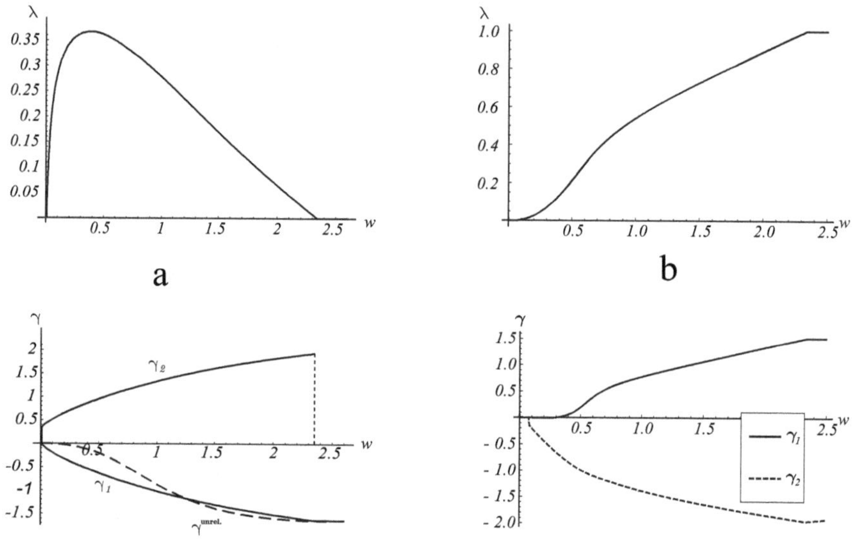

Fig. 3.10 Volume fraction λ of domain 2 and plastic slips γ_1 and γ_2 of both domains for: (a) Rate-independent approach, (b) viscous approach, plastic slip $\gamma^{\text{unrel.}}$ for homogeneous evolution.

approaches is shown in Fig. 3.10. For the rate-independent approach the volume fraction of the second domain begins almost directly to evolve and reaches a maximum value of $\lambda = 0.36$ at $w = 0.4$. Then the volume fraction decreases again until $\lambda = 0$ at $w = 2.35$, from thereon the material is homogeneous again. For the viscous model lamination will set in at approximately $w = 0.10$ when the volume fraction of domain two starts to evolve. Therefore in comparison to the rate-independent model, the initiation is delayed. This had to be expected because the plastic slip γ_2 has to evolve continuously now. At approximately $w = 2.35$, a uniform microstructure has established that only consists of the second domain of the laminate. At approximately $w = 2.35$, a uniform microstructure has established that only consists of the second domain of the laminate.

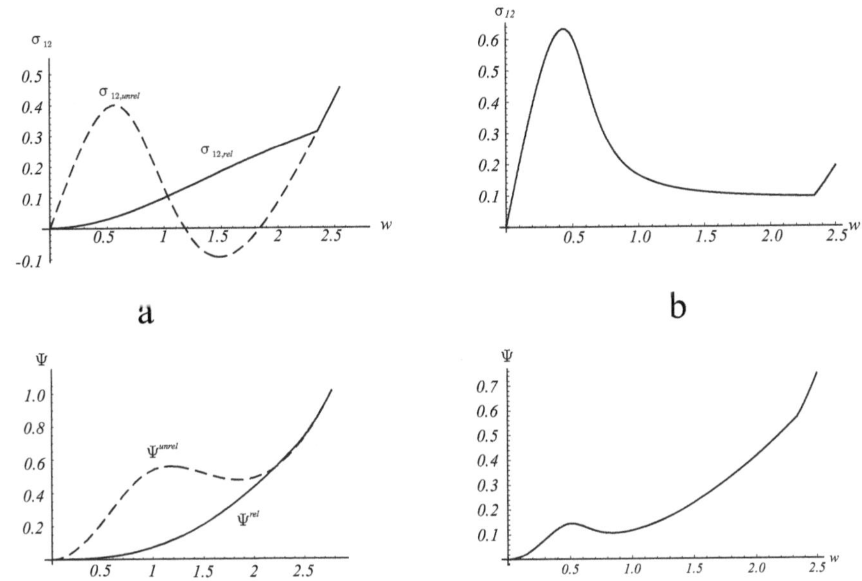

Fig. 3.11 Cauchy shear stress and free energy for: (a) rate-independent approach, (b) viscous approach

Simultaneously to the evolution of the volume fraction, plastic slip evolves in both domains of the laminate. Both models lead to similar results: For the rate-independent model, the plastic slip in both domains starts to evolve directly. The plastic slip in domain one decreases from zero to $\gamma_1 = -1.6$. The plastic slip in domain two jumps to $\gamma_2 = 0.4$ and then increases further to $\gamma_2 = 2.0$. For the viscous model, the evolution of the plastic slip in domain one is drastically delayed due to the viscous effects and starts at $w = 0.35$. Thereafter, the plastic slip increases to $\gamma_2 = 1.4$, then the material consists only of the first domain. Except the delay at the onset of the microstructure, the values of the plastic slip of the first domain correspond to the absolute values resulting for the rate-independent model, only the sign differs. The plastic slip in domain two is negative and jumps from zero to $\gamma_2 = -0.15$. During further loading, its absolute value increases slowly to $|\gamma_2| = 1.9$, which is again similar to the absolute value of the results of the rate-independent model.

Along with the evolving microstructure, the resulting Cauchy stresses

$$\sigma = \mu \left(\mathbf{F}_e \mathbf{F}_e^{-T} - \mathbf{I} \right) \tag{3.60}$$

can be calculated for both models. In the upper part of Fig. 3.11 the Cauchy shear stresses, determined (a) with the rate-independent and (b) with the viscous approach, are presented. On the left side, the stresses of the rate-independent model are shown. While the microstructure evolves, the resulting stress increases due to the hardening slowly until the crystal is homogeneous again. From there on the stress increases with a larger slope. On the right hand side, the stresses of the viscous approach are presented: during the first part of the shearing, the stress increases with slightly decreasing slope. This follows from the non-linear neo-Hookean energy that we choose. When the first domain of the laminate has evolved, hence plastic slip occurs, the slope of the stress is reduced but still increasing. The total amount of plastic deformation during this period remains still too small to influence the stresses.

However, when the plastic slip also evolves in domain one (in this case the larger domain), the stress drops quite drastically. This surprising effect can be explained in the following way. In the rate-independent model the laminate is initiated at a load when the relaxed energy starts to differ from the original one. In the viscosity limit, this happens essentially when the energy looses ellipticity, i.e. much later. This means, that the energy will then increase with a smaller rate during laminate evolution to compensate for the higher starting value, resulting in smaller stresses. This phenomenon is strong enough to counter even hardening effects. A similar effect has been observed in [YBG11] using a related model based on energy relaxation.

At a loading larger than $w > 2.35$, there exists no further material that could transform to domain two. Consequently, stress starts to increase again. Summarized, the evolutions of the shear stresses for the two models are not very closely related, as the precise values of the microstructure parameters have a great impact.

The free energies for both approaches are displayed in the lower part of Fig. 3.11. The graph for the rate-independent model is convex. Due to the viscous delay of the microstructure, the energy for the viscous model still exhibits a non convexity at the onset and the ending of the microstructure. In addition, the energy obtained by the viscous model is smaller than that for the rate-independent one which is physically expected due to the additional term in the dissipation potential.

As a second example we investigate the microstructure evolution for a plane-strain tension-compression test with the macroscopic deformation gradient

$$\mathbf{F} = \begin{pmatrix} 1+w & 0 & 0 \\ 0 & 1/(1+w) & 0 \\ 0 & 0 & 1 \end{pmatrix}. \tag{3.61}$$

The loading is illustrated in Fig. 3.12. Computations were carried out with $\kappa = 0.01\,\text{GPa}$ and the rate-independent model is tested with constant load increments $\Delta w = 2 \cdot 10^{-4}$ up to the maximum load of $w_{\text{max}} = 2.5$. For the viscous model, the parameters coincide with those of the shear test. In both models, the slip system is oriented under an angle of $\bar{\phi} = 70°$. Again, due to the loss of rank-one convexity, the homogeneous deformation state becomes unstable and decomposes into a laminate. The microstructure evolution, including the evolution of plastic slips and the volume fractions, is summarized in Fig. 3.13. On the left side, the presented evolution is based on the rate-independent ansatz, on the right side, the evolution is resulting from the viscous ansatz.

In the upper part of Fig. 3.13 the evolution of the volume fraction of domain two is presented for both models. Both results show similar characteristics: already for a very small amount of loading a microstructure is established and the second domain

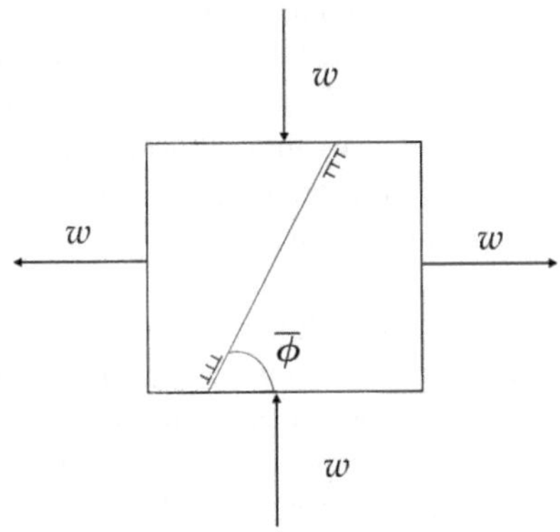

Fig. 3.12 Tension-compression test

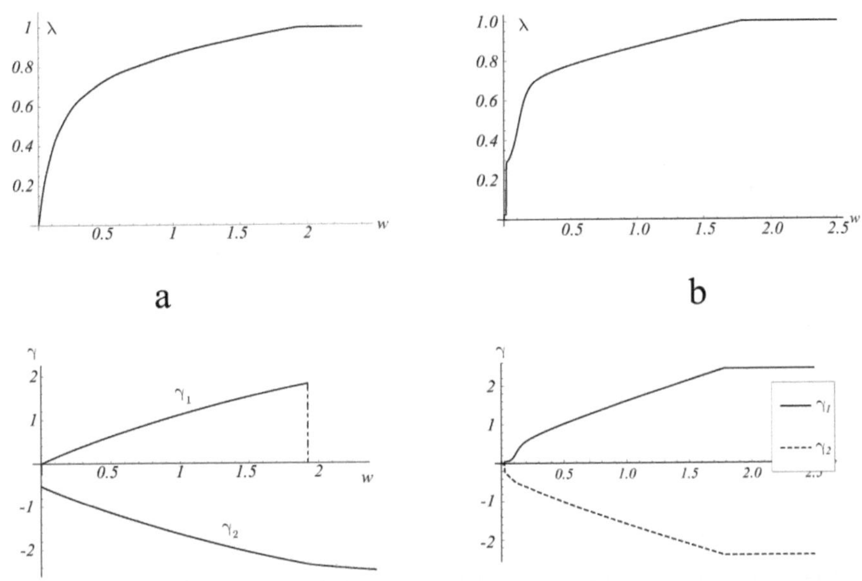

Fig. 3.13 Volume fraction λ of domain 2 and plastic slips γ_1 and γ_2 of both domains for: (a) rate-independent approach, (b) viscous approach.

increases monotonically from 0 to 1. From there on, the crystal is homogeneous again, consisting only of the second domain. A closer observation reveals slight differences, the volume fraction in the rate-independent model (left side of Figure 3.13) increases monotonously to one (reached at $w = 1.9$), while for the viscous approach (right side of Figure 3.13), the volume fraction increases rapidly up to a value of 30% before the slope is less pronounced. At $w = 1.75$, the process is completed and $\lambda = 1$.

The evolution of the corresponding plastic slips in both domains for the outlined approaches is presented in the lower part of Fig. 3.13. The plastic slip for the rate-independent ansatz is shown on the left hand side of Figure 3.13. The plastic slip in domain one is positive and increases monotonously from zero to $\gamma_1 = 1.9$. The plastic slip of the second domain jumps from zero to -0.6 and then decreases further to $\gamma_2 = -2.4$. Meanwhile the plastic slip in the viscous model (lower part on the right hand side of Figure 3.13) evolves similar: in domain one, the plastic slip increases, after a small delay, up to $\gamma_1 = 2.4$, which is rather similar to the results of the rate-independent ansatz. The plastic slip of domain two jumps from zero to -0.15, then it decreases further to -2.4 during formation of laminate two. Except the reduced jump in the beginning, this also corresponds to the evolution obtained by the rate-independent ansatz. Once again, the observed delays in the viscous model can be explained by the viscous regularization of the microstructure initiation.

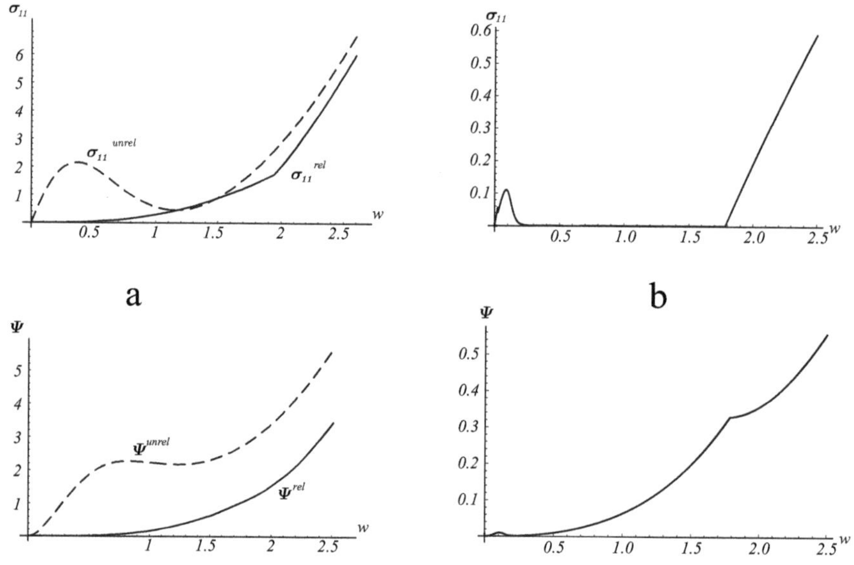

Fig. 3.14 Cauchy normal stress and free energy for: (a) rate-independent approach, (b) viscous approach

The Cauchy normal stress σ_{11} and the free energy are presented in Figure 3.14. The stress which is exhibited by the rate-independent approach (on the left hand in the upper part of Fig. 3.14) increases linearly as the microstructure evolves. When the crystal is homogeneous again, the slope increases. The normal stress of the viscous model is presented on the right hand side of Figure 3.14. During the first loading steps, the stress increases up to a maximum value of 0.12 GPa. In these first loading steps, the volume fraction of domain two is negligible: even though there is a large amount of plastic slip in domain one, the total amount of plastic slip is too small to significantly influence the stress.

With the increase of volume fraction and the plastic slip in domain one, the stress drops drastically (as in the case of shear loading) to 0.0 GPa and then remains constant. At first glance, the zero stresses during the evolution of microstructure are surprising since hardening is present. However, the hardening can be observed in the individual domains only. Since plastic slip is identical in both domains, except for the sign, the stresses will increase in every domain due to hardening, but cancel each other effectively. No macroscopic hardening can be observed. While this behavior is very interesting, our simulations indicate, that it is not generic but occurs only for specific parameter values.

The graph of the energy of the rate-independent ansatz (on the left hand in the lower part of Fig. 3.14) appears convex. The energy of the viscous approach is shown on the right hand in the lower part of Fig. 3.14. The graph of the energy displays two characteristic kinks. This is once again due to the delayed initiation and disappearance of microstructure caused by the viscous regularization.

3.9 Conclusion and Discussion

We have presented two strategies to model the time-continuous evolution of laminate microstructures in finite-strain plasticity. For both studies we have in great detail studied the aspects of laminate initiation, laminate rotation, and continuous evolution, which all have to be treated in a different manner. We employed partially relaxed approximations of the nonconvex potentials involved. For an incompressible neo-Hookean material we have derived a closed-form, partially-relaxed energy which corresponds to a laminate of first order. The relaxation of the laminate energy is carried out only with respect to those variables that change purely elastically, in our case the amplitudes of the deformation gradient in each domain. The course of all remaining unknowns is determined from dissipative evolution equations, where we employ the principle of the minimum of the dissipation potential.

The first approach is based on a dissipation potential leading to rate-independent evolution. While the results obtained this way are very instructive, the numerical effort is rather high and the approach has a tendency to instabilities. Thus it is doubtful that it can be applied to more complicated situations than treated here. Therefore, the model has been modified by adding a quadratic term to the dissipation potential leading to viscous behavior. Now the evolution equations no longer require any global minimization. They can be evaluated directly with standard approaches for numerical integration. For the comparison of both approaches, a shear test and a tension-compression test were performed. Both tests show good agreement between the numerical schemes, except a delayed occurrence and vanishing of microstructure in the viscous case. This behavior, however, had to be expected due to the local nature of all equations in the latter case.

While this paper concentrates more on the principal aspects of laminate evolution, the focus of future work will lie on the extension of the present models to physically more realistic ones, involving for example several slip systems, on the solution of full boundary-value problems, and on comparison with experimental data.

Acknowledgements. This work was partially supported by the Deutsche Forschungsgemeinschaft through Forschergruppe 797 "Analysis and computation of microstructure in finite plasticity".

References

[AD14] Anguige, K., Dondl, P.W.: Relaxation of the single-slip condition in strain-gradient plasticity. Proceedings of the Royal Society of London A: Mathematical, Physical and Engineering Sciences 470(2169) (2014)

[BC10] Boiger, W., Carstensen, C.: On the strong convergence of gradients in stabilised degenerate convex minimisation problems. SIAM J. Numer. Anal. 47(6), 4569–4580 (2010)

[BCC+06] Bartels, S., Carstensen, C., Conti, S., Hackl, K., Hoppe, U., Orlando, A.: Relaxation and the computation of effective energies and microstructures in solid mechanics. In: Mielke, A. (ed.) Analysis, Modeling and Simulation of Multiscale Problems, pp. 197–224. Springer (2006)

[BCHH04] Bartels, S., Carstensen, C., Hackl, K., Hoppe, U.: Effective relaxation for microstructure simulations: algorithms and applications. Computer Methods in Applied Mechanics and Engineering 193(48-51), 5143–5175 (2004). Advances in Computational Plasticity

[Bha03] Bhattacharya, K.: Microstructure of Martensite. Why it forms and how it gives rise to the shape-memory effect. Oxford University Press (2003)

[BJ87] Ball, J., James, R.: Fine phase mixtures as minimizers of energy. Arch. Rat. Mech. Anal. 100, 13–52 (1987)

[CDK09] Conti, S., Dolzmann, G., Klust, C.: Relaxation of a class of variational models in crystal plasticity. Proceedings of the Royal Society of London A: Mathematical, Physical and Engineering Sciences 465(2106), 1735–1742 (2009)

[CDK11] Conti, S., Dolzmann, G., Kreisbeck, C.: Asymptotic behavior of crystal plasticity with one slip system in the limit of rigid elasticity. SIAM Journal on Mathematical Analysis 43(5), 2337–2353 (2011)

[CDK13] Conti, S., Dolzmann, G., Kreisbeck, C.: Relaxation and microstructure in a model for finite crystal plasticity with one slip system in three dimensions. Discrete and Continuous Dynamical Systems - Series S 6(1), 1–16 (2013)

[CHHO06] Carstensen, C., Hackl, K., Hoppe, U., Orlando, A.: Computational microstructures in phase transition solids and finite-strain elastoplasticity. GAMM-Mitteilungen 29(2), 215–246 (2006)

[CKM02] Carstensen, C., Hackl, K., Mielke, A.: Non-convex potentials and microstructures in finite-strain plasticity. Proc. R.Soc. London A458, 299–317 (2002)

[CM95] Christian, J., Mahajan, S.: Deformation twinning. Progress in Materials Science 39(1-2), 1–157 (1995)

[CO05] Conti, S., Ortiz, M.: Dislocation microstructures and the effective behavior of single crystals. Archive for Rational Mechanics and Analysis 176(1), 103–147 (2005)

[CSMC05] Canadinc, D., Sehitoglu, H., Maier, H., Chumlyakov, Y.: Strain hardening behavior of aluminum alloyed hadfield steel single crystals. Acta Materialia 53(6), 1831–1842 (2005)

[CSO08] Carstensen, C., Conti, S., Orlando, A.: Mixed analytical-numerical relaxation in finite single-slip crystal plasticity. Cont. Mech. Thermodyn. 20, 275–301 (2008)

[CT05] Conti, S., Theil, F.: Single-slip elastoplastic microstructures. Arch. Rat. Mech. Anal. 178, 125–148 (2005)

[Dac82] Dacorogna, B.: Quasiconvexity and relaxation of nonconvex problems in the calculus of variations. J. Funct. Anal. 46, 102–118 (1982)

[DDMD09] Dmitrieva, O., Dondl, P., Müller, S., Raabe, D.: Lamination microstructure in shear deformed copper single crystals. Acter Mater. 57, 3439–3449 (2009)

[HF08] Hackl, K., Fischer, F.: On the relation between the principle of maximum dissipation and inelastic evolution given by dissipation potentials. Proc. R. Soc. A 464(2089), 117–132 (2008)

[HHK14] Hackl, K., Hoppe, U., Kochmann, D.M.: Variational modeling of microstructures in plasticity. In: Schröder, J., Hackl, K. (eds.) Plasticity and Beyond. CISM International Centre for Mechanical Sciences, vol. 550, pp. 65–129. Springer Vienna (2014)

[HHM12] Hackl, K., Heinz, S., Mielke, A.: A model for the evolution of laminates in finite-strain elastoplasticity. ZAMM-Journal of Applied Mathematics and Mechanics/Zeitschrift für Angewandte Mathematik und Mechanik 92(11-12), 888–909 (2012)

[HK08] Hackl, K., Kochmann, D.: Relaxed potentials and evolution equations for inelastic microstructures. In: Theoretical, Computational and Modelling Aspects of Inelastic Media. Springer, Berlin (2008)

[HK09] Hackl, K., Kochmann, D.: Time-continuous evolution of microstructures in finite plasticity. In: Variational Concepts with Applications to the Mechanics of Materials. Springer, Berlin (2009)

[HK10] Hackl, K., Kochmann, D.: Time-continuous evolution of microstructures in finite plasticity. In: IUTAM Symposium on Variational Concepts with Applications to the Mechanics of Materials, vol. 21 (2010)

[HK11] Hackl, K., Kochmann, D.: An incremental strategy for modeling laminate microstructures in finite plasticity–energy reduction, laminate orientation and cyclic behavior. In: Multiscale Methods in Computational Mechanics, pp. 117–134. Springer (2011)

[KH10] Kochmann, D., Hackl, K.: Influence of hardening on the cyclic behavior of laminate microstructures in finite crystal plasticity. Technische Mechanik 30, 387–400 (2010)

[KK11] Kochmann, D., Hackl, K.: The evolution of laminates in finite crystal plasticity: a variational approach. Continuum Mechanics and Thermodynamics 23, 63–85 (2011)

[MCJ03] Lambrecht, M., Miehe, C., Dettmar, J.: Energy relaxation of non-convex incremental stress potentials in a strain-softening elastic-plastic bar. Int. J. Solids Struct. 40, 1369–1391 (2003)

[MGB07] Miehe, C., Gürses, E., Birkle, M.: A computational framework of configurational-force-driven brittle fracture based on incremental energy minimization. Int. J. Frac. 145, 245–259 (2007)

[Mie02] Mielke, A.: Finite elastoplasticity, Lie groups and geodesics on SL(d). Springer, Berlin (2002)

[Mie03] Mielke, A.: Energetic formulation of multiplicative elasto-plasticity using dissipation distances. Cont. Mech. Thermodyn. 15, 351–382 (2003)

[Mie04] Mielke, A.: Deriving new evolution equations for microstructures via relaxation of variational incremental problems. Computer Methods in Applied Mechanics and Engineering 193(48-51), 5095–5127 (2004)

[Mie14] Miehe, C.: Variational gradient plasticity at finite strains. part i: Mixed potentials for the evolution and update problems of gradient-extended dissipative solids. Computer Methods in Applied Mechanics and Engineering 268, 677–703 (2014)

[ML03] Miehe, C., Lambrecht, M.: Analysis of microstructure development in shearbands by energy relaxation of incremental stress potentials: Large-strain theory for standard dissipative solids. International Journal for Numerical Methods in Engineering 58(1), 1–41 (2003)

[MLG04] Miehe, C., Lambrecht, M., Gürses, E.: Analysis of material instabilities in inelastic solids by incremental energy minimization and relaxation methods: evolving deformation microstructures in finite plasticity. J. Mech. Phys. Solids 52, 2725–2769 (2004)

[MM09] Mainik, A., Mielke, A.: Global existence for rate-independent gradient plasticity at finite strain. Journal of Nonlinear Science 19(3), 221–248 (2009)

[MRF10] Miehe, C., Rosato, D., Frankenreiter, I.: Fast estimates of evolving orientation microstructures in textured bcc polycrystals at finite plastic strains. Acta Materialia 58(15), 4911–4922 (2010)

[MS08] Mielke, A., Stefanelli, U.: A discrete variational principle for rate-independent evolution. Advances in Calculus of Variations 1(4), 399–431 (2008)

[OR99] Ortiz, M., Repetto, E.: Nonconvex energy minimization and dislocation structures in ductile single crystals. J. Mech. Phys. Solids 47, 397–462 (1999)

[YBG11] Yalcinkaya, T., Brekelmans, W., Geers, M.: Deformation patterning driven by rate dependent non-convex strain gradient plasticity. Journal of the Mechanics and Physics of Solids 59(1), 1–17 (2011)

Chapter 4
Variational Gradient Plasticity: Local-Global Updates, Regularization and Laminate Microstructures in Single Crystals

Steffen Mauthe and Christian Miehe

Abstract. This work summarizes recent results on the formulation and numerical implementation of gradient plasticity based on incremental variational potentials as outlined in a recent sequence of work [Mie14, MMH14, MWA14, MAM13]. We focus on variational gradient crystal plasticity and outline a formulation and finite element implementation of micromechanically-motivated multiplicative gradient plasticity for single crystals. In order to partially overcome the complexity of full multislip scenarios, we suggest a new viscous regularized formulation of rate-independent crystal plasticity, that exploits in a systematic manner the long- and short-range nature of the involved variables. To this end, we outline a multifield scenario, where the macro-deformation and the plastic slips on crystallographic systems are the primary fields. We then define a long-range state related to the primary fields and in addition a short-range plastic state for further variables describing the plastic state. The evolution of the short-range state is fully determined by the evolution of the long-range state, which is systematically exploited in the algorithmic treatment. The model problem under consideration accounts in a canonical format for basic effects related to statistically stored and geometrically necessary dislocation flow, yielding micro-force balances including non-convex cross-hardening, kinematic hardening and size effects. Further key ingredients of the proposed algorithmic formulation are geometrically exact updates of the short-range state and a distinct regularization of the rate-independent dissipation function that preserves the range of the elastic domain. The model capability and algorithmic performance is shown in a first multislip scenario in an fcc crystal. A second example presents the prediction of formation and evolution of laminate microstructure.

Steffen Mauthe · Christian Miehe
Institute of Applied Mechanics (CE),
University of Stuttgart, Pfaffenwaldring 7, 70569 Stuttgart, Germany
e-mail: steffen.mauthe@mechbau.uni-stuttgart.de

S. Conti and K. Hackl (eds.), *Analysis and Computation of Microstructure in Finite Plasticity*,
Lecture Notes in Applied and Computational Mechanics 78, DOI: 10.1007/978-3-319-18242-1_4

4.1 Introduction

With the ongoing trend of miniaturization and nanotechnology, the predictive modeling of size effects play an increasingly important role in metal plasticity. In this context, two specific phenomena have recently received growing attention: plastic laminate microstructure stemming from latent hardening and size effects stemming from geometrically necessary dislocations (GNDs). In this work, we propose a gradient plasticity model based on the framework of gradient crystal plasticity that approximates the laminate interfaces in a phasefield like manner and incorporates the effects of GNDs by use of the dislocation density tensor. Our considerations are based on the work of Miehe et al. [MMH14] and inspired by ideas presented in Hildebrand & Miehe [HM12]. A key challenge of the numerical implementation of gradient crystal plasticity is the complexity within full multislip scenarios, in particular in context of rate-independent settings. We outline variational-based formulations and efficient numerical implementation of gradient crystal plasticity based Nye's dislocation tensor in the free energy, which is well suited for large-scale computations.

The physically-based phenomenological description of macroscopic plastic strains in metallic single crystals has been guided by the pioneering works Taylor [Tay34, Tay38], Schmid & Boas [SB35], Nye [Nye53] and Kröner [Kro60]. The mechanism of inelastic distortion in ductile single crystals was found to be governed by plastic slip on a certain set of crystallographic systems where the shear stresses reach critical values. Mathematical continuum descriptions of elastic-plastic deformations in crystals have been developed by Hill [Hil66] in the small-strain format and by Rice [Ric71], Hill & Rice [HR72], Mandel [Man72], Teodosiu [Teo70], Kröner & Teodosiu [KT72] and Asaro [Asa83a] in the context of finite strains, see also the reviews Lin [Lin71], Asaro [Asa83b], Havner [Hav92] and Bassani [Bas93]. The geometric basis of what is often called the *macroscopic continuum slip theory* of finite crystal elastoplasticity is a multiplicative decomposition of the local deformation gradient into a plastic part solely due to multislip on given crystallographic planes and a remaining part which describes elastic distortions and rigid rotations of the lattice, yielding the standard definition $\boldsymbol{F}^e := \boldsymbol{F}\boldsymbol{F}^{p-1}$. The constitutive equations which govern the slip resistance and the slip evolution can be motivated by micromechanical investigations of defects in crystals as reviewed in Cottrell [Cot53], Seeger [See58], Nabarro [Nab67], Hirth & Lothe [HL68], Mura [Mur87], Hull & Bacon [HB84] and Phillips [Phi01]. The formulation of micromechanically motivated hardening laws for multislip is a cornerstone of the continuum slip theory. We refer to the reviews Kocks [Koc66], Kocks et al. [KAA75], Asaro [Asa83b], Perzyna [Per88], Havner [Hav92], Bassani [Bas93], Cuitiño & Ortiz [CnO92] and references cited therein. Algorithmic representations of *local* finite crystal plasticity models suitable for the numerical simulation of initial boundary value problems in context with the application of finite element methods were developed by Peirce et al. [PAN82], Needleman et al. [NALP85], Cuitiño & Ortiz [CnO92], Anand & Kothari [AK96], Miehe [Mie96], Kothari & Anand [KA98], Ortiz & Stainier [OS99] and Miehe et al. [MSS99, MSL02], who discuss alternative numerical schemes for the updates of the stresses and the active set of slip systems in the multisurface frameworks of rate-dependent and rate-independent crystal plasticity. A current status of *computational local crystal plasticity* at finite strains is outlined in Miehe & Schotte [MS04].

However, the above mentioned *local approach* to crystal plasticity is not able to describe size effects. Experimental evidence for nano indentation tests of single crystals, torsion tests of microwires and microbending tests of thin films show that inhomogeneous plastic flow at small scales is inherently size dependent, with '*smaller being stronger*', see Hall [Hal51], Petch [Pet53], Fleck et al. [FMAH94] and Stölken & Evans [SE98], among others. As a consequence, an important field of research in crystal plasticity treats advanced formulations that allow to model size and boundary layer effects accompanied with additional hardening. Micromechanical basis of such investigations are so-called *geometrically necessary dislocations* (GNDs) occuring in inhomogeneous plastic deformations, which supplement *statistically stored dislocations* (SSDs) governing homogeneous plastic deformations. Basic definitions for the phenomenological quantification of GNDs can be traced back to Nye [Nye53], Kröner [Kro60], Ashby [Ash70], Fleck & Hutchinson [FH97], Nix & Gao [NG98] and Arzt [Arz98]. A key ingredient here is the definition of the *dislocation density tensor*, which can be related to *Cartans torsion tensor*. This owes mainly to the description of defect distributions by objects of differential geometry, as outlined in the classical works Kondo [Kon52], Bilby et al. [BBS55], Kröner & Seeger [KS59] and Kröner [Kro60] which identifed the continuum-mechanical theory of dislocations with a Cartan geometry. This geometric aspect was elaborated in Steinmann [Ste96] with a view on multiplicative plasticity.

Continuum strain-gradient theories for single crystal plasticity which follow this geometric viewpoint were outlined in Menzel & Steinmann [MS00] in the small-strain context and in Gurtin [Gur00, Gur02] and Cermelli & Gurtin [CG01] for finite deformations. Here, the central ingredient leading to size effects was to make the free energy dependent of the lattice-curvature-based *dislocation density tensor* related in a geometrically consistent format to the curl of the plastic deformation map Curl$\boldsymbol{F}^p$. It was shown in Mielke & Müller [MM06] that such an ansatz allows the proof of existence theorems in finite plasticity. Similar but more physically-based continuum models of gradient crystal plasticity based on dislocation densities defined on crystallographic slip systems and their interactions are outlined in Arsenlis & Parks [AP99], Arsenlis et at. [APBB04], Evers et al. [EBG04a, EBG04b], Gurtin [Gur08], Kuroda & Tvergaard [KT08] and Ertürk et al. [EvG09]. Detailed comparative reviews on different modeling approaches to gradient crystal plasticity provide Svendsen [Sve02] and Svendsen & Bargmann [SB10]. However, large scale three-dimensional *numerical implementations* of gradient crystal plasticity are missing in the literature, partly due to the complexity of the above mentioned models. This is in particular evident with respect to the modeling of size effects in *polycrystals*. Here, the current status of numerical implementation is often restricted to simplified gradient models and idealized slip models for planar double slip, such as the works Ekh et al. [EGRS07], Bargmann et al. [BERS10] and Wulfinghoff & Böhlke [WB12].

The formation of laminate deformation microstructure during plastic deformation of single crystals has been experimentally observed e.g. by Dmitrieva et al. [DDMR09] and can have a strong effect on the hardening properties of the material. Mathematically, the formation of laminate microstructure is triggered by a non-convexity of the latent hardening function that favors an inhomogeneous state of laminate microstructure of alternating states with fewer active slip systems over a homogeneous multi-slip state for certain slip combinations. Such a dominance of latent hardening over self hardening has been observed e.g. by Kocks [Koc64] and Franciosi et al. [FBZ80]. One approach to the modeling of plastic laminates is that of a minimization of the condensed non-convex free energy function with respect to a parametrized laminate microstructure. This approach is referred to as relaxation and restricts the possible microstructure. Examples for application of relaxation to single crystal plasticity can be found in Ortiz & Repetto [OR99], Carstensen et al. [CHM02], Miehe et al. [MLG04], Kochmann [Koc09], Kochmann & Hackl [KH11] and Conti et al. [CDK09, CDK13]. Since these approaches do not fully resolve the microstructure, an incorporation of size effects based on GNDs is not directly possible. An alternative approach is obtained by a suitable slip gradient regularization together with a nonconvex hardening function in the slips. A small strain single slip example for this approach in one dimension with non-convex self hardening is given in Yalcinkaya et al. [YBG11] and very similarly in Klusemann et al. [KBS12]. A two-dimensional small strain double slip approach with non-convex latent hardening and an isotropic gradient term is outlined in Yalcinkaya et al. [YBG12]. A model framework which combines strain-gradient plasticity and lamination by partially relaxing the free energy function and hence restoring existence and uniqueness of solutions has been recently proposed by Anguige & Dondl [AD14].

Following Miehe [Mie14] and Miehe et al. [MMH14], the aim of this paper is to outline an efficient implementation of finite gradient plasticity suitable for large-scale problems, that is in a rigorous format based on variational principles. To this end, we outline a multifield scenario where the macro-deformation φ and the plastic slips γ^α on crystallographic systems are the *primary fields*. The first ingredient is a variational-based formulation of gradient crystal plasticity that exploits in a geometrically consistent format the dislocation density tensor $\boldsymbol{A} := \boldsymbol{F}^p \mathrm{Curl} \boldsymbol{F}^p / \det[\boldsymbol{F}^p]$ defined in Cermelli & Gurtin [CG02]. To overcome the problem of active set searches in rate-independent multislip scenarios, we apply a particular *viscous regularized technique*. This regularization recasts the non-smooth setting of rate-independent crystal plasticity in a natural way into a close related smooth setting. The proposed regularization ansatz ensures the *resolved shear stresses to be bounded* by the critical Schmid stress. This is in sharp contrast to standard regularization techniques based on power laws with high exponents, used in the classical literature of crystal plasticity. The next ingredient of the proposed formulation is a systematic differentiation between long-range and short-range states. This differentiation is based purely on the chosen algorithmic structure and does not imply any physical properties. Directly related to the primary fields, we define as the *long-range state* $\mathfrak{G} := \{\nabla\varphi, \gamma^1, ..., \gamma^m, \nabla\gamma^1, ..., \nabla\gamma^m\}$ the deformation gradient, the plastic slips and their gradients. We then introduce as the *short-range plastic state* $\mathfrak{L} := \{\boldsymbol{F}^p, \boldsymbol{A}^T, \gamma^{+1}, ..., \gamma^{+m}\}$ the plastic deformation map, the dislocation density tensor and scalar hardening parameters associated with the slip systems. It is shown that the evolution of the short-range state is fully determined by the evolution of the long-range state. This separation into long- and short-range states is systematically exploited in the algorithmic treatment by a new update structure, where the short-range variables play the role of a local history base. The evolution equations of the proposed model of finite gradient crystal plasticity are governed by a rate-type variational principle, which represents a canonical two-field minimization structure as outlined in Miehe [Mie14]. Following the general concept outlined in this paper, we formulate a rate-potential that governs the minimization principle. The Euler equations of the minimization principle are the quasi-static *equilibrium condition* and the *coupled micro-balances* which govern the slip on the crystallographic systems. The formulation is specified for a simple model problem with isotropic constitutive functions, that demonstrate a particular back stress response on the slip systems due to the geometric dislocation density tensor. The algorithmic implementation uses a particular update for the short-range plastic deformation map $\boldsymbol{F}^p$, governed by an approximation of the exponential map suggested in Miehe & Schotte [MS04] that satisfies exactly the plastic incompressibility constraint. In contrast, the dislocation density tensor $\boldsymbol{A}^T$ as well as the hardening parameters $\gamma^{+1}, ..., \gamma^{+m}$ are updated in a *semi-explicit manner* based on a frozen sensitivity $\mathfrak{L}_{,\mathfrak{G}}$. This is a key ingredient with regard to a geometrically consistent but efficient numerical implementation. The computational formulation is governed by an incremental potential that is spatially discretized by finite elements with nodal degrees for the deformation map as well for the plastic slip.

The paper is structured as follows. We outline in Section 4.2 the theoretical formulation, including the particular regularization technique and the concept of separation into long- and short-range states. Section 4.3 contains the algorithmic formulation, including the definition of incremental potentials which govern the geometrically consistent implicit-explicit updates of the short-range state variables and the finite element design. We then demonstrate in Section 4.4 the modeling capabilities and algorithmic performance is demonstrated for a fcc crystal with 12 slip systems, i.e. $3+12=15$ nodal degrees of freedom. Section 4.5 then shows the capability of the presented model to predict formation and evolution of laminate microstructure stemming from latent hardening in crystals.

4.2 A Multifield Formulation of Gradient Crystal Plasticity

4.2.1 Introduction of Long-Range Field Variables

Let $\mathcal{B} \subset \mathcal{R}^d$ be the reference configuration of the crystal with dimension $d \in [2,3]$ in space and $\partial\mathcal{B} \subset \mathcal{R}^{d-1}$ its surface with normal $\boldsymbol{n}$ as depicted in Figure 4.1. We study the deformation of the crystal under mechanical loading in the range $\mathcal{T} \subset \mathcal{R}$ of time and are interested in predicting the *macro-motion field* at material points

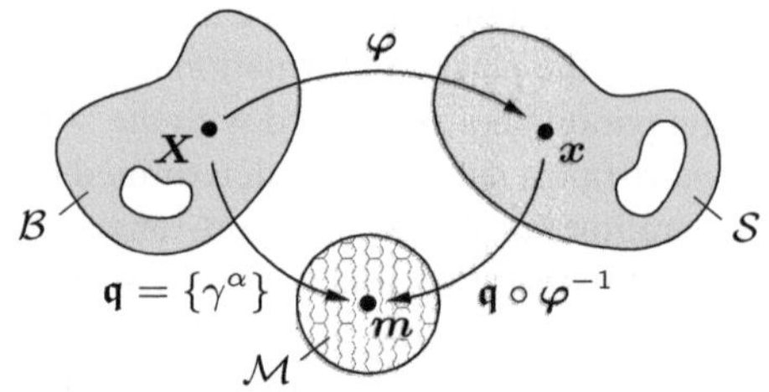

Fig. 4.1 Geometry of map-type finite plasticity. The reference configuration $\mathcal{B} \in \mathcal{R}^3$ and the current configuration $\mathcal{S} \in \mathcal{R}^3$ are considered as differentiable manifolds. $\boldsymbol{\varphi} : \mathcal{B} \times \mathcal{T} \to \mathcal{S}$ is the nonlinear *deformation map* that determines the current position $\boldsymbol{x} \in \mathcal{S}$. $\mathfrak{q} : \mathcal{B} \times \mathcal{T} \to \mathcal{M}$ assembles $\alpha = 1 \dots m$ *plastic slips* γ^α.

$\boldsymbol{X} \in \mathcal{B}$ at time $t \in \mathcal{T}$. Microstructural plastic slip mechanisms of the crystal are described by *micro-motion fields*, which generalize the classical notion of locally evolving internal variables to global fields driven by additional micro-balances. In what follows, $\nabla(\cdot) := \partial_{\boldsymbol{X}}(\cdot)$ and $\dot{(\cdot)} := \partial_t(\cdot)$ denote the material gradient and the time derivative of the field $(\cdot)$, respectively.

4.2.1.1 Macro-motion Field

In the large strain context, we describe the macroscopic motion of the body by the *macro-motion field*

$$\boldsymbol{\varphi} : \begin{cases} \mathcal{B} \times \mathcal{T} \to \mathcal{R}^d \\ (\boldsymbol{X}, t) \mapsto \boldsymbol{\varphi}(\boldsymbol{X}, t) , \end{cases} \tag{4.1}$$

which maps at time $t \in \mathcal{T}$ points $\boldsymbol{X} \in \mathcal{B}$ of the reference configuration $\mathcal{B}$ onto points $\boldsymbol{x} = \boldsymbol{\varphi}(\boldsymbol{X}, t)$ of the current configuration $\boldsymbol{\varphi}_t(\mathcal{B})$. The material *deformation gradient* $\boldsymbol{F} := \nabla\boldsymbol{\varphi}(\boldsymbol{X}, t)$ is constrained by the condition $J := \det[\nabla\boldsymbol{\varphi}] > 0$ on its Jacobian. The exterior surface of the reference configuration is decomposed via $\partial\mathcal{B} = \partial\mathcal{B}_{\varphi} \cup \partial\mathcal{B}_P$ into a part $\partial\mathcal{B}_{\varphi}$, where the macro-motion is prescribed by Dirichlet-type boundary conditions $\boldsymbol{\varphi}(\boldsymbol{X}, t) = \boldsymbol{\varphi}_D(\boldsymbol{X}, t)$ at $\boldsymbol{X} \in \partial\mathcal{B}_{\varphi}$ and a part $\partial\mathcal{B}_P$, where a macro-traction $\bar{\boldsymbol{t}}_N(\boldsymbol{X}, t)$ is prescribed by Neumann-type boundary conditions as outlined below. Clearly, we have $\partial\mathcal{B}_{\varphi} \cap \partial\mathcal{B}_P = 0$.

4.2.1.2 Micro-motion Fields

For multislip crystal plasticity with m slip systems $\alpha = 1, ..., m$, we introduce the *micro-motion fields*

$$\gamma^\alpha : \begin{cases} \mathcal{B} \times \mathcal{T} \to \mathcal{R} \\ (\boldsymbol{X}, t) \mapsto \gamma^\alpha(\boldsymbol{X}, t) , \end{cases} \tag{4.2}$$

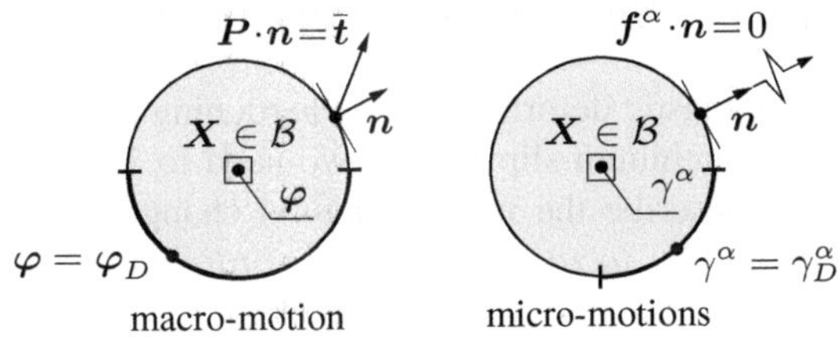

Fig. 4.2 Primary fields. The basic fields of the coupled problem are the deformation map $\boldsymbol{\varphi}$ and the plastic slips γ^α defined on the domain $\mathcal{B} \subset \mathcal{R}^d$ of the reference configuration.

characterizing the *plastic slip* on the m slip systems $\alpha = 1, ..., m$. They describe in a homogenized sense the dislocation movement through the crystal due to structural changes of the lattice on lower scales. The gradients of these fields $\boldsymbol{g}^\alpha := \nabla\gamma^\alpha(\boldsymbol{X}, t)$ govern the compatibility features of gradient-type crystal-plasticity. Their presence in the energy storage and dissipation potential functions induce the balance-type structure of the *partial differential equations* for the evolution of the plastic slips. With respect to *each* plastic slip field γ^α, we decompose the surface of the solid via $\partial\mathcal{B} = \partial\mathcal{B}_{\gamma^\alpha} \cup \partial\mathcal{B}_{H^\alpha}$ into a part $\partial\mathcal{B}_{\gamma^\alpha}$, as depicted in Figure 4.2, where the plastic slip is prescribed by Dirichlet-type boundary conditions $\gamma^\alpha(\boldsymbol{X}, t) = \gamma^\alpha_D(\boldsymbol{X}, t)$ at $\boldsymbol{X} \in \partial\mathcal{B}_{\gamma^\alpha}$ and a part $\partial\mathcal{B}_{H^\alpha}$, where *micro-tractions* $\bar{\boldsymbol{t}}^\alpha_N(\boldsymbol{X}, t)$ are assumed as Neumann-type boundary conditions. We have $\partial\mathcal{B}_{\gamma^\alpha} \cap \partial\mathcal{B}_{H^\alpha} = 0$. The Dirichlet and Neumann conditions allow the specification of microstructural constraints for the plastic slips, for example at grain boundaries of polycrystals.

4.2.1.3 Long-Range Constitutive State Variables

With regard to the subsequent specification of *stored energy* and *dissipation potential functionals*, we introduce the set of *global* constitutive state variables

$$\mathfrak{G} := \{\boldsymbol{F}, \gamma^1, ..., \gamma^m, \boldsymbol{g}^1, ..., \boldsymbol{g}^m\} \tag{4.3}$$

associated with the fields introduced above. Hence, the constitutive response functions of the crystal are assumed to depend on the deformation gradient $\boldsymbol{F}$, the plastic slips $\{\gamma^\alpha\}_{\alpha=1...m}$ on the m systems and their spatial gradients $\{\boldsymbol{g}^\alpha\}_{\alpha=1...m}$. The introduction of these variables is consistent with a *simple material of the grade one*, where the constitutive functions depend on both the macro- and micro-motions and their first gradients. Note that the dependence on the macro-motion $\boldsymbol{\varphi}$ drops out due to the argument of *material frame invariance*, which also restricts the dependence on the deformation gradient $\boldsymbol{F}$ in (4.3) to a dependence on the right Cauchy-Green tensor $\boldsymbol{C}(\boldsymbol{X}, t) := \boldsymbol{F}^T(\boldsymbol{X}, t)\boldsymbol{F}(\boldsymbol{X}, t)$.

4.2.2 Introduction of Short-Range Field Variables

In order to describe the plastic deformation and hardening of single crystals within the framework of the continuum slip theory, we need to introduce several short-range variables. They describe the microstructural changes with spatially short-range actions and evolve by *local* equations, i.e. *ordinary differential equations*, driven by the long-range variables introduced above.

4.2.2.1 The Plastic Deformation Map

The most important internal variable in the continuum slip theory of plasticity is the *plastic deformation map*

$$\boldsymbol{F}^p : \begin{cases} \mathcal{B} \times \mathcal{T} \to \mathcal{R}^{d\times d} \\ (\boldsymbol{X}, t) \mapsto \boldsymbol{F}^p(\boldsymbol{X}, t) \ , \end{cases} \tag{4.4}$$

that determines the superimposed plastic deformation of the crystal solely due to *plastic shearing* on the m slip systems $\alpha = 1, ..., m$. The evolution of $\boldsymbol{F}^p$ in time is determined by the *local evolution equation*

$$\frac{d}{dt}\boldsymbol{F}^p = \sum_{\alpha=1}^{m} \boldsymbol{F}^p_{,\gamma^\alpha} \dot{\gamma}^\alpha \quad \text{with} \quad \boldsymbol{F}^p_{,\gamma^\alpha} := \mathfrak{M}^\alpha \boldsymbol{F}^p \tag{4.5}$$

in terms of the *sensitivities* $\boldsymbol{F}^p_{,\gamma^\alpha}$ with respect to the long-range fields $\{\gamma^\alpha\}_{\alpha=1...m}$ and with the initial condition $\boldsymbol{F}^p(\boldsymbol{X}, 0) = \boldsymbol{1}$. Here, $\mathfrak{M}^\alpha := \mathfrak{s}^\alpha \otimes \mathfrak{n}^\alpha$ are *constant* second-order structural tensors, where the orthogonal unit vectors $\mathfrak{n}^\alpha$ and $\mathfrak{s}^\alpha$ define the slip plane normal and the slip direction of the slip system α. The basic assumption that plastic flow occurs by simple shearing induces the plastic deformation map $\boldsymbol{F}^p \in SL(3) = \{\boldsymbol{F}^p | \det[\boldsymbol{F}^p] = 1\}$ as an element of the special linear group $SL(3)$ of unimodular second-order tensors. This so-called *plastic incompressibility constraint* is a central ingredient of the Schmid-type continuum theory of crystals.

4.2.2.2 The Geometric Dislocation Density Tensor

The next variable needed in a physically-based theory of crystal plasticity is the geometrically consistent *dislocation density tensor*

$$\boldsymbol{A}^T : \begin{cases} \mathcal{B} \times \mathcal{T} \to \mathcal{R}^{d\times d} \\ (\boldsymbol{X}, t) \mapsto \boldsymbol{A}^T(\boldsymbol{X}, t) \end{cases} \tag{4.6}$$

that depends on the plastic deformation map $\boldsymbol{F}^p$ and its material curl via the definition

$$\boldsymbol{A}^T := \frac{1}{\det[\boldsymbol{F}^p]}(\text{Curl}^T \boldsymbol{F}^p)\boldsymbol{F}^{pT} \ , \tag{4.7}$$

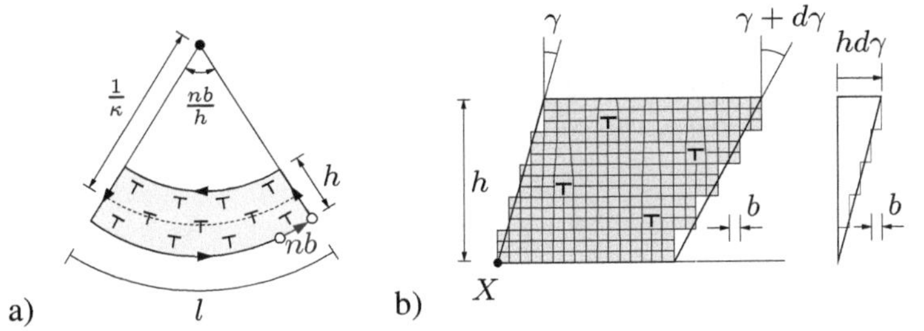

Fig. 4.3 Geometric necessary dislocations. The storage of GNDs is induced by a lattice curvature. a) GNDs in bending test with density $\rho = \kappa/b$, considered by Ashby and Nye. b) GNDs induced by inhomogeneous plastic deformation with density $\rho = 1/b \; \partial\gamma/\partial X$.

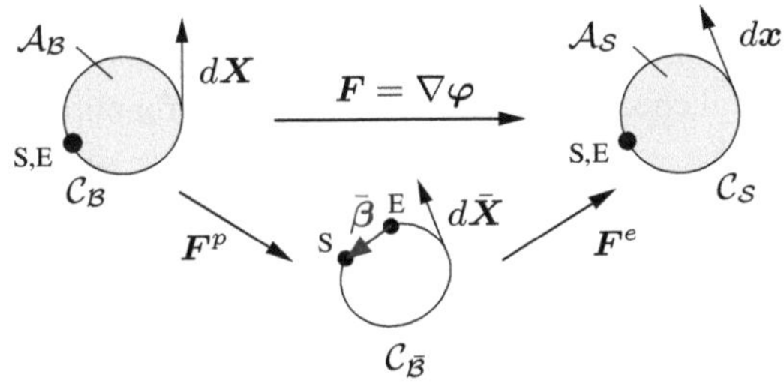

Fig. 4.4 Dislocation density tensor. Geometrically consistent measures for the incompatibility of the intermediate configuration. The closure failure of a line integral in the intermediate configuration corresponds to the macroscopic Burgers vector $\bar{\boldsymbol{\beta}}$.

see Cermelli & Gurtin [CG01]. It plays a crucial role as an incompatibility measure of the plastic deformation. The tensor is needed as a macroscopic measure for the amount of geometrically necessary dislocations (GNDs). The geometrical relation between a plastic lattice curvature and the amount of GNDs goes back to Nye [Nye53], see Figure 4.3. Taking the time-derivative of (4.7), we obtain the *local evolution equation*

$$\frac{d}{dt}\boldsymbol{A}^T = \sum_{\alpha=1}^{m} \{ \, \boldsymbol{A}^T_{,\gamma^\alpha}\dot{\gamma}^\alpha + \boldsymbol{A}^T_{,\boldsymbol{g}^\alpha} \cdot \dot{\boldsymbol{g}}^\alpha \, \} \tag{4.8}$$

with initial condition $\boldsymbol{A}^T(\boldsymbol{X}, 0) = \boldsymbol{0}$ and with the *sensitivities* with respect to the plastic slips $\{\gamma^\alpha\}_{\alpha=1\ldots m}$ and their gradients $\{\boldsymbol{g}^\alpha\}_{\alpha=1\ldots m}$

$$\boldsymbol{A}^T_{,\gamma^\alpha} := \mathfrak{M}^\alpha \boldsymbol{A}^T + \boldsymbol{A}^T \mathfrak{M}^{\alpha T} \quad \text{and} \quad \boldsymbol{A}^T_{,\boldsymbol{g}^\alpha} := -\mathfrak{s}^\alpha \otimes [\boldsymbol{F}^p \times (\boldsymbol{F}^{pT}\mathfrak{n}^\alpha)] \, . \tag{4.9}$$

4.2.2.3 Hardening Parameters

Finally, to be able to model hardening effects on all slip systems, we introduce m *short-range accumulated plastic slip fields* associated with the slips γ^α as

$$\gamma^{+\alpha} : \begin{cases} \mathcal{B} \times \mathcal{T} \to \mathcal{R} \\ (\boldsymbol{X}, t) \mapsto \gamma^{+\alpha}(\boldsymbol{X}, t) \end{cases} \tag{4.10}$$

for $\alpha = 1, ..., m$. They are needed for the description of the typical latent hardening of single crystals due to statistically stored dislocations (SSDs). We assume these hardening variables to be governed by the *evolution equations*

$$\frac{d}{dt}\gamma^{+\alpha} = f'(\dot{\gamma}^\alpha)\dot{\gamma}^\alpha \quad \text{with} \quad \gamma^{+\alpha}(\boldsymbol{X}, 0) = 0 \tag{4.11}$$

in terms of the long-range fields $\{\gamma^\alpha\}_{\alpha=1...m}$ introduced in (4.2), with the *sensitivity* f'. Here, the canonical choice is the non-smooth norm function

$$f(\dot{\gamma}^\alpha) = |\dot{\gamma}^\alpha| \quad \text{and} \quad f'(\dot{\gamma}^\alpha) = \text{sign}(\dot{\gamma}^\alpha) \, , \tag{4.12}$$

which characterizes the sensitivity in (4.11) to be the signum function. Hence, the hardening parameter $\gamma^{+\alpha}$ accumulates the absolute amount of slip activity on slip system α. In what follows, we consider *regularizations* of the non-smooth norm function, which are convenient for numerical implementations. A particular choice is the regularized function

$$f(\dot{\gamma}^\alpha) = \dot{\gamma}_0 \ln[\cosh(\dot{\gamma}^\alpha/\dot{\gamma}_0)] \quad \text{and} \quad f'(\dot{\gamma}^\alpha) = \tanh(\dot{\gamma}^\alpha/\dot{\gamma}_0) \, , \tag{4.13}$$

where the the signum function is approximated by a tanh function in terms of the regularization parameter $\dot{\gamma}_0$. A visualization is given in Figure 4.5.

4.2.2.4 Short-Range Constitutive State Variables

We denote the internal variables introduced above as the set of *local* constitutive state variables

$$\mathfrak{L} := \{\boldsymbol{F}^p, \boldsymbol{A}^T, \gamma^{+1}, ..., \gamma^{+m}\} \, , \tag{4.14}$$

which enter the stored energy and dissipation potential functionals besides the *global* constitutive variables $\mathfrak{G}$. Observe from (4.5), (4.8) and (4.11) that the evolution of these local variables is essentially determined by the evolution of the global variables, such that we can write

$$\frac{d}{dt}\mathfrak{L} = \mathfrak{L}_{,\mathfrak{G}} \cdot \dot{\mathfrak{G}} \, , \tag{4.15}$$

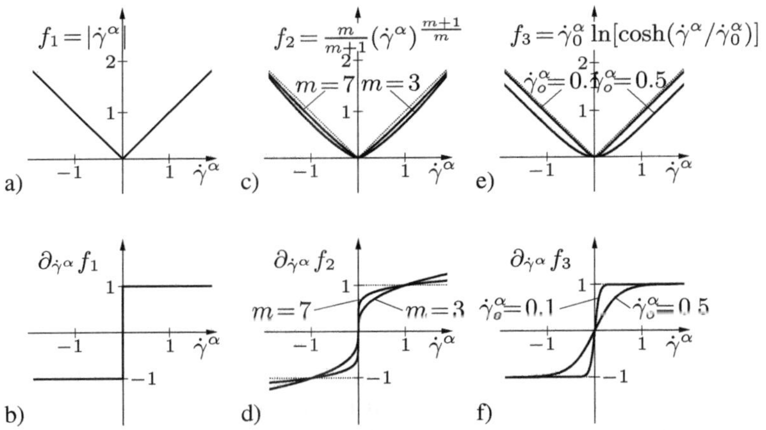

Fig. 4.5 Regularization technique. The non-smooth function $f_1 = |\dot{\gamma}^\alpha|$ in a) has the subdifferential in b). A common approximation is $f_2 = \frac{m}{m+1}(\dot{\gamma}^\alpha)^{\frac{m+1}{m}}$ in c) with derivative in d). We propose the regularization $f_3 = \dot{\gamma}_0^\alpha \ln[\cosh(\dot{\gamma}^\alpha/\dot{\gamma}_o^\alpha)]$ in e) with derivative in f).

where $\mathfrak{L}_{,\mathfrak{G}}$ contains the state-dependent *sensitivities* as introduced in (4.5), (4.9) and (4.12) or (4.13). This property is exploited in the proposed computational setting of gradient crystal plasticity, where the variables $\boldsymbol{F}^p$, $\boldsymbol{A}^T$ and $\gamma^{+\alpha}$ form a *local history*.

4.2.3 Energy Storage, Dissipation Potential and Load Functionals

4.2.3.1 Stored Energy Functional

The energy stored in a crystal due to macroscopic lattice distortions and microscopic stress fields takes the form

$$E(\varphi, \gamma^1, ..., \gamma^m, \mathfrak{L}) = \int_{\mathcal{B}} \psi(\mathfrak{G}, \mathfrak{L})\, dV \, . \tag{4.16}$$

The energy functional depends on the long-range macro- and micro-motion fields φ and γ^α through the global constitutive variables $\mathfrak{G}$. The energy density $\psi = \psi(\mathfrak{G}, \mathfrak{L})$ describes the stored energy per unit volume of the crystal and is a constitutive function of both the global as well as the local state variables. We assume a split of this constitutive function into three parts

$$\psi = \psi_{LAT}(\boldsymbol{F}, \boldsymbol{F}^p) + \psi_{SSD}(\gamma^{+1}, ..., \gamma^{+m}) + \psi_{GND}(\boldsymbol{A}^T) \, . \tag{4.17}$$

The first term ψ_{LAT} characterizes the energy storage due to macroscopic elastic distortion of the crystal *lattice*. Following a standard argument of crystal plasticity that assumes invariance with respect to previous plastic deformations, this energy has the form $\psi_{LAT}(\boldsymbol{F}, \boldsymbol{F}^p) = \psi_{LAT}(\boldsymbol{F}^e)$ and depends solely on the elastic deformation map $\boldsymbol{F}^e := \boldsymbol{F}\boldsymbol{F}^{p-1}$. A simple example of this function is provided by the isotropic Neo-Hookean function

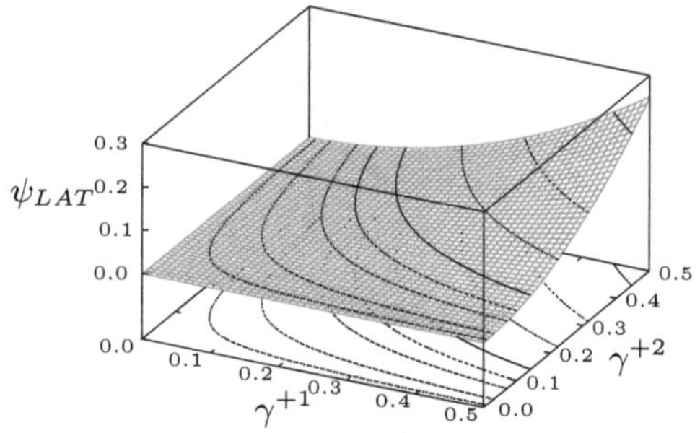

Fig. 4.6 Plot of the energy ψ_{SSD} due to latent hardening as defined in (4.19) with material parameters $\hat{\tau} = 4.0$ N/m^2, $\hat{\gamma} = 1.0$ and $b = 0$. For the specific choice $b = 0$, ψ_{SSD} does not include self hardening, only cross-hardening. Further note that ψ_{LAT} is a *nonconvex* function.

$$\psi_{LAT}(\boldsymbol{F}^e) = \frac{\mu}{2}\left[\mathrm{tr}(\boldsymbol{F}^{eT}\boldsymbol{F}^e) - 3\right] + \frac{\mu}{\beta}\left[\det(\boldsymbol{F}^{eT}\boldsymbol{F}^e)^{\frac{\beta}{2}} - 1\right], \tag{4.18}$$

where $\mu > 0$ and $\beta > 0$ are material parameters specifying the elastic response.

The second term ψ_{SSD} in (4.17) describes the stored energy due to micro-stress fields caused by statistically stored dislocations (SSDs). It models hardening effects on the slip systems due to forest dislocations with the typical latent hardening phenomena. Here, we propose an ansatz

$$\psi_{SSD}(\gamma^{+1}, ..., \gamma^{+m}) = \frac{1}{2}\hat{\tau}\,\hat{\gamma}\left[\sum_{\alpha=1}^{m}\sum_{\beta=1}^{m}\left(\frac{\gamma^{+\alpha}}{\hat{\gamma}}\right)^n a^{\alpha\beta}\left(\frac{\gamma^{+\beta}}{\hat{\gamma}}\right)^n\right], \tag{4.19}$$

where $\hat{\tau} > 0$ and $\hat{\gamma} > 0$ are a reference resolved shear stress and a reference slip strain, and $a^{\alpha\beta}$ is a constant interaction matrix $a^{\alpha\beta} = \left[q + (b - q)\delta^{\alpha\beta}\right]$. For $n = b = 1$, this formulations yields the bilinear ansatz of HUTCHINSON [Hut76] or KOCKS [Koc64]. For $q = b = n = 1$, (4.19) results in isotropic Taylor-type hardening with the linear hardening modulus $h = \hat{\tau}/\hat{\gamma} > 0$, see TAYLOR [Tay38], whereas $q > b$ leads to a non-positive definite interaction matrix $a^{\alpha\beta}$ and describes off-diagonal-dominant latent hardening of the slip systems. This induces the typical laminate microstructures in plasticity, see below.

Finally, the third term ψ_{GND} in (4.17) describes the stored energy due to micro-stress fields caused by geometrically necessary dislocations (GNDs). It is assumed to have the isotropic form

$$\psi_{GND}(\boldsymbol{A}^T) = \frac{\lambda l^2}{2}\|\boldsymbol{A}^T\|^2 \tag{4.20}$$

of the transposed geometric dislocation density tensor introduced in (4.7), where the material parameter l constitutes a length-scale, see e.g. Gurtin [Gur02] and where λ is another material parameter.

With the above constitutive functions at hand, we may define the *rate of energy functional*

$$
\begin{aligned}
\mathcal{E}(\dot{\boldsymbol{\varphi}}, \dot{\gamma}^1, ..., \dot{\gamma}^m) &:= \frac{d}{dt} E \\
&= \int_{\mathcal{B}} [\, \partial_{\mathfrak{G}} \psi \cdot \dot{\mathfrak{G}} + \partial_{\mathfrak{L}} \psi \cdot \dot{\mathfrak{L}} \,] \, dV \\
&= \int_{\mathcal{B}} [\delta_{\boldsymbol{\varphi}} \psi \cdot \dot{\boldsymbol{\varphi}} + \sum_{\alpha=1}^{m} \delta_{\gamma^\alpha} \psi \, \dot{\gamma}^\alpha] dV + \int_{\partial \mathcal{B}_P} [(\partial_{\boldsymbol{F}} \psi) \cdot \boldsymbol{n}] \cdot \dot{\boldsymbol{\varphi}} dA \\
&\quad + \sum_{\alpha=1}^{m} \int_{\partial \mathcal{B}_{H^\alpha}} [(\partial_{\boldsymbol{g}^\alpha} \psi) \cdot \boldsymbol{n}] \dot{\gamma}^\alpha dA
\end{aligned}
\tag{4.21}
$$

at a given state $\{\mathfrak{G}, \mathfrak{L}\}$. Hence, we have introduced the variational or functional derivatives $\delta_{\boldsymbol{\varphi}} \psi := -\text{Div}[\partial_{\boldsymbol{F}} \psi]$ and $\delta_{\gamma^\alpha} \psi := \partial_{\gamma^\alpha} \psi - \text{Div}[\partial_{\boldsymbol{g}^\alpha} \psi]$.

4.2.3.2 Dissipation Potential Functional

We consider the evolution of the plastic slips to be governed by a dissipation potential functional

$$
D(\dot{\gamma}^1, ..., \dot{\gamma}^m) = \int_{\mathcal{B}} \phi(\dot{\gamma}^1, ..., \dot{\gamma}^m) \, dV \ , \tag{4.22}
$$

where the dissipation potential function $\phi = \phi(\dot{\gamma}^1, ..., \dot{\gamma}^m)$ is assumed to be a function of the slip rates $\dot{\gamma}^\alpha$ only and to be *convex*, which guarantees the consistency of the proposed formulation with the second axiom of thermodynamics. A decoupled representation is given by

$$
\phi = \sum_{\alpha=1}^{m} [\, \tau_c \, f(\dot{\gamma}^\alpha) + \frac{\eta}{2} (\dot{\gamma}^\alpha)^2 \,] \ , \tag{4.23}
$$

consisting of a rate-independent part governed by the constant critical resolved shear stress $\tau_c > 0$ and a viscous regularization term governed by the viscosity parameter $\eta > 0$. The function f is identical with the one defined in (4.12) or (4.13). For the canonical non-smooth rate-independent response, we have $f(\dot{\gamma}^\alpha) = |\dot{\gamma}^\alpha|$. In analogy to (4.12), we consider *regularizations* of these non-smooth norm functions, which are convenient for numerical implementation. Particularly, as given in (4.13) we choose $f(\dot{\gamma}^\alpha) = \dot{\gamma}_0 \ln[\, \cosh(\dot{\gamma}^\alpha / \dot{\gamma}_0) \,]$.

4.2.3.3 Load Functional

We assume an external loading of the crystal due to a body force field $\overline{\gamma}$ per unit volume of the reference domain $\mathcal{B}$ and a macro traction field $\overline{\boldsymbol{t}}_N$ per unit of the surface. Hence, the loading power functional is a function of the rate $\dot{\boldsymbol{\varphi}}$, particularly

$$P_{ext}(\dot{\boldsymbol{\varphi}}, t) := \int_{\mathcal{B}} \overline{\gamma}(\boldsymbol{X}, t) \cdot \dot{\boldsymbol{\varphi}}\, dV + \int_{\partial\mathcal{B}_P} \overline{\boldsymbol{t}}_N(\boldsymbol{X}, t) \cdot \dot{\boldsymbol{\varphi}}\, dA + \sum_{\alpha} \int_{\partial\mathcal{B}_{H^\alpha}} \tilde{t}_N^\alpha \dot{\gamma}^\alpha\, dA\,. \tag{4.24}$$

4.2.4 Rate-Type Variational Principle and Euler Equations

Based on the rate of energy, dissipation potential and load functionals $\mathcal{E} := \frac{d}{dt}E$, D and P_{ext} introduced in (4.21), (4.22) and (4.24) and conceptually drawing from Miehe [Mie11, Mie14], we define the *rate-type potential*

$$\underbrace{\Pi(\dot{\boldsymbol{\varphi}}, \dot{\gamma}^1, ..., \dot{\gamma}^m)}_{potential} := \underbrace{\mathcal{E}(\dot{\boldsymbol{\varphi}}, \dot{\gamma}^1, ..., \dot{\gamma}^m)}_{rate\ of\ energy} + \underbrace{D(\dot{\gamma}^1, ..., \dot{\gamma}^m)}_{dissipation} - \underbrace{P_{ext}(\dot{\boldsymbol{\varphi}}; t)}_{external\ power} \tag{4.25}$$

at a given state $\{\mathfrak{G}, \mathfrak{L}\}$ at time t. The rate potential takes the explicit form

$$\Pi = \int_{\mathcal{B}} \pi\, dV + \int_{\partial\mathcal{B}_P} [(\partial_{\boldsymbol{F}}\psi) \cdot \boldsymbol{n} - \overline{\boldsymbol{t}}_N] \cdot \dot{\boldsymbol{\varphi}}\, dA + \sum_{\alpha=1}^{m} \int_{\partial\mathcal{B}_{H^\alpha}} [(\partial_{\boldsymbol{g}^\alpha}\psi) \cdot \boldsymbol{n}] \cdot \dot{\gamma}^\alpha\, dA \tag{4.26}$$

in terms of the rate bulk potential per unit volume

$$\pi(\dot{\boldsymbol{\varphi}}, \dot{\gamma}^1, ..., \dot{\gamma}^m) = [\delta_{\boldsymbol{\varphi}}\psi - \bar{\gamma}] \cdot \boldsymbol{\varphi} + \sum_{\alpha=1}^{m} \delta_{\gamma^\alpha}\psi\, \dot{\gamma}^\alpha + \phi(\dot{\gamma}^1, ..., \dot{\gamma}^m)\,. \tag{4.27}$$

Note that the only nonlinear entry occurs through the dissipation function ϕ. All other terms are linear in the rates $\{\dot{\boldsymbol{\varphi}}, \dot{\gamma}^1, ..., \dot{\gamma}^m\}$ of the macro- and micro-motion fields. We assume that the rates of the macro- and micro-deformation fields at a given state are governed by the variational principle

$$\{\dot{\boldsymbol{\varphi}}, \dot{\gamma}^1, ..., \dot{\gamma}^m\} = \text{Arg}\{ \inf_{\dot{\boldsymbol{\varphi}}} \inf_{\dot{\gamma}^\alpha} \Pi(\dot{\boldsymbol{\varphi}}, \dot{\gamma}^1, ..., \dot{\gamma}^m) \}\,. \tag{4.28}$$

Hence, the constitutive setting of gradient plasticity for single crystals has an underlying rate-variational structure. Taking the variation of the rate potential (4.26) with arbitrary rates $\dot{\boldsymbol{\varphi}} \in \mathcal{W}^0_{\boldsymbol{\varphi}}$ and $\dot{\gamma^\alpha} \in \mathcal{W}^0_{\gamma^\alpha}$, where $\mathcal{W}^0_{\boldsymbol{\varphi}}$ and $\mathcal{W}^0_{\gamma^\alpha}$ are the admissible rates that are zero on the Dirichlet boundary, we obtain as Euler equations the coupled field equations of gradient plasticity

$$\text{Div}[\boldsymbol{P}] + \bar{\gamma} = \boldsymbol{0} \quad \text{in } \mathcal{B} \quad \text{and} \quad \psi_{,\gamma^\alpha} - \text{Div}[\psi_{,\boldsymbol{g}^\alpha}] + \phi_{,\dot{\gamma}^\alpha} = 0 \quad \text{in } \mathcal{B}\,, \tag{4.29}$$

for $\alpha = 1, ..., m$, where $\boldsymbol{P} = \delta_{\boldsymbol{F}}\psi$ is the first Piola or nominal stress tensor and where we have assumed ϕ as smooth. They determine the evolution of the plastic slip and through (4.5), (4.8) and (4.11) the evolution of the plastic map $\boldsymbol{F}^p$, $\text{Curl}^T \boldsymbol{F}^p$ and the hardening parameters $\gamma^{+\alpha}$. Furthermore, we obtain the Neumann-type boundary conditions

$$(\partial_{\boldsymbol{F}}\psi) \cdot \boldsymbol{n} = \bar{\boldsymbol{t}}_N \text{ on } \partial\mathcal{B}_P \quad \text{and} \quad (\partial_{\boldsymbol{g}^\alpha}\psi) \cdot \boldsymbol{n} = \tilde{t}_N^\alpha \text{ on } \partial\mathcal{B}_{H^\alpha} \,. \tag{4.30}$$

4.2.5 Explicit Form of the Micro-force Balance Equations

We now outline the explicit form of the micro-balance equations $(4.29)_2$ based on the specific constitutive functions introduced above. With the additive split (4.17) and the constitutive functions (4.18), (4.19), (4.20) and (4.23) we get

$$(\tau^\alpha - \tau_B^\alpha) = [\tau_c + \tilde{h}^\alpha] f'(\dot{\gamma}^\alpha) + \eta\dot{\gamma}^\alpha + \text{Div}[\boldsymbol{\kappa}^\alpha] \tag{4.31}$$

in $\mathcal{B}$ for $\alpha = 1, ..., m$. Here, we introduced the hardening contribution $\tilde{h}^\alpha := \partial_{\gamma^{+\alpha}}\psi_{SSD}$, the back stress $\tau_B^\alpha := \partial_{\gamma^\alpha}\psi_{GND}$ and the nonlocal contribution $\kappa^\alpha := \partial_{\boldsymbol{g}^\alpha}\psi_{GND}$ which can be evaluated as

$$\tau_B^\alpha = \lambda l^2 \boldsymbol{A}^T : \left[\mathfrak{M}^\alpha \boldsymbol{A}^T + \boldsymbol{A}^T \mathfrak{M}^{\alpha T}\right] \quad \text{and} \quad \boldsymbol{\kappa}^\alpha = -\lambda l^2 \mathfrak{s}^\alpha \cdot [\boldsymbol{A}^T \boldsymbol{F}^p \times (\boldsymbol{F}^{pT} \mathfrak{n}^\alpha)] \,. \tag{4.32}$$

First, we observe the *forest hardening* of the slip systems related to statistically stored dislocations (SSDs), governed by the not positive definite interaction matrix $a^{\alpha\beta}$. Next, we recognize a *back stress hardening* response due to contribution τ_B^α, related to the geometrically necessary dislocations (GNDs). Hence, the effective Schmid stress on the slip systems α is determined by $\tau_{eff}^\alpha := \tau^\alpha - \tau_B^\alpha$. The *divergence term* contains the length-scale-dependent contribution resulting from the GNDs. Finally, we have a viscous regularization term governed by the viscosity η.

4.3 Algorithmic Formulation of Gradient Crystal Plasticity

4.3.1 Time-Discrete Field Variables in Incremental Setting

The variational structure outlined above is of great importance with regard to the time-discrete setting of gradient-type dissipative solids, see Miehe [Mie11, Mie14]. To show this, we consider solutions of the field variables at the discrete times $0, \ldots, t_n, t_{n+1}, \ldots, T$ of the process interval $[0, T]$. In order to advance the solution within a typical time step, we focus on the finite time increment $[t_n, t_{n+1}]$, where

$$\tau_{n+1} := t_{n+1} - t_n > 0 \tag{4.33}$$

denotes the step size. In the subsequent treatment, all field variables at time t_n are assumed to be *known*. The goal then is to determine the fields at time t_{n+1} based on variational principles valid for the time increment under consideration. In order to obtain a compact notation, we drop in what follows the subscript $n+1$ and consider all variables without subscript to be evaluated at time t_{n+1}.

4.3.2 Update Algorithms for the Short-Range Field Variables

4.3.2.1 The Plastic Deformation Map

A geometrically consistent integration of the evolution equation (4.5) for the plastic deformation map $\boldsymbol{F}^p$ that satisfies the plastic incompressibility constraint $\boldsymbol{F}^P \in SL(3)$ in an algorithmically exact manner is provided by the exponential map, see e.g. Miehe [Mie96]. It is given by

$$\boldsymbol{F}^p = \exp[\boldsymbol{N}]\boldsymbol{F}^p_n \quad \text{with} \quad \boldsymbol{N} := \sum_{\alpha=1}^{m}(\gamma^\alpha - \gamma^\alpha_n)\,\mathfrak{M}^\alpha \tag{4.34}$$

where γ^α are the current accumulated slip on systems $\alpha = 1...m$. The algorithm preserves $\boldsymbol{F}^p \in SL(3)$ due to the property $\det\{\exp[\boldsymbol{N}]\} = \exp\{\mathrm{tr}[\boldsymbol{N}]\}$ and the deviatoric nature of $\mathfrak{M}^\alpha$. The efficient approximation of this exponential algorithm proposed in Miehe & Schotte [MS04] has the two-step multiplicative format

$$\boldsymbol{F}^p = \left(\det[\bar{\boldsymbol{F}}^p]\right)^{-1/3}\bar{\boldsymbol{F}}^p \quad \text{with} \quad \bar{\boldsymbol{F}}^p := [\boldsymbol{1} - \xi\boldsymbol{N}]^{-1}[\boldsymbol{1} + (1-\xi)\boldsymbol{N})]\boldsymbol{F}^p_n \,. \tag{4.35}$$

The first step $(4.35)_2$ considers a family of algorithms parameterized by the integration parameter $\xi \in [0,1]$. For $\xi = 0$ this includes a *fully explicit* update of $\boldsymbol{F}^p$ and for $\xi = 1$ a *fully implicit* Euler update of $\boldsymbol{F}^p$. $\xi = 1/2$ recovers the second-order accurate trapezoidal rule, which can also be considered as the Padé approximation of the exponential map (4.34). Clearly, this first step does not preserve the group structure $SL(3)$. This is achieved by the second step $(4.35)_1$, which enforces the unimodular characteristic of the plastic map.

4.3.2.2 The Geometric Dislocation Density Tensor

The integration of the evolution equation (4.8) for the transposed dislocation density tensor $\boldsymbol{A}^T$ does not need to satisfy any geometric constraint. We hence use the explicit update equation

$$\boldsymbol{A}^T = \boldsymbol{A}^T_n + \sum_{\alpha=1}^{m}\{\ \boldsymbol{S}^\alpha_n(\gamma^\alpha - \gamma^\alpha_n) + \boldsymbol{\mathcal{S}}^\alpha_n \cdot (\boldsymbol{g}^\alpha - \boldsymbol{g}^\alpha_n)\ \} \tag{4.36}$$

in terms of the sensitivities (4.9) which we rename as

$$\boldsymbol{S}^\alpha_n := \mathfrak{M}^\alpha\boldsymbol{A}^T_n + \boldsymbol{A}^T_n\mathfrak{M}^{\alpha T} \quad \text{and} \quad \boldsymbol{\mathcal{S}}^\alpha_n := -\mathfrak{s}^\alpha \otimes [\boldsymbol{F}^p_n \times (\boldsymbol{F}^{pT}_n\mathfrak{n}^\alpha)] \tag{4.37}$$

evaluated in terms of the density tensor $\boldsymbol{A}_n^T$ and the plastic deformation map $\boldsymbol{F}_n^p$ at time t_n. This linear structure guarantees the identical form of the variational-based micro-balance equation as in the continuous setting. Note that the algorithmic sensitivities of $\boldsymbol{A}^T$ with respect to γ^α and $\boldsymbol{g}^\alpha$ are determined by the *constant* tensors $\boldsymbol{S}_n^\alpha$ and $\boldsymbol{\mathcal{S}}_n^\alpha$. Hence, the second-order sensitivities vanish.

4.3.2.3 Hardening Parameters

Finally, the integration of the evolution equation (4.11) for the hardening parameters $\gamma^{+\alpha}$ is assumed to have the form

$$\gamma^{+\alpha} = \gamma_n^{+\alpha} + \tau f([\gamma^\alpha - \gamma_n^\alpha]/\tau) \tag{4.38}$$

in term of the norm function (4.12) or its regularization (4.13). The first- and second-order sensitivities read

$$\gamma_{,\alpha}^{+\alpha} = f'([\gamma^\alpha - \gamma_n^\alpha]/\tau) \quad \text{and} \quad \gamma_{,\alpha\beta}^{+\alpha} = f''([\gamma^\alpha - \gamma_n^\alpha]/\tau)/\tau\delta_{\alpha\beta} \ , \tag{4.39}$$

where we introduced the Kronecker delta $\delta_{\alpha\beta}$.

4.3.3 *Time-Discrete Incremental Variational Principle*

Associated with the time interval (4.33), we define the incremental energy functional of the crystal due to the coupled macro-micro-motion by

$$E^\tau(\varphi, \gamma^1, ..., \gamma^m) \ := \ \int_{\mathcal{B}} \{ \ \psi(\mathfrak{G}, \widehat{\mathfrak{L}}^\tau(\mathfrak{G})) - \psi_n \ \} \, dV \tag{4.40}$$

at given constitutive state $\{\mathfrak{G}_n, \mathfrak{L}_n\}$, where E is the energy functional defined in (4.16). It is governed by the free energy function ψ. Next, we define associated with the time interval (4.33) an *algorithmic* expression for an incremental dissipation potential functional

$$D^\tau(\gamma^1, ..., \gamma^m) \ := \ \tau \int_{\mathcal{B}} \phi([\gamma^1 - \gamma_n^1]/\tau, ..., [\gamma^m - \gamma_n^m]/\tau) \, dV \ , \tag{4.41}$$

where ϕ is the dissipation function defined in (4.23). Finally, we consider the incremental *algorithmic* form of the load functional (4.24)

$$P_{ext}^\tau(\varphi) = \int_{\mathcal{B}} \overline{\gamma} \cdot (\varphi - \varphi_n) dV + \int_{\partial\mathcal{B}_P} \bar{t}_N \cdot (\varphi - \varphi_n) dA + \sum_\alpha \int_{\partial\mathcal{B}_{H^\alpha}} \tilde{t}_N^\alpha (\gamma^\alpha - \gamma_n^\alpha) dA \tag{4.42}$$

at given state φ_n and γ_n^α. Then, the time-discrete counterpart of the rate-potential Π in (4.25) takes the algorithmic form

$$\underbrace{\Pi^\tau(\varphi,\gamma^1,...,\gamma^m)}_{potential} := \underbrace{E^\tau(\varphi,\gamma^1,...,\gamma^m)}_{energy} + \underbrace{D^\tau(\gamma^1,...,\gamma^m)}_{dissipation} - \underbrace{P^\tau_{ext}(\varphi)}_{work} . \quad (4.43)$$

We may write this incremental potential as

$$\begin{aligned}\Pi^\tau = &\int_{\mathcal{B}} \{\, \pi^\tau(\mathfrak{G}) - \overline{\gamma}\cdot(\varphi-\varphi_n)\,\}\, dV \\ &- \int_{\partial\mathcal{B}_P} \bar{t}_N\cdot(\varphi-\varphi_n)\, dA - \sum_\alpha \int_{\partial\mathcal{B}_{H^\alpha}} \tilde{t}_N^\alpha(\gamma^\alpha-\gamma_n^\alpha)\, dA\, ,\end{aligned} \quad (4.44)$$

where we call π^τ the *incremental internal work density*. It has the representation

$$\pi^\tau(\mathfrak{G}) = \psi(\mathfrak{G},\widehat{\mathfrak{L}}^\tau(\mathfrak{G})) - \psi_n + \tau\phi\left([\gamma^1-\gamma_n^1]/\tau,...,[\gamma^m-\gamma_n^m]/\tau\right) \quad (4.45)$$

in terms of the free energy function ψ and dissipation potential function ϕ, respectively. Then, the finite-step-sized incremental minimization principle

$$\{\varphi,\gamma^1,...,\gamma^m\} = \text{Arg}\{\inf_\varphi \inf_{\gamma^\alpha} \Pi^\tau(\varphi,\gamma^1,...,\gamma^m)\} \quad (4.46)$$

determines the macro- and micro-motion fields φ and $\gamma^1,...,\gamma^m$ at the current time t_{n+1} as the *minimum* of the incremental functional Π^τ.

4.3.4 Space-Time-Discrete Incremental Variational Principle

We now consider the spatial discretization of the coupled problem of gradient-extended crystal plasticity by a finite element method. Let $\mathfrak{T}^h$ denote a finite element triangulation of the solid domain $\mathcal{B}$. The index h indicates a typical mesh size based on E^h finite element domains $\mathcal{B}_e^h \in \mathfrak{T}^h$ and N^h global nodal points. Associated with the triangulation $\mathfrak{T}^h$, we write the finite element interpolations of the long-range constitutive state (4.3)

$$\mathfrak{G}^h = \boldsymbol{B}(\boldsymbol{X})\,\boldsymbol{d} \quad (4.47)$$

in terms of the *global nodal state vector* $\mathbf{d} \in \mathcal{R}^{(d+m)N^h}$, which contains the current macro-motion φ and the m scalar micro-motion fields $\gamma^1,...,\gamma^m$ at all nodal points of the finite element mesh. $\boldsymbol{B}$ is a symbolic representation of a global interpolation matrix for the coupled problem. We use *quadratic interpolations* for both the macro- as well as the micro-motion fields, in what follows denoted as Q2-Q2-elements. Clearly, this global array is never formulated explicitly, but represents symbolically the interpolations on all finite element domains $\mathcal{B}_e^h \in \mathfrak{T}^h$. It governs the space-discrete representation of the potential (4.44)

$$\Pi^{\tau h}(\boldsymbol{d}) = \int_{\mathcal{B}} \pi^{\tau}(\boldsymbol{B}\boldsymbol{d})\, dV \; + l.t.\,, \tag{4.48}$$

where the loading terms $(l.t.)$ are not displayed in order to keep the notation compact. Then, the space-time-discrete minimization principle

$$\boldsymbol{d} = \text{Arg}\{ \inf_{\boldsymbol{d}} \Pi^{\tau h}(\boldsymbol{d}) \} \tag{4.49}$$

determines the global nodal state vector $\boldsymbol{d}$ of the finite element mesh at the current time t_{n+1}. The necessary condition of the discrete variational problem (4.49) reads

$$\Pi^{\tau h}_{,\boldsymbol{d}} = \int_{\mathcal{B}} \boldsymbol{B}^T [\partial_{\mathfrak{G}} \pi^{\tau}]\, dV - \boldsymbol{L} = \boldsymbol{0} \tag{4.50}$$

and provides a nonlinear algebraic system for the determination of the nodal state vector $\boldsymbol{d}$. $\boldsymbol{L}$ is the nodal load vector due to body forces and Neumann loads. A standard Newton-type iteration scheme is applied to the fully coupled nonlinear algebraic system (4.50) updates the global nodal state by the algorithm

$$\boldsymbol{d} \Leftarrow \boldsymbol{d} - [\,\Pi^{\tau h}_{,\boldsymbol{dd}}\,]^{-1}[\Pi^{\tau h}_{,\boldsymbol{d}}] \quad \text{with} \quad \Pi^{\tau h}_{,\boldsymbol{dd}} := \int_{\mathcal{B}} \boldsymbol{B}^T [\partial^2_{\mathfrak{G}\mathfrak{G}} \pi^{\tau}] \boldsymbol{B}\, dV \tag{4.51}$$

in terms of the monolithic tangent matrix of the coupled problem $\Pi^{\tau h}_{,\boldsymbol{dd}}$. Observe the *symmetry of the tangent matrix* induced by the incremental variational structure of the coupled problem. The update (4.51) is performed until convergence is achieved in the sense $|\Pi^{\tau h}_{,\boldsymbol{d}}| < tol$.

The finite element residual and tangent are governed by the *generalized stress and tangent arrays* $\mathfrak{S} := \partial_{\mathfrak{G}} \pi^{\tau}(\mathfrak{G})$ and $\mathfrak{C} := \partial^2_{\mathfrak{G}\mathfrak{G}} \pi^{\tau}(\mathfrak{G})$, i.e. the first and second derivatives of the incremental internal work density π^{τ} defined in (4.45) by the

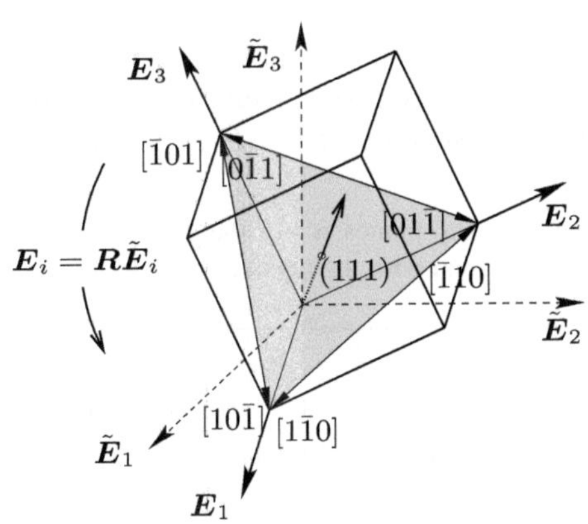

Fig. 4.7 Orientation of the fcc unit cell. The standard cartesian base $\{\tilde{\boldsymbol{E}}_i\}_{i=1,3}$ is rotated to the base $\{\boldsymbol{E}_i\}_{i=1,2,3}$ aligned to the fcc crystal. The (111) slip plane of the fcc crystal is marked by the shading.

discretized constitutive state vector $\mathfrak{G}$ at current time t_{n+1}. The partitioned structure of the generalized stress array takes the form

$$\mathcal{S} = \partial_{\mathfrak{G}} \pi^{\tau}(\mathfrak{G}) \begin{bmatrix} \boldsymbol{P}^{e} \boldsymbol{F}^{p-T} \\ -(\tau^{\alpha} - \tau^{\alpha}_{B}) + [\, \tau_{c} + \tilde{h}^{\alpha} \,] f'(\dot{\gamma}^{\alpha}) + \eta \dot{\gamma}^{\alpha} \\ \boldsymbol{\kappa}^{\alpha} \end{bmatrix} \tag{4.52}$$

for $\alpha = 1 \dots m$ with the algorithmic expression $\dot{\gamma}^{\alpha} = [\gamma^{\alpha} - \gamma^{\alpha}_{n}]/\tau$.

4.4 Example 1: Analysis of an F.C.C. Crystal Grain Aggregate

4.4.1 Slip Systems and Euler Angles

With the space-time discrete algorithmic formulation of our gradient crystal plasticity model, we are now in the position to carry out numerical simulations. We restrict our attention to the example of copper, which has an f.c.c. crystal structure that is characterized by four $\{111\}$ slip planes defined by slip normals $\tilde{\mathfrak{n}}$ and three $\langle 110 \rangle$ slip directions $\tilde{\mathfrak{s}}$ on each plane, yielding $m = 12$ slip systems $\{\tilde{\mathfrak{s}}^{\alpha}, \tilde{\mathfrak{n}}^{\alpha}\}$ in total, see Figure 4.7 and Table 4.1, where the slip directions and slip normals of all slip systems are given with respect to an orthogonal frame $\{\tilde{\boldsymbol{E}}_{A}\}_{A=1,2,3}$ aligned to the crystal axes. A relative rotation of the crystal and thereby $\{\tilde{\boldsymbol{E}}_{A}\}_{A=1,2,3}$ with respect to the fixed orthogonal frame $\{\boldsymbol{E}_{A}\}_{A=1,2,3}$ can be described by $\boldsymbol{E}_{A} = \boldsymbol{R}\, \tilde{\boldsymbol{E}}_{A}$ with $\boldsymbol{R} = \boldsymbol{Q}_{3}(\vartheta_{3}) \boldsymbol{Q}_{2}(\vartheta_{2}) \boldsymbol{Q}_{1}(\vartheta_{1}) \in SO(3)$ and where the matrices $\boldsymbol{Q}_{1}, \boldsymbol{Q}_{2}$ and $\boldsymbol{Q}_{3}$ are defined by the explicit expressions

$$\boldsymbol{Q}_{1} = \begin{bmatrix} \cos\vartheta_{1} & -\sin\vartheta_{1} & 0 \\ \sin\vartheta_{1} & \cos\vartheta_{1} & 0 \\ 0 & 0 & 1 \end{bmatrix}, \quad \boldsymbol{Q}_{2} = \begin{bmatrix} 1 & 0 & 0 \\ 0 & \cos\vartheta_{2} & -\sin\vartheta_{2} \\ 0 & \sin\vartheta_{2} & \cos\vartheta_{2} \end{bmatrix}, \quad \boldsymbol{Q}_{3} = \begin{bmatrix} \cos\vartheta_{3} & -\sin\vartheta_{3} & 0 \\ \sin\vartheta_{3} & \cos\vartheta_{3} & 0 \\ 0 & 0 & 1 \end{bmatrix},$$

in terms of the Euler angles ϑ_{1}, ϑ_{2} and ϑ_{3} that parametrize rotations about the Z-axis, the X-axis and again the Z-axis, respectively. This also governs the rotation of the slip directions and normals, such that $\mathfrak{s}^{\alpha} = \boldsymbol{R}\, \tilde{\mathfrak{s}}^{\alpha}$ and $\mathfrak{n}^{\alpha} = \boldsymbol{R}\, \tilde{\mathfrak{n}}^{\alpha}$.

Table 4.1 The slip directions $\tilde{\mathfrak{s}}^{\alpha}$ and planes $\tilde{\mathfrak{n}}^{\alpha}$ of the 12 slip systems of f.c.c. crystals in terms of a basis $\{\tilde{\boldsymbol{E}}_{A}\}_{A=1,2,3}$ aligned with the crystal axes

	$\sqrt{2}\tilde{\mathfrak{s}}$	$\sqrt{3}\tilde{\mathfrak{n}}$		$\sqrt{2}\tilde{\mathfrak{s}}$	$\sqrt{3}\tilde{\mathfrak{n}}$		$\sqrt{2}\tilde{\mathfrak{s}}$	$\sqrt{3}\tilde{\mathfrak{n}}$
A2	$[0\,1\,\bar{1}]$	$(\bar{1}\,1\,1)$	A3	$[1\,0\,1]$	$(\bar{1}\,1\,1)$	A6	$[\bar{1}\,\bar{1}\,0]$	$(\bar{1}\,1\,1)$
B2	$[0\,1\,\bar{1}]$	$(1\,1\,1)$	B4	$[\bar{1}\,0\,1]$	$(1\,1\,1)$	B5	$[1\,\bar{1}\,0]$	$(1\,1\,1)$
C1	$[0\,1\,1]$	$(1\,1\,\bar{1})$	C3	$[1\,0\,1]$	$(1\,1\,\bar{1})$	C5	$[1\,\bar{1}\,0]$	$(1\,1\,\bar{1})$
D1	$[0\,1\,1]$	$(\bar{1}\,1\,\bar{1})$	D4	$[1\,0\,\bar{1}]$	$(\bar{1}\,1\,\bar{1})$	D6	$[\bar{1}\,\bar{1}\,0]$	$(\bar{1}\,1\,\bar{1})$

Table 4.2 Material parameters for Copper

No.	Parameter	Name	Value	Unit
1.	μ	shear modulus	54135.0	MPa
2.	β	exponent	2.125	-
3.	τ_0	initial Schmid stress	1.0	MPa
4.	q	off-diagonal component	1.4	-
5.	b	diagonal component	1.0	-
6.	n	exponent	1	-
7.	$\hat{\tau}$	reference shear stress	0.0	MPa
8.	$\hat{\gamma}$	reference slip strain	1.0	-
9.	λ	energy density	54135.0	MPa
10.	η	viscosity	0.1	MPa s
11.	$\dot{\gamma}_0$	rate regularization	0.1	s^{-1}
12.	l	GND length scale	$0.1 - 2.0$	m

4.4.2 Voronoi-Tessellated Unit Cell under Shear

Consider the numerical example of a Voronoi-tessellated unit cell under shear deformation. The unit cell consists of five grains and has a periodic structure in all three space dimensions. The unit cell has a dimension of $10 \times 10 \times 10$ mm. In each grain the initial orientation of the crystal lattice is supposed to be different. At the boundaries of the cell $\partial\mathcal{B}$, homogeneous boundary conditions are applied

$$\varphi(\boldsymbol{X}) = \bar{\boldsymbol{F}}(\Gamma)\boldsymbol{X} , \quad \text{with} \quad \bar{\boldsymbol{F}}(\Gamma) = \boldsymbol{1} + \Gamma\, \boldsymbol{e}_2 \otimes \boldsymbol{E}^3 , \tag{4.53}$$

Here, $\boldsymbol{e}_i$ for $i = 1, 2, 3$ are the covariant basis vectors of the spatial configuration, $\boldsymbol{E}^i$ for $i = 1, 2, 3$ the contravariant basis vectors of the Lagrangian configuration. The prescribed shear Γ is applied with a rate of $\dot{\Gamma} = 1.0\ \mathrm{s}^{-1}$. Furthermore, at the grain boundaries, "micro-clamped" boundary conditions for the plastic slips γ^α are applied $\gamma^\alpha = 0$ for $\boldsymbol{X} \in \partial\mathcal{B}_{\text{grain}}$. The geometric setup is depicted in Figure 4.8 and the material parameters are given in Table 4.2. The Voronoi-tessellated unit cell is discretized by 23.210 quadratic tetrahedral elements with 33.871 nodes. In the numerical simulation, a time step of $\Delta t = 10^{-6}$ s is used. The unit cell is deformed up to a shear of $\Gamma = 20.0$ %. The numerical results obtained for two different length scale parameters are shown in Figures 4.9 and 4.10. Figure 4.10 shows the accumulated plastic slip $\bar{\gamma} = \sum_\alpha \gamma^\alpha$. Observe that the smaller length scale l leads to a more distinct transition zone between the micro-clamped grain boundaries and the interior of the grains. Furthermore, the smaller length scale yield a higher amount of plastic

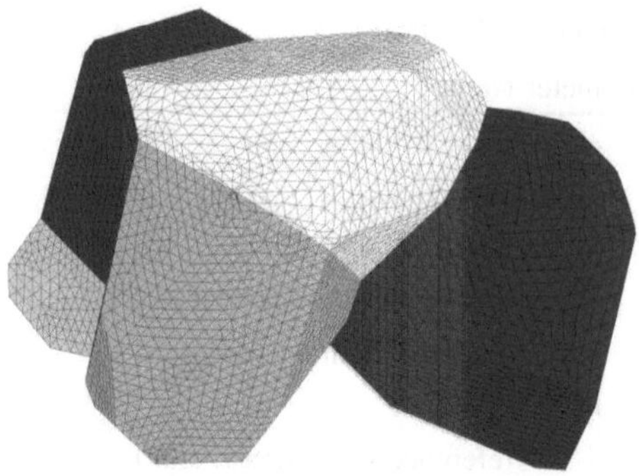

Fig. 4.8 Three-dimensional Voronoi-tessellated unit cell under shear: Geometric setup

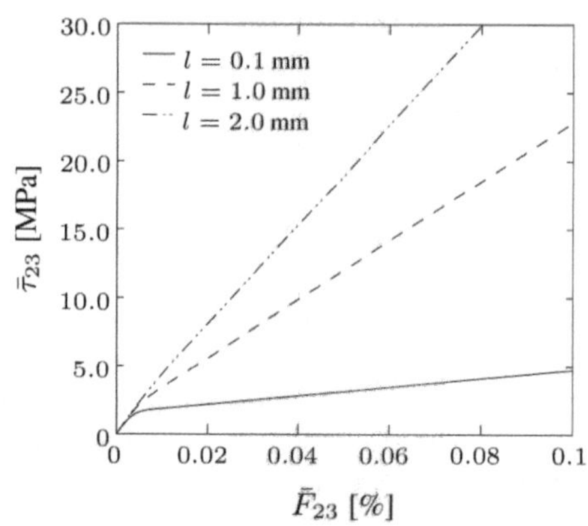

Fig. 4.9 Three-dimensional Voronoi-tessellated unit cell under shear: Resultant homogenized shear stress $\bar{\tau}_{23}$ over the homogeneus strain $\bar{F}_{23}$ for different length scales l

slip activity $\bar{\gamma}$ inside the grains. By means of homogenization we observe a stiffer macroscopic behavior for the unit cell with higher length scale. This is visualized in Figure 4.9 for three different length scales.

4.5 Example 2: Laminate Microstructure in Single Crystals

In the next example, we want to demonstrate the capability of our gradient crystal plasticity model to predict the formation and evolution of plastic deformation microstructure in f.c.c. Copper with same plane double slip.

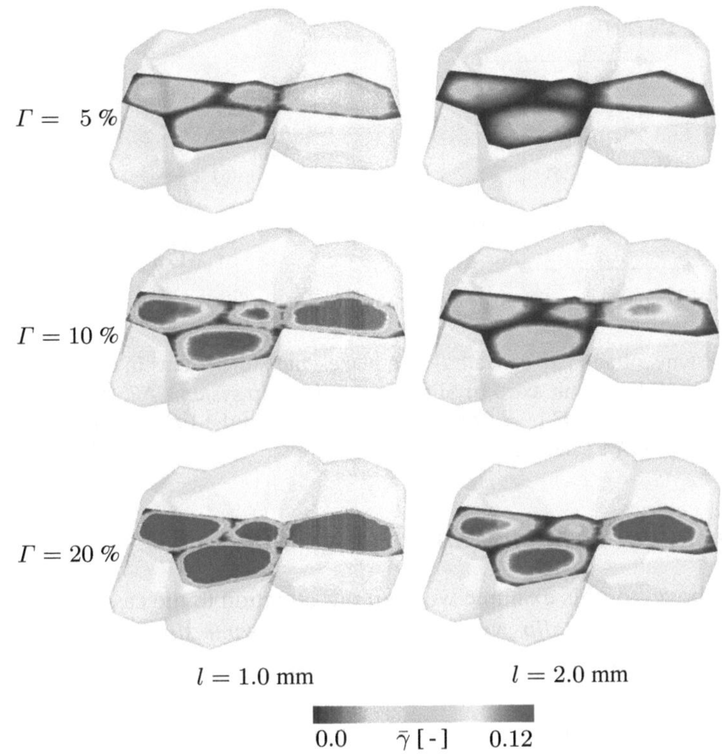

Fig. 4.10 Three-dimensional Voronoi-tessellated unit cell under shear: Evolution of accumulated plastic slip $\bar{\gamma}$ for different length scales and at different deformation states

4.5.1 Double Slip Systems

For the subsequent considerations, we restrict our attention to two slip systems with common slip normal. Particularly, we choose the systems A2 and A6, see Table 4.1. Assuming a relative rotation $\boldsymbol{R}$ of $\vartheta_1 = -30.00°$, $\vartheta_2 = 54.74°$ and $\vartheta_3 = -45.00°$ between the local crystal basis $\{\tilde{\boldsymbol{E}}_A\}$ and the global referential basis $\{\boldsymbol{E}_A\}$, we obtain the referential slip directions and slip normals as

$$\mathfrak{s}^1 = \frac{1}{2}\begin{bmatrix} -\sqrt{3} \\ 1 \\ 0 \end{bmatrix} \boldsymbol{E}_A\,, \quad \mathfrak{s}^2 = \frac{1}{2}\begin{bmatrix} \sqrt{3} \\ 1 \\ 0 \end{bmatrix} \boldsymbol{E}_A \quad \text{and} \quad \mathfrak{n} = \begin{bmatrix} 0 \\ 0 \\ 1 \end{bmatrix} \boldsymbol{E}_A\,. \tag{4.54}$$

Figure 4.11b shows the orientation of the slip systems in the frame $\{\boldsymbol{E}_A\}$.

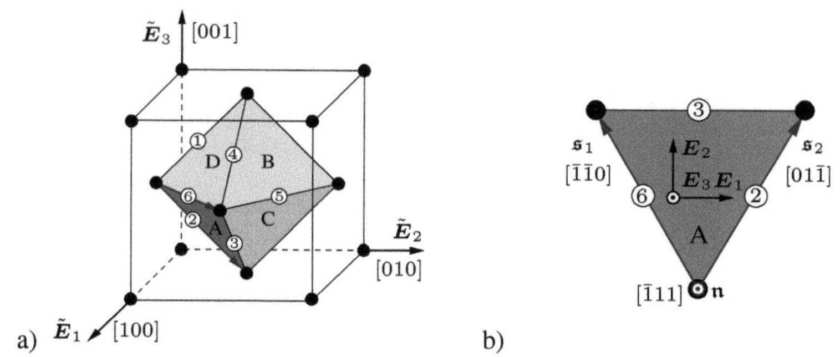

Fig. 4.11 Slip systems of an f.c.c. crystal: a) Slip planes and slip systems in the coordinate system $\tilde{\boldsymbol{E}}_A$ aligned with the unit axes. b) Slip plane A and slip systems A2, A3 and A6 in the rotated coordinate system $\boldsymbol{E}_A = \boldsymbol{R}\tilde{\boldsymbol{E}}_A$ after rotation with the Euler rotation $\boldsymbol{R}$ with $\vartheta_1 = -30.00°$, $\vartheta_2 = 54.74°$ and $\vartheta_3 = -45.00°$.

4.5.2 *Implications of Same Plane Double Slip*

As mentioned above, for this example we limit our attention to the case of a single crystal with *only two active slip systems that lie on the same slip plane* such that $m = 2$ and $\mathfrak{n}^1 = \mathfrak{n}^2 = \mathfrak{n}$.

4.5.2.1 Plastic Deformation

Making use of the assumption $m = 2$ and $\mathfrak{n} = \mathfrak{n}^1 = \mathfrak{n}^2$, equation (4.5) can be directly integrated and leads to

$$\boldsymbol{F}^p(\gamma^\alpha) = \boldsymbol{1} + \sum_{\alpha=1}^{2} \gamma^\alpha \mathfrak{s}^\alpha \otimes \mathfrak{n} \quad \Rightarrow \quad \boldsymbol{F}^{p\,-1}(\gamma^\alpha) = \boldsymbol{1} - \sum_{\alpha=1}^{2} \gamma^\alpha \mathfrak{s}^\alpha \otimes \mathfrak{n}\,. \tag{4.55}$$

Unlike general multi-slip, the assumption of same plane double slip thus allows the derivation of an explicit relation between plastic deformation and slips as $\boldsymbol{F}^p(\gamma^\alpha)$.

4.5.2.2 Dislocation Density Tensor

For same plane double slip, (4.7) with (4.55) yields the simplified relation

$$\boldsymbol{A}(\nabla\gamma^\alpha) = \sum_{\alpha=1}^{2} \mathfrak{s}^\alpha \otimes \left[(\nabla\gamma^\alpha \cdot \mathfrak{s}^\alpha)\mathfrak{t}^\alpha - (\nabla\gamma^\alpha \cdot \mathfrak{t}^\alpha)\mathfrak{s}^\alpha \right]. \tag{4.56}$$

Note from (4.56) that the dislocation density tensor $\boldsymbol{A}$ and hence also the measure $||\boldsymbol{A}||^2$ are both independent of slip gradient components in the normal direction $\mathfrak{n}$.

4.5.2.3 Stored Energy Functional

For the double slip problem with $m = 2$, with the explicit dislocation density tensor (4.56) at hand, and introducing $\boldsymbol{b}^{\alpha\beta} = (1 - \delta^{\alpha\beta})\frac{\mathfrak{s}^{\alpha}\cdot\mathfrak{s}^{\beta}}{2}\left[(\mathfrak{s}^{\alpha}\cdot\mathfrak{s}^{\beta})\mathfrak{t}^{\alpha}\otimes\mathfrak{t}^{\beta} - (\mathfrak{t}^{\alpha}\cdot\mathfrak{s}^{\beta})\mathfrak{s}^{\alpha}\otimes\mathfrak{t}^{\beta} + (\mathfrak{t}^{\alpha}\cdot\mathfrak{t}^{\beta})\mathfrak{s}^{\alpha}\otimes\mathfrak{s}^{\beta}\right] + \delta^{\alpha\beta}\left[\mathfrak{t}^{\alpha}\otimes\mathfrak{t}^{\beta} + \mathfrak{s}^{\alpha}\otimes\mathfrak{s}^{\beta}\right]$ we can rewrite the energy of GNDs ψ_{GND} defined in (4.20) as

$$\psi_{GND}(\nabla\gamma^{\alpha}) = \frac{\lambda}{2}\, l_{GND}^{2} \sum_{\alpha=1}^{2}\sum_{\beta=1}^{2} \nabla\gamma^{\alpha}\cdot(\boldsymbol{b}^{\alpha\beta}\nabla\gamma^{\beta})\,. \tag{4.57}$$

Note that due to the structure of the tensors $\boldsymbol{b}^{\alpha\beta}$ derived from the definition of $\boldsymbol{A}$ and defined solely by the slip directions $\mathfrak{s}^{\alpha}$ and the transverse slip directions $\mathfrak{t}^{\alpha}$, the gradient energy ψ_{GND} as given in (4.57) punishes *all but* the $\mathfrak{n}$-components of the gradients $\nabla\gamma^{\alpha}$. However, the $\mathfrak{n}$-direction is exactly the direction in which we expect the formation of plastic laminate interfaces. To see this, consider two adjacent domains with equal elastic deformation $\boldsymbol{F}^{e}$ but with distinct plastic slip states $\{\check{\gamma}^{1}, \check{\gamma}^{2}\}$ and $\{\tilde{\gamma}^{1}, \tilde{\gamma}^{2}\}$ and hence distinct total deformations

$$\{\check{\gamma}^{1}, \check{\gamma}^{2}\} \Rightarrow \check{\boldsymbol{F}} = \boldsymbol{F}^{e}\check{\boldsymbol{F}}^{p} = \boldsymbol{F}^{e}(\boldsymbol{1} + (\check{\gamma}^{1}\mathfrak{s}^{1} + \check{\gamma}^{2}\mathfrak{s}^{2})\otimes\mathfrak{n})\,, \tag{4.58}$$
$$\{\tilde{\gamma}^{1}, \tilde{\gamma}^{2}\} \Rightarrow \tilde{\boldsymbol{F}} = \boldsymbol{F}^{e}\tilde{\boldsymbol{F}}^{p} = \boldsymbol{F}^{e}(\boldsymbol{1} + (\tilde{\gamma}^{1}\mathfrak{s}^{1} + \tilde{\gamma}^{2}\mathfrak{s}^{2})\otimes\mathfrak{n})\,. \tag{4.59}$$

By subtracting (4.58) from (4.59), one can see that the resulting deformation gradients $\check{\boldsymbol{F}}$ and $\tilde{\boldsymbol{F}}$ of such domains are rank one connected by an interface with reference normal $\mathfrak{n}$,

$$\check{\boldsymbol{F}} - \tilde{\boldsymbol{F}} = \left\{\boldsymbol{F}^{e}[(\check{\gamma}^{1} - \tilde{\gamma}^{1})\mathfrak{s}^{1} + (\check{\gamma}^{2} - \tilde{\gamma}^{2})\mathfrak{s}^{2}]\right\}\otimes\mathfrak{n}\,. \tag{4.60}$$

We hence see that the material can form laminate deformation microstructure with coherent sharp interfaces with normals $\mathfrak{n}$ between adjacent layers of (any two) distinct slip states. To regularize the resulting sharp interfaces, we add a gradient term that energetically punishes *only* the $\mathfrak{n}$-components of $\nabla\gamma^{\alpha}$,

$$\psi_{LAM}(\nabla\gamma^{\alpha}) = \frac{\lambda}{2}\, l_{\Gamma}^{2} \sum_{\alpha=1}^{2} \nabla\gamma^{\alpha}(\mathfrak{n}\otimes\mathfrak{n})\nabla\gamma^{\alpha}\,, \tag{4.61}$$

where $\lambda > 0$ is the material parameter associated with the energy density of lattice defects already introduced in (4.20) and $l_{\Gamma} > 0$ is proportional to the length-scale of the regularized transition between states of different plastic slip at the interfaces. Introducing the set of structural tensors $c^{\alpha\beta} = \delta^{\alpha\beta}(\mathfrak{n}\otimes\mathfrak{n})$, reformulating the energy ψ_{LAM} in the spirit of (4.57), we can introduce the gradient energy $\psi_{GRAD} = \psi_{GND} + \psi_{LAM}$ given by

$$\psi_{GRAD}(\nabla\gamma^{\alpha}) = \frac{\lambda}{2} \sum_{\alpha=1}^{2}\sum_{\beta=1}^{2} \nabla\gamma^{\alpha}\cdot\left[(l_{GND}^{2}\,\boldsymbol{b}^{\alpha\beta} + l_{\Gamma}^{2}\,\boldsymbol{c}^{\alpha\beta})\nabla\gamma^{\beta}\right] \tag{4.62}$$

Table 4.3 Material parameters for Copper , taken from Dmitrieva et al. [DDMR09] to fit the experiment

No.	Parameter	Name	Value	Unit
1.	μ	shear modulus	48000.0	MPa
2.	β	exponent	2.125	-
3.	τ_0	initial Schmid stress	100.0	MPa
4.	q	off-diagonal component	1.0	-
5.	b	diagonal component	0.0	-
6.	n	exponent	2	-
7.	$\hat{\tau}$	reference shear stress	400.0	MPa
8.	$\hat{\gamma}$	reference slip strain	0.01	-
9.	λ	energy density	10000.0	MPa
10.	η	viscosity	100.0	MPa s
11.	$\dot{\gamma}_0$	rate regularization	0.001	s^{-1}
12.	l	GND length scale	0.5	m
13.	l_Γ	regularization length	0.05	m

that punishes *all* contributions of the slip gradients $\nabla\gamma^\alpha$. Choosing $l_\Gamma \ll l_{GND}$ leads to an anisotropic gradient energy $\psi_{GRAD}(\nabla\gamma^1 = \|\nabla\gamma^1\|\mathfrak{m}^1, \nabla\gamma^2 = \|\nabla\gamma^2\|\mathfrak{m}^2)$ with $\psi_{GRAD}(\alpha\mathfrak{n}, \alpha\mathfrak{n}) < \psi_{GRAD}(\alpha\mathfrak{m}^1, \alpha\mathfrak{m}^2) \forall \mathfrak{m}^1, \mathfrak{m}^2 \neq \pm\mathfrak{n} \in \mathcal{R}^3\ \forall\alpha \in \mathcal{R}$. This referential gradient anisotropy is somewhat similar to that introduced in the context of martensitic laminates with a known rank-one direction in Hildebrand & Miehe [HM12].

4.5.3 Laminate Deformation Microstructure in Single Crystal Copper

The two numerical simulations are inspired by the shear experiments described in Dmitrieva et al. [DDMR09], where deformation laminate microstructure is observed in a Copper single crystal with the orientation depicted in Figure 4.11b. The material parameters are taken from Dmitrieva et al. [DDMR09] to fit the experiment, and given in Table 4.3. With this in mind, we consider a cubic domain with length $L=1$ mm, which we discretize by $40\times40\times40$ finite elements. We prescribe homogeneous boundary conditions

$$\varphi(\boldsymbol{X}) = \bar{\boldsymbol{F}}(\Gamma)\boldsymbol{X}\,, \quad \text{with} \quad \bar{\boldsymbol{F}}(\Gamma) = \boldsymbol{1} + \Gamma\,\bar{\mathfrak{s}} \otimes \mathfrak{n}\,, \tag{4.63}$$

where $\bar{\mathfrak{s}} = (\mathfrak{s}^2 - \mathfrak{s}^1)/|\mathfrak{s}^2 - \mathfrak{s}^1|$ and where $\dot{\Gamma} = 0.25\ s^{-1}$. Note that (4.63) is a time-dependent simple shear that cannot be accommodated by single slip on one of the two slip systems. In all simulations, we use a time step of $\Delta t = 0.0001$s. With this essential deformation boundary condition, we carry out two simulations

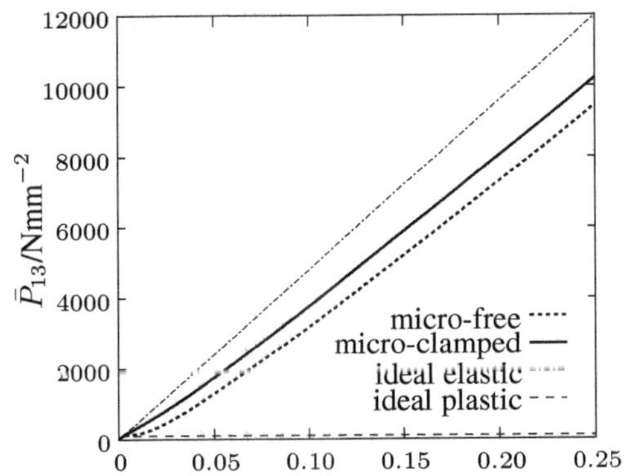

Fig. 4.12 The homogenized stress-strain response $\bar{P}_{13}(\bar{F}_{13} = \Gamma)$ for laminate deformation microstructure formation and evolution during micro-free and micro-clamped homogeneous simple shear. The GND-governed boundary layer of the plastic slips in the micro-clamped simulations yields a stiffer response.

with different micro-boundary conditions for the plastic slips. In one simulation, we use natural micro-boundary condition of the form $\tilde{\boldsymbol{t}}_N^\alpha = 0$ on $\partial\mathcal{B}$ and in the other, we use essential micro-boundary condition of the form $\gamma^\alpha = 0$ on $\partial\mathcal{B}$. These boundary conditions are sometimes referred to as *micro-free* and *micro-clamped* boundary conditions, see e.g. Svendsen et al. [SBER10]. The results of the simulations are given in Figures 4.12–4.14. Under micro-free boundary conditions, our model predicts that the plastic deformation state in $\mathcal{B}$ is homogeneous up to $-\gamma^1 = \gamma^2 \approx 0.015$, see Figure 4.13a, which depicts a state slightly past this point. The non-convexity of the latent hardening contribution $\hat{\psi}_{SSD}$ then triggers the formation of laminate microstructure with interface normal $\mathbf{n}$ as predicted, where in one layer, $\gamma^1 \approx -0.015$ stays constant while γ^2 accommodates further deformation, while in the other $\gamma^2 \approx 0.015$ stays constant while γ^1 accommodates further deformation, see Figures 4.13b-e. Note from the figures that the width of the smooth transition between the states $\{-0.015, \gamma^2\}$ and $\{\gamma^1, 0.015\}$ is governed by the regularization length scale l_Γ and that (4.13) very well approximates the accumulated slips $\gamma^{+\alpha} = |\gamma^\alpha|$. Under micro-clamped boundary conditions, our model predicts that the plastic deformation state in $\mathcal{B}$ is homogeneous besides a boundary layer up to $-\gamma^1 = \gamma^2 \approx 0.0175$, see Figure 4.14a, which depicts a state slightly before this point. Then, again, formation of microstructure starts, see Figures 4.14b-e. Note from the figures that the width of the smooth transition between the states $\{-0.0175, \gamma^2\}$ and $\{\gamma^1, 0.0175\}$ in the direction $\mathbf{n}$ is again governed by the regularization length scale l_Γ while the width of the boundary layer at the boundaries in $\boldsymbol{E}_1$ and $\boldsymbol{E}_2$-direction is now governed by the GND length scale l. To show the effect of the boundary conditions on the homogenized overall behavior, Figure 4.12 depicts the homogenized stress-strain relation $\bar{P}_{13}(\bar{F}_{13})$. The figure nicely shows the more pronounced hardening under micro-clamped boundary conditions caused

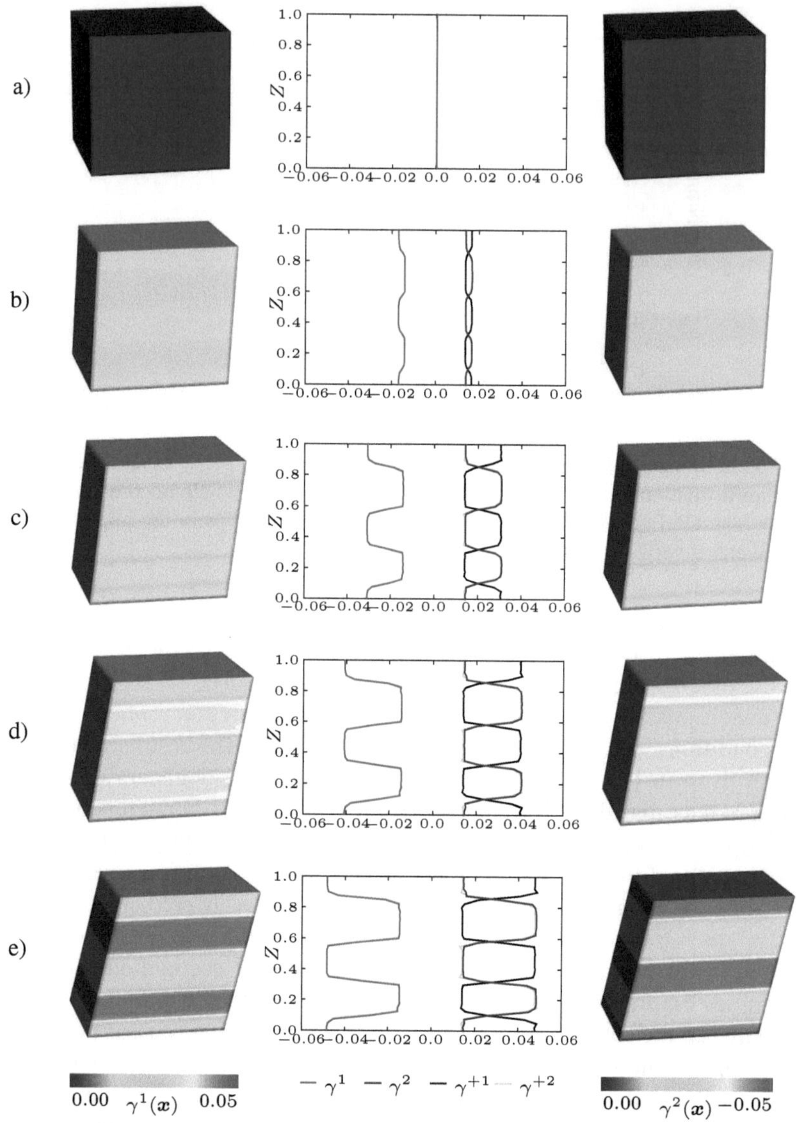

Fig. 4.13 Snapshots of the distributions of $\gamma^1(\boldsymbol{x})$ and $\gamma^2(\boldsymbol{x})$ in $\mathcal{S}$ and of $\gamma^1(Z)$, $\gamma^2(Z)$ and projected values of $\gamma^{+1}(Z)$, $\gamma^{+2}(Z)$ along the referential line connecting the points $(\frac{1}{2}L, \frac{1}{2}L, 0)$ and $(\frac{1}{2}L, \frac{1}{2}L, L)$ during *micro-free* homogeneous shear of all points on $\partial\mathcal{B}_\varphi = \partial\mathcal{B}$ with $\varphi(\boldsymbol{X},t) = \bar{\varphi}(\boldsymbol{X},t) = (\boldsymbol{1} + \Gamma(t)\boldsymbol{e}_1 \otimes \boldsymbol{E}_3)\boldsymbol{X}$: a) $t = 0.0$s, $\Gamma = 0.0$, b) $t = 0.25$s, $\Gamma = 0.0625$, c) $t = 0.5$s, $\Gamma = 0.125$, d) $t = 0.75$s, $\Gamma = 0.1875$ and e) $t = 1.0$s, $\Gamma = 0.25$.

by the inhibition of plastic slip in the boundary layer with width proportional to l_{GND}. Comparing the stress strain curves to the ones measured by Dmitrieva et al. [DDMR09], we see from the slope of the curve $\bar{P}_{13}(\bar{F}_{13})$ that our model dras-

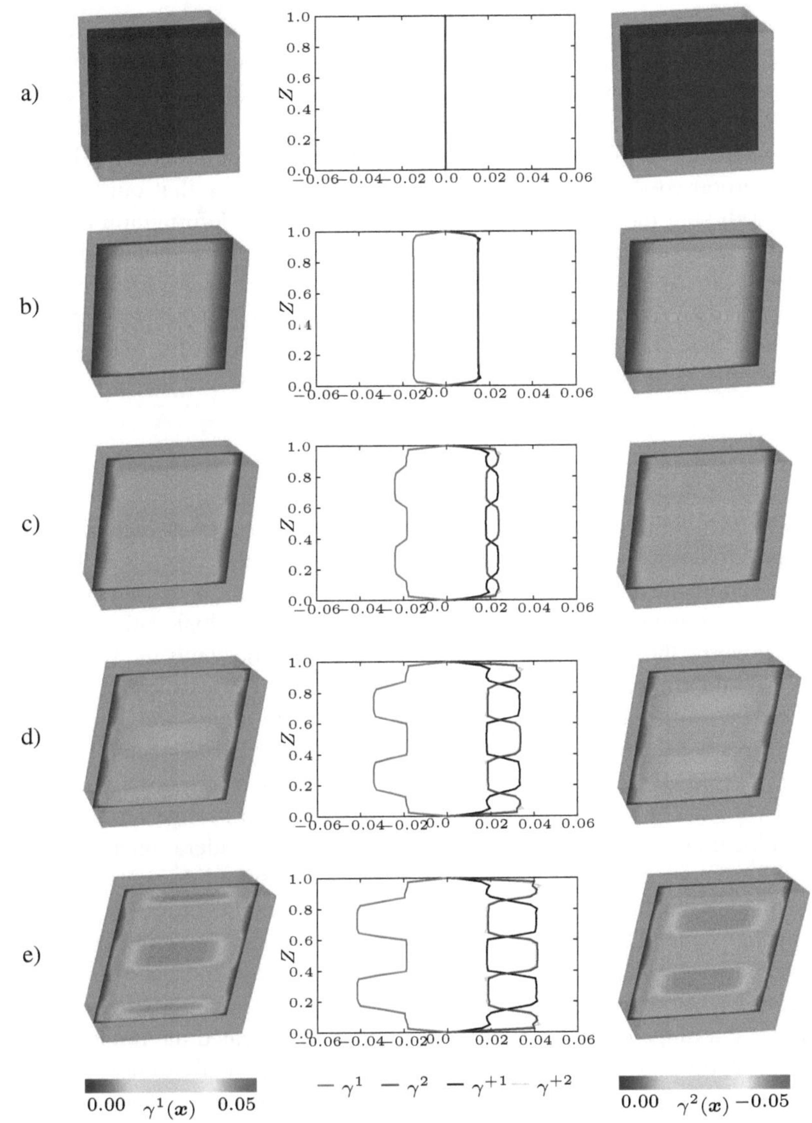

Fig. 4.14 Snapshots of the distributions of $\gamma^1(\boldsymbol{x})$ and $\gamma^2(\boldsymbol{x})$ in $\mathcal{S}$ and of $\gamma^1(Z)$, $\gamma^2(Z)$ and projected values of $\gamma^{+1}(Z)$, $\gamma^{+2}(Z)$ along the referential line connecting the points $(\frac{1}{2}L, \frac{1}{2}L, 0)$ and $(\frac{1}{2}L, \frac{1}{2}L, L)$ during *micro-clamped* homogeneous shear of all points on $\partial\mathcal{B}_\varphi = \partial\mathcal{B}$ with $\boldsymbol{\varphi}(\boldsymbol{X}, t) = \bar{\boldsymbol{\varphi}}(\boldsymbol{X}, t) = (\boldsymbol{1} + \Gamma(t)\boldsymbol{e}_1 \otimes \boldsymbol{E}_3)\boldsymbol{X}$: a) $t = 0.0$s, $\Gamma = 0.0$, b) $t = 0.25$s, $\Gamma = 0.0625$, c) $t = 0.5$s, $\Gamma = 0.125$, d) $t = 0.75$s, $\Gamma = 0.1875$ and e) $t = 1.0$s, $\Gamma = 0.25$.

tically overestimates the hardening and from the form of $\bar{P}_{13}(\bar{F}_{13})$ that the chosen bi-quadratic hardening function $\hat{\psi}_{SSD}$ is not completely in agreement with experimental results. Also, we note that the predicted deformation laminate microstructure with alternating activities of the A2 and the A6 slip systems is *not* in agreement to that observed by Dmitrieva et al. [DDMR09], where alternating activities of the A2 and the A6 slip systems are expected, but where it is concluded that in disagreement with this expectation, the laminate consists of regions with A2/A6 activities alternating with regions of C1/C5 activity. Such behavior can of course not be captured in the proposed double slip model. In summary, we see that our model is capable of predicting the formation and evolution of plastic deformation laminate microstructure in combination with length scale effects due to geometrically necessary dislocations. The hardening parameters necessary to trigger the formation of laminates lead to an overestimation of the overall hardening. However, due to its strong physical basis and its canonical form, it is an ideal basis for extensions to multi slip and also quantitatively realistic latent hardening laws.

4.6 Conclusion

We outlined a formulation and finite element implementation of micromechanically-motivated multiplicative gradient plasticity for single crystals, which may be considered as a canonically simple setting. The central ingredient is a contribution to the free energy storage in terms of the geometrically consistent dislocation gradient tensor, that governs the GND-based kinematic hardening mechanisms. In order to partially overcome the complexity of full multislip scenarios, we suggested a new *viscous regularized concept* of rate-independent crystal plasticity, and exploited in a systematic manner the long- and short-range nature of the variables involved, see also Miehe et al. [MMH14]. The separation into long- and short-range states was algorithmically treated by a new update structure, where the short-range variables play the role of a local history base. The model problem under consideration accounts in a canonical format for basic effects related to statistically stored and geometrically necessary dislocation flow, yielding micro-force balances including non-convex cross-hardening, kinematic hardening and size effects. Further key ingredients of the proposed algorithmic formulation are geometrically exact updates of the short-range state and a distinct regularization of the rate-independent dissipation function that preserves the range of the elastic domain. We demonstrated the modeling capabilities and algorithmic performance by a large-scale numerical example for multislip scenarios of f.c.c. single crystals. Furthermore, the capability of the model to predict the formation and evolution of plastic laminate microstructure due to latent hardening has been shown in a reduced double-slip setting. We have demonstrated this under micro-free and micro-clamped boundary conditions in combination with length scale effects caused by GNDs. More detailed results on computational gradient plasticity are outlined in the recent works [Mie14, MMH14, MWA14, MAM13].

Acknowledgements. Support for this research was provided by the German Science Foundation (Deutsche Forschungsgemeinschaft), as part of the research group "MICROPLAST" (FOR797) on *Analysis and Computation of Microstructures in Finite Plasticity*, project Mi295/12-2. We also thank Felix Hildebrand for his contribution on laminate microstructure.

References

[AD14] Anguige, K., Dondl, P.: Relaxation of the single-slip condition in strain-gradient plasticity. Proceedings of the Royal Society A 470(2169), 1–15 (2014)

[AK96] Anand, L., Kothari, M.: A computational procedure for rate-independent crystal plasticity. Journal of the Mechanics and Physics of Solids 44, 525–558 (1996)

[AP99] Arsenlis, A., Parks, D.M.: Crystallographic aspects of geometrically-necessary and statistically-stored dislocation density. Acta Materialia 47, 1597–1611 (1999)

[APBB04] Arsenlis, A., Parks, D.M., Becker, R., Bulatov, V.V.: On the evolution of crystallographic dislocation density in non-homogeneously deforming crystals. Journal of the Mechanics and Physics of Solids 52, 1213–1246 (2004)

[Arz98] Arzt, E.: Size effects in materials due to microstructural and dimensional constraints: a comparative review. Acta Materialia 46, 5611–5626 (1998)

[Asa83a] Asaro, R.: Crystal plasticity. Journal of Applied Mechanics 50, 921–934 (1983)

[Asa83b] Asaro, R.: Micromechanics of crystals and polycrystals. Advances in Applied Mechanics 23, 1–115 (1983)

[Ash70] Ashby, M.F.: The deformation of plastically non-homogeneous materials. The Philosophical Magazine A 21, 399–424 (1970)

[Bas93] Bassani, J.L.: Plastic flow of crystals. Advances in Applied Mechanics 30, 191–258 (1993)

[BBS55] Bilby, B.A., Bullough, R., Smith, E.: Continuous distributions of dislocations: a new application of the methods of non-riemannian geometry. Proceedings of the Royal Society London A 231, 263–273 (1955)

[BERS10] Bargmann, S., Ekh, M., Runesson, K., Svendsen, B.: Modeling of polycrystals with gradient crystal plasticity - a comparison of strategies. Philosophical Magazine 90, 1263–1288 (2010)

[CDK09] Conti, S., Dolzmann, G., Klust, C.: Relaxation of a class of variational models in crystal plasticity. Proceedings of the Royal Society A 465, 1735–1742 (2009)

[CDK13] Conti, S., Dolzmann, G., Kreisbeck, C.: Relaxation of a model in finite plasticity with two slip systems. Mathematical Models and Methods in Applied Sciences 23(11), 2111–2128 (2013)

[CG01] Cermelli, P., Gurtin, M.E.: On the characterization of geometrically necessary dislocations in finite plasticity. Journal of the Mechanics and Physics of Solids 49, 1539–1568 (2001)

[CG02] Cermelli, P., Gurtin, M.E.: Geometrically necessary dislocations in viscoplastic single crystalls and bicrystals undergoing small deformations. International Journal of Solids and Structures 39, 6281–6309 (2002)

[CHM02] Carstensen, C., Hackl, K., Mielke, A.: Nonconvex potentials and microstructures in finite-strain plasticity. Proceedings of the the Royal Society of London, Series A 458, 299–317 (2002)

[CnO92] Cuitiño, A.M., Ortiz, M.: Computational modelling of single crystals. Modelling and Simulation in Materials Science and Engineering 1, 225–263 (1992)

[Cot53] Cottrell, A.H.: Dislocations and plastic flow of crystals. Oxford University Press, London (1953)

[DDMR09] Dmitrieva, O., Dondl, P., Müller, S., Raabe, D.: Lamination microstructure in shear deformed copper single crystals. Acta Materialia 57, 3439–3449 (2009)

[EBG04a] Evers, L.P., Brekelmans, W.A.M., Geers, M.G.D.: Non-local crystal plasticity model with intrinsic ssd and gnd effects. Journal of the Mechanics and Physics of Solids 52, 2379–2401 (2004)

[EBG04b] Evers, L.P., Brekelmans, W.A.M., Geers, M.G.D.: Scale dependent crystal plasticity framework with dislocation density and grain boundary effects. International Journal of Solids and Structures 41, 5209–5230 (2004)

[EGRS07] Ekh, M., Grymer, M., Runesson, K., Svedberg, T.: Gradient crystal plasticity as part of the computational modelling of polycrystals. International Journal for Numerical Methods in Engineering 72, 197–220 (2007)

[EvG09] Ertürk, I., van Dommelen, J.A.W., Geers, M.G.D.: Energetic dislocation interactions and thermodynamical aspects of strain gradient crystal plasticity theories. Journal of the Mechanics and Physics of Solids 57, 1801–1814 (2009)

[FBZ80] Franciosi, P., Berveiller, M., Zaoui, A.: Latent hardening in copper and aluminium single crystals. Acta Metallurgica 28, 273–283 (1980)

[FH97] Fleck, N.A., Hutchinson, J.W.: Strain gradient plasticity. Advances in Applied Mechanics 33, 295–362 (1997)

[FMAH94] Fleck, N.A., Müller, G.M., Ashby, M.F., Hutchinson, J.: Strain gradient plasticity: theory and experiment. Acta Materialia 42, 475–487 (1994)

[Gur00] Gurtin, M.E.: On the plasticity of single crystals: free energy, microforces, plastic-strain gradients. Journal of the Mechanics and Physics of Solids 48, 989–1036 (2000)

[Gur02] Gurtin, M.E.: A gradient theory of single-crystal viscoplasticity that accounts for geometrically necessary dislocations. Journal of the Mechanics and Physics of Solids 50, 5–32 (2002)

[Gur08] Gurtin, M.E.: A finite-deformation, gradient theory of single-crystal plasticity with free energy dependent on densities of geometrically necessary dislocations. International Journal of Plasticity 24, 702–725 (2008)

[Hal51] Hall, E.: The deformation and ageing of mild steel: III discussion of results. Proceedings of the Physical Society. Section B 64, 747–753 (1951)

[Hav92] Havner, K.S.: Finite plastic deformation of crystalline solids. Cambridge University Press, Cambridge (1992)

[HB84] Hull, D., Bacon, D.J.: Introduction to dislocations. Pergamon Press, Oxford (1984)

[Hil66] Hill, R.: Generalized constitutive relations for incremental deformation of metal crystals by multislip. Journal of the Mechanics and Physics of Solids 14, 95–102 (1966)

[HL68] Hirth, J.P., Lothe, J.: Theory of dislocations. McGraw-Hill, London (1968)

[HM12] Hildebrand, F., Miehe, C.: A phase field model for the formation and evolution of martensitic laminate microstructure at finite strains. Philosophical Magazine 92, 4250–4290 (2012)

[HR72] Hill, R., Rice, J.: Constitutive analysis of elastic-plastic crystals at arbitrary strain. Journal of the Mechanics and Physics of Solids 20, 401–413 (1972)

[Hut76] Hutchinson, J.W.: Bounds and self-consistent estimates for creep of polycrystalline materials. Proceedings of the the Royal Society of London, Series A 348, 101–127 (1976)

[KA98] Kothari, M., Anand, L.: Elasto-viscoplastic constitutive equations for polycrystalline metals: application to tantalum. Journal of the Mechanics and Physics of Solids 46, 51–83 (1998)

[KAA75] Kocks, U.F., Argon, A.S., Ashby, M.F.: Thermodynamics and kinetics of slip. Progress in Materials Science 19, 141–145 (1975)

[KBS12] Klusemann, B., Bargmann, S., Svendsen, B.: Two models for gradient inelasticity based on non-convex energy. Computational Materials Science 64, 96–100 (2012)

[KH11] Kochmann, D.M., Hackl, K.: The evolution of laminates in finite crystal plasticity: a variational approach. Continuum Mechanics and Thermodynamics 23, 63–85 (2011)

[Koc64] Kocks, U.F.: Latent hardening and secondary slip in aluminium and silver. Transactions of the Metallurgical Society of AIME 230, 1160–1167 (1964)

[Koc66] Kocks, U.F.: A statistical theory of flow stress and work-hardening. The Philosophical Magazine 13, 541–566 (1966)

[Koc09] Kochmann, D.: Mechanical Modeling of Microstructures in Elasto-Plastically Deformed Crystalline Solids. PhD thesis, Ruhr-Universität Bochum (2009)

[Kon52] Kondo, K.: On the geometrical and physical foundations of the theory of yielding. Proceedings Japan National Congress of Applied Mechanics 2, 41–47 (1952)

[Kro60] Kroener, E.: Allgemeine Kontinuumstheorie der Versetzungen und Eigenspannungen. Archive for Rational Mechanics and Analysis 4, 273–334 (1960)

[KS59] Kröner, E., Seeger, A.: Nicht-lineare Elastizitätstheorie der Versetzungen und Eigenspannungen. Archive for Rational Mechanics and Analysis 3, 97–119 (1959)

[KT72] Kröner, E., Teodosiu, C.: Lattice defect approach to plasticity and viscoplasticity. In: Sawzuk, A. (ed.) Problems in Plasticity. Nordhoff International Publishing (1972)

[KT08] Kuroda, M., Tvergaard, V.: A finite deformation theory of higher-order gradient-crystal plasticity. Journal of the Mechanics and Physics of Solids 56, 2573–2584 (2008)

[Lin71] Lin, T.H.: Physical theory of plasticity. Advances in Applied Mechanics 11, 255–311 (1971)

[MAM13] Miehe, C., Aldakheel, F., Mauthe, S.: Mixed variational principles and robust finite element implementations of gradient plasticity at small strains. International Journal for Numerical Methods in Engineering 94, 1037–1074 (2013)

[Man72] Mandel, J.: Plasticité clasique et viscoplasticité. CISM Courses and Lectures No.97. Springer (1972)

[Mie96] Miehe, C.: Exponential map algorithm for stress updates in anisotropic elastoplasticity at large strains for single crystals. International Journal Numerical Methods in Engineering 39, 3367–3390 (1996)

[Mie11] Miehe, C.: A multi-field incremental variational framework for gradient-extended standard dissipative solids. Journal of the Mechanics and Physics of Solids 59, 898–923 (2011)

[Mie14] Miehe, C.: Variational gradient plasticity at finite strains. Part I: Mixed potentials for the evolution and update problems of gradient-extended dissipative solids. Computer Methods in Applied Mechanics and Engineering 268, 677–703 (2014)

[MLG04] Miehe, C., Lambrecht, M., Gürses, E.: Analysis of material instabilities in inelastic solids by incremental energy minimization and relaxation methods: Evolving deformation microstructures in finite plasticity. Journal of the Mechanics and Physics of Solids 52, 2725–2769 (2004)

[MM06] Mielke, A., Müller, S.: Lower semicontinuity and existence of minimizers in incremental finite-strain plasticity. Zeitschrift für angewandte Mathematik und Mechanik 86, 233–250 (2006)

[MMH14] Miehe, C., Mauthe, S., Hildebrand, F.: Variational gradient plasticity at finite strains. Part III: Local-global updates and regularization techniques in multiplicative plasticity for single crystals. Computer Methods in Applied Mechanics and Engineering 268, 735–762 (2014)

[MS00] Menzel, A., Steinmann, P.: On the continuum formulation of higher gradient plasticity for single and polycrystals. Journal of the Mechanics and Physics of Solids 48, 1777–1796 (2000)

[MS04] Miehe, C., Schotte, J.: Anisotropic finite elastoplastic analysis of shells: Simulation of earing in deep-drawing of single- and polycrystalline sheets by taylor-type micro-to-macro transitions. Computer Methods in Applied Mechanics and Engineering 193, 25–57 (2004)

[MSL02] Miehe, C., Schotte, J., Lambrecht, M.: Homogenization of inelastic solid materials at finite strains based on incremental minimization principles. Journal of the Mechanics and Physics of Solids 50, 2123–2167 (2002)

[MSS99] Miehe, C., Schröder, J., Schotte, J.: Computational homogenization analysis in finite plasticity. simulation of texture development in polycrystalline materials. Computer Methods in Applied Mechanics and Engineering 171, 387–418 (1999)

[Mur87] Mura, T.: Micromechanics of defects in solids. Martinus Nijhoff Publishers, Dordrecht (1987)

[MWA14] Miehe, C., Welschinger, F., Aldakheel, F.: Variational gradient plasticity at finite strains. Part II: Local-global updates and mixed finite elements for additive plasticity in the logarithmic strain space. Computer Methods in Applied Mechanics and Engineering 268, 704–734 (2014)

[Nab67] Nabarro, F.R.N.: Theory of crystal dislocations. Oxford University Press, London (1967)

[NALP85] Needleman, A., Asaro, R.J., Lemonds, J., Peirce, D.: Finite element analysis of crystalline solids. Computer Methods in Applied Mechanics and Engineering 52, 689–708 (1985)

[NG98] Nix, W.D., Gao, H.: Indentation size effects in crystalline materials: a law for strain gradient plasticity. Journal of the Mechanics and Physics of Solids 46, 411–425 (1998)

[Nye53] Nye, J.F.: Some geometrical relations in dislocated crystals. Acta Metallurgica 1, 153–162 (1953)

[OR99] Ortiz, M., Repetto, E.A.: Nonconvex energy minimization and dislocation structures in ductile single crystals. Journal of the Mechanics and Physics of Solids 47, 397–462 (1999)

[OS99] Ortiz, M., Stainier, L.: The variational formulation of viscoplastic constitutive updates. Computer Methods in Applied Mechanics and Engineering 171, 419–444 (1999)

[PAN82] Peirce, D., Asaro, R., Needleman, A.: An analysis of nonuniform and localized deformation in ductile single crystals. Acta Metallurgica 30, 1087–1119 (1982)

[Per88] Perzyna, P.: Temperature and rate dependent theory of plasticity of crystalline solids. Revue de Physique Appliquée 23, 445–459 (1988)

[Pet53] Petch, N.J.: The cleavage strength of polycrystals. Journal of the Iron and Steel Institute 174, 25–28 (1953)

[Phi01] Phillips, R.: Crystals, defects and microstructures. Cambridge University Press, Cambridge (2001)

[Ric71] Rice, J.R.: Inelastic constitutive relations for solids: an internal-variable theory and its application to metal plasticity. Journal of the Mechanics and Physics of Solids 19, 433–455 (1971)

[SB35] Schmid, E., Boas, W.: Kristallplastizität. Springer (1935)

[SB10] Svendsen, B., Bargmann, S.: On the continuum thermodynamic rate variational formulation of models for extended crystal plasticity at large deformation. Journal of the Mechanics and Physics of Solids 58, 1253–1271 (2010)

[SBER10] Svendsen, B., Bargmann, S., Ekh, M., Runesson, K.: Modeling of polycrystals with gradient crystal plasticity: A comparison of strategies. Philosophical Magazine 90, 1263–1288 (2010)

[SE98] Stölken, J., Evans, A.: A microbend test method for measuring the plasticity length scale. Acta Materialia 46, 5109–5115 (1998)

[See58] Seeger, A.: Theorie der Gitterfehlstellen. In: Fluegge, S. (ed.) Handbuch der Physik, vol. VII/1. Springer (1958)

[Ste96] Steinmann, P.: Views on multiplicative elastoplasticity and the continuum theory of dislocations. International Journal of Engineering Science 34, 1717–1735 (1996)

[Sve02] Svendsen, B.: Continuum thermodynamic models for crystal plasticity including the effects of geometrically-necessary dislocations. Journal of the Mechanics and Physics of Solids 50, 1297–1329 (2002)

[Tay34] Taylor, G.I.: The mechanism of plastic deformation of crystals. Proceedings of the Royal Society London A 145, 362–387 (1934)

[Tay38] Taylor, G.I.: Plastic strain in metals. Journal of the Institute of Metals 62, 307–324 (1938)

[Teo70] Teodosiu, C.: A dynamic theory of dislocations and its applications to the theory of the elastic-plastic continuum, Fundamental Aspects of Dislocation Theory. National Bureau of. Standards (U.S.) Special Publication, vol. II, pp. 837–876 (1970)

[WB12] Wulfinghoff, S., Böhlke, T.: Equivalent plastic strain gradient enhancement of single crystal plasticity: theory and numerics. Proceedings of the Royal Society London A 468, 2682–2703 (2012)

[YBG11] Yalcinkaya, T., Brekelmans, W.A.M., Geers, M.G.D.: Deformation patterning driven by rate dependent non-convex strain gradient plasticity. Journal of the Mechanics and Physics of Solids 59, 1–17 (2011)

[YBG12] Yalçinkaya, T., Brekelmans, W., Geers, M.: Non-convex rate dependent strain gradient crystal plasticity and deformation patterning. International Journal of Solids and Structures 49, 2625–2636 (2012)

Chapter 5
Variational Approaches and Methods for Dissipative Material Models with Multiple Scales

Alexander Mielke

Abstract. In a first part we consider evolutionary systems given as generalized gradient systems and discuss various variational principles that can be used to construct solutions for a given system or to derive the limit dynamics for multiscale problems via the theory of evolutionary Gamma-convergence. On the one hand we consider a family of viscous gradient system with quadratic dissipation potentials and a wiggly energy landscape that converge to a rate-independent system. On the other hand we show how the concept of Balanced-Viscosity solution arise in the vanishing-viscosity limit.

As applications we discuss, first, the evolution of laminate microstructures in finite-strain elastoplasticity and, second, a two-phase model for shape-memory materials, where H-measures are used to construct the mutual recovery sequences needed in the existence theory.

5.1 Introduction

This work shows how methods from abstract evolutionary systems can be employed for the study of material models which allow for small or finite-strain elastic deformation y and are characterized by further internal or dissipative variables z which may describe damage, plastic deformations, magnetization, polarization, or phase transformations. The common feature of all models considered is their description in terms of an energy functional $\mathscr{E}$ and a dissipation potential $\mathscr{R}$. Hence the evolution of the state $q = (y,z)$ can be described by a generalized force balance, namely

$$0 \in \partial_{\dot{q}}\mathscr{R}(q(t),\dot{q}(t)) + \mathrm{D}_q\mathscr{E}(t,q(t)). \tag{5.1}$$

Alexander Mielke
Weierstraß-Institut für Angewandte Analysis und Stochastik,
Mohrenstraße 39, 10117 Berlin, Germany
e-mail: `alexander.mielke@wias-berlin.de`

S. Conti and K. Hackl (eds.), *Analysis and Computation of Microstructure in Finite Plasticity*,
Lecture Notes in Applied and Computational Mechanics 78, DOI: 10.1007/978-3-319-18242-1_5

Here $\partial_{\dot{q}}\mathscr{R}(q,\dot{q})$ denotes the convex subdifferential of the dissipation potential $\mathscr{R}$, where for each state q the function $\mathscr{R}(q,\cdot)$ is nonnegative, convex, and lower semi-continuous and satisfies $\mathscr{R}(q,0)=0$. Thus, the possibly set-valued subdifferential $\partial_{\dot{q}}\mathscr{R}(q,\dot{q})$ contains the dissipative forces generated by the rate $\dot{q}$ if the system is in the state q. These forces have to be balanced by the potential restoring forces $-\mathrm{D}_q\mathscr{E}(t,q)$.

The formulation of material models in terms of the functionals $\mathscr{E}$ and $\mathscr{R}$ instead of general PDEs shows additional physical structure that can be exploited mathematically. In particular, one can employ the rich theory of the calculus of variations, even for evolutionary systems. As a first case, we see that a very useful time discretization of (5.1) can be obtained by the time-incremental minimization problem

$$q_{k+1}\in \operatorname*{Arg\,min}_{q}\Big(\mathscr{E}(t_{k+1},q)+(t_{k+1}-t_k)\mathscr{R}\big(q_{k+\theta},\tfrac{1}{t_{k+1}-t_k}(q-q_k)\big)\Big). \tag{5.2}$$

In the context of abstract evolutionary systems this scheme relates to De Giorgi's theory of minimizing movements, and one way of obtaining solutions is via De Giorgi's $(\mathscr{R},\mathscr{R}^*)$-principle, also called the energy-dissipation principle (EDP), which is given by the simple variational characterization via

$$\mathscr{E}(T,q(T))+\int_0^T \mathscr{R}(q,\dot{q})+\mathscr{R}^*(q,-\mathrm{D}\mathscr{E}(t,q))\,\mathrm{d}t\le \mathscr{E}(0,q(0))+\int_0^T \partial_t\mathscr{E}(t,q)\,\mathrm{d}t.$$

This principle and its equivalence to (5.1) will be discussed in Section 5.2.1.

The EDP is also extremely useful for studying multiscale problems given in terms of generalized gradient systems $(\boldsymbol{X},\mathscr{E}_\varepsilon,\mathscr{R}_\varepsilon)$, where $\varepsilon\in[0,1]$ is a small parameter. The major question is under what conditions the solutions $q_\varepsilon:[0,T]\to\boldsymbol{X}$ for $(\boldsymbol{X},\mathscr{E}_\varepsilon,\mathscr{R}_\varepsilon)$ converge to a solution $q_0:[0,T]\to\boldsymbol{X}$ for $(\boldsymbol{X},\mathscr{E}_0,\mathscr{R}_0)$ in the limit $\varepsilon\to 0$. If this holds and additionally the energies converge, i.e. $\mathscr{E}_\varepsilon(t,q_\varepsilon(t))\to\mathscr{E}_0(t,q_0(t))$ we call this *evolutionary Γ-convergence.* In general, the Γ-convergences $\mathscr{E}_\varepsilon\xrightarrow{\Gamma}\mathscr{E}_0$ and $\mathscr{R}_\varepsilon\xrightarrow{\Gamma}\mathscr{R}_0$ are not enough. We discuss some of the results from [Mie14] and give applications to models with wiggly energies, where for $\varepsilon>0$ the dissipation potentials $\mathscr{R}_\varepsilon(q,v)=\frac{1}{2}\langle v,\mathbb{G}_\varepsilon(q)v\rangle$ are quadratic and satisfy $\mathscr{R}_\varepsilon\to 0$, but the limiting dissipation potential $\mathscr{R}_0$ is 1-homogeneous, such that $(\boldsymbol{X},\mathscr{E}_0,\mathscr{R}_0)$ is a rate-independent system (RIS), such as linearized elastoplasticity, see Section 5.4.2.

Moreover, the vanishing-viscosity limit $\varepsilon\to 0$ of generalized gradient systems $(\boldsymbol{X},\mathscr{E},\mathscr{R}_\varepsilon)$, where the "small-viscosity dissipation potential" has the form $\mathscr{R}_\varepsilon(q,v)=\Psi(q,v)+\frac{\varepsilon}{2}\langle v,\mathbb{G}v\rangle$, can also be studied efficiently using a reparametrized version of the EDP, see Section 5.4.3. This leads to the notion of *balanced-viscosity solutions* (also called BV solutions) for RIS $(\boldsymbol{X},\mathscr{E},\Psi,\mathbb{G})$, where $\mathbb{G}$ indicates the additional viscosity structure which determines the jump behavior.

For purely rate-independent models it is advantageous to replace the infinitesimal dissipation metric Ψ by the dissipation distance $\mathscr{D}(q_0,q_1)$ between two states z_0 and z_1. This leads to the notion of *energetic rate-independent systems* (ERIS). In particular, the time-incremental minimization (5.2) does not depend on the time step and can be replaced by

$$q^{k+1} \in \operatorname*{Arg\,min}_{q \in \boldsymbol{X}} \Big(\mathscr{E}(t_{k+1}, q) + \mathscr{D}(q_k, q) \Big). \tag{5.3}$$

It was observed in [MTL02] that all accumulation points of the piecewise interpolants of the solutions of (5.3) are so-called *energetic solutions*, see (5.5) for the purely energetic definition of this solution concept.

A corresponding notion of evolutionary Γ-convergence for ERIS $(\boldsymbol{X}, \mathscr{E}_\varepsilon, \mathscr{R}_\varepsilon)$ was developed in [MRS08], see also [MR15] for more details. Using this approach and the general existence theory for finite-strain elastoplasticity from [MM09, Mie10] it was shown in [MS13] that linearized elastoplasticity can be derived as the evolutionary Γ-limit of finite-strain elastoplasticity, if the yields stress is tending to 0, see Section 5.3.2.

In Section 5.5 we discuss two rate-independent material models that describe the evolution of microstructures. The first one is a mathematical version of the model proposed in [KH11], where laminates are considered as dissipative internal variables and equipped with a physically motivated dissipation distance, see Section 5.5.1 and [HHM12]. In Section 5.5.2 the two-phase model introduced in [MTL02] is reconsidered using a new construction for mutual recovery sequences, which allows us to generalize the original existence proof considerably.

In Table 5.1 and 5.2 we give an overview of the discussed topics in this work.

Table 5.1 A survey on the existence results discussed here via variational formulations

Topic	Section	Reference
Energetic solutions for finite-strain elastoplasticity	Sec. 5.2.2	[MM09, Mie10, MR15]
Energetic solutions for laminate evolution	Sec. 5.5.1	[HHM12]
Energetic solutions for a two-phase SM material	Sec. 5.5.2	[MTL02, HM15]
Balanced-Viscosity solutions	Sec. 5.4.3	[MRS09, MZ14, MRS14b]
Finite-strain viscoplasticity	Sec. 5.2.1	[MRS15]
One-dimensional finite-strain viscoelasticity via EVI	Sec. 5.2.1	[MOŞ14]

Table 5.2 A survey on the results on evolutionary Γ-convergence discussed in this work

ε-dependent multiscale problem	limit problem	Section	Reference
wiggly-energy ODE	scalar dry-friction model	Sec. 5.4.1	[PT02, Mie12]
system of bistable, viscous springs	1D pseudo-elasticity	Sec. 5.4.2	[MT12]
finite-strain elastoplasticity	linearized elastoplasticity	Sec. 5.3.2	[MS13]

5.2 Variational Formulations for Evolution

A main point of looking in different variational principles lies in the fact that theses principles lead to different mathematical formulations. For instance, when looking to global existence results for material models allowing for finite strains and the associated geometric nonlinearities, it is highly desirable to use minimization principles on the energy such that the rich theory of direct methods from the calculus of variations are applicable, such as weak lower semicontinuity, existence of minimizers, Γ-convergence, and relaxation techniques.

5.2.1 Generalized Gradient Systems and the Energy-Dissipation Principle

We now convert the formal ideas from the introduction into rigorous mathematical statements. We call a triple $(\boldsymbol{X},\mathscr{E},\mathscr{R})$ a *generalized gradient system (gGS)*, if $\boldsymbol{X}$ is a Banach space, $\mathscr{E}:[0,T]\times\boldsymbol{X}\to\mathbb{R}_\infty:=\mathbb{R}\cup\{\infty\}$ is an energy functional, and $\mathscr{R}:\boldsymbol{X}\times\boldsymbol{X}\to[0,\infty]$ is a *dissipation potential*, which means that for all $q\in\boldsymbol{X}$ the functional $\mathscr{R}(q,\cdot):\boldsymbol{X}\to\mathbb{R}_\infty$ is lower semicontinuous, nonnegative, convex, and satisfies $\mathscr{R}(q,0)=0$. We speak of a classical gradient system, or simply a gradient system, if $\mathscr{R}(q,\cdot)$ is quadratic, i.e. there exists a (viscosity) operator $\mathbb{G}(q)=\mathbb{G}(q)^*\geq 0$ such that $\mathscr{R}(q,v)=\frac{1}{2}\langle\mathbb{G}(q)v,v\rangle$. However, plasticity requires non-quadratic dissipation potentials, e.g. of the form $\mathscr{R}(\dot\pi)=\sigma_{\mathrm{yield}}\|\dot\pi\|_{\mathrm{L}^1}+\frac{1}{2}\mu_{\mathrm{visc}}\|\dot\pi\|^2_{\mathrm{L}^2}$. In particular, the rate-independent case requires $\mathscr{R}(q,\lambda v)=\lambda\mathscr{R}(q,v)$ for all $\lambda>0$, which is incompatible with a quadratic form.

The following proposition from convex analysis shows that there are several completely equivalent formulations of the generalized force balance (5.1). The equivalences of the points (ii) to (iv) are also called the Fenchel equivalences, cf. [Fen49]. The essential tools is the Fenchel-Legendre transform $\Psi^*:\boldsymbol{X}^*\to\mathbb{R}_\infty$ of a convex function $\Psi:\boldsymbol{X}\to\mathbb{R}_\infty$ defined via

$$\Psi^*(\xi):=\sup\{\,\langle\xi,v\rangle-\Psi(v)\,|\,v\in\boldsymbol{X}\,\}.$$

Note that in a reflexive Banach space we have $(\Psi^*)^*=\Psi$.

Proposition 5.2.1 (Equivalent formulations). *Let $\boldsymbol{X}$ be a reflexive Banach space and $\Psi:\boldsymbol{X}\to\mathbb{R}_\infty$ be proper, convex, and lower semicontinuous. Then, for every $\xi\in\boldsymbol{X}^*$ and every $v\in\boldsymbol{X}$ the following five statements are equivalent:*

(i) $v\in\operatorname*{Arg\,min}_{w\in\boldsymbol{X}}\big(\Psi(w)-\langle\xi,w\rangle\big)$; *(ii)* $\xi\in\partial\Psi(v)$;

(iii) $\Psi(v)+\Psi^*(\xi)=\langle\xi,v\rangle$;

(iv) $v\in\partial\Psi^*(\xi)$; *(v)* $\xi\in\operatorname*{Arg\,min}_{\eta\in\boldsymbol{X}^*}\big(\Psi^*(\eta)-\langle\eta,v\rangle\big)$.

Note that the definition of Ψ^* immediately implies the Young-Fenchel inequality $\Psi(w)+\Psi^*(\eta)\geq\langle\eta,w\rangle$ for all w and η. Thus, (iii) expresses an optimality as well.

Defining the dual dissipation potential $\mathscr{R}^*$ via $\mathscr{R}^*(q,\cdot):=(\mathscr{R}(q,\cdot))^*$ we can apply these equivalences to reformulate (5.1) in the following ways:

(I) Rayleigh principle [Ray71]

$$\text{(HRP)}\quad \dot q\in\operatorname{Arg\,min}\Big(\mathscr{R}(q,v)-\langle\mathrm{D}\mathscr{E}(t,q),v\rangle\Big);$$

(II) Force balance in $\boldsymbol{X}^*$ Rayleigh-Biot equation [Ray71, Bio55]

$$\text{(FB)} \quad 0 \in \partial_{\dot{q}}\mathscr{R}(q,\dot{q}) + \mathrm{D}\mathscr{E}(t,q) \ \in \boldsymbol{X}^*;$$

(III) Power balance in $\mathbb{R}$ De Giorgi's $(\mathscr{R},\mathscr{R}^*)$ formulation [DMT80]

$$\text{(PB)} \quad \mathscr{R}(q,\dot{q}) + \mathscr{R}^*(q,-\mathrm{D}\mathscr{E}(t,q)) = -\langle \mathrm{D}\mathscr{E}(t,q),\dot{q}\rangle;$$

(IV) Rate equation in $\boldsymbol{X}$ Onsager equation [Ons31]

$$\text{(RE)} \quad \dot{q} \in \partial_\xi \mathscr{R}^*(q,-\mathrm{D}\mathscr{E}(t,q)) \ \in \boldsymbol{X};$$

(V) Maximum dissipation principle cf. e.g. [HF08]

$$\text{(MDP)} \quad \mathrm{D}\mathscr{E}(t,q) \in \operatorname{Arg\,max}\Big(\langle \xi,\dot{q}\rangle - \mathscr{R}^*(q,\xi)\Big).$$

In fact, [Ray71, Eqn. (26)] also includes the kinetic energy $\mathscr{K}$, which we omit in our quasistatic approximation, namely $\frac{\mathrm{d}}{\mathrm{d}t}\big(\mathrm{D}_{\dot{q}}\mathscr{K}(q,\dot{q})\big) + \mathrm{D}_{\dot{q}}\mathscr{R}(q,\dot{q}) + \mathrm{D}_q\mathscr{E}(t,q) = 0$.

Note that we have changed the sign in (V) to justify the name of (MDP). The reason for this will become apparent in the rate-independent setting where $\mathscr{R}^*$ only takes the two values 0 and ∞, see (5.4) and [HF08].

Before returning to the general situation, we highlight the three different cases (II)–(IV) for the classical viscous dissipation, i.e. $\mathscr{R}(u,v) = \frac{1}{2}\langle \mathbb{G}v,v\rangle$ and $\mathscr{R}^*(u,\xi) = \frac{1}{2}\langle \xi,\mathbb{K}\xi\rangle$ with $\mathbb{K} = \mathbb{G}^{-1}$. Then, we have

$$\text{(FB)} \quad \mathbb{G}\dot{u} = -\mathrm{D}\mathscr{E}(u) \qquad \text{(RE)} \quad \dot{u} = -\mathbb{K}\mathrm{D}\mathscr{E}(u) = -\nabla_{\mathbb{G}}\mathscr{E}(u)$$

$$\text{(PB)} \quad \frac{1}{2}\langle \mathbb{G}\dot{u},\dot{u}\rangle + \frac{1}{2}\langle \mathrm{D}\mathscr{E}(u),\mathbb{K}\mathrm{D}\mathscr{E}(u)\rangle = -\langle \mathrm{D}\mathscr{E}(u),\dot{u}\rangle,$$

where (RE) can be seen as a "gradient evolution", as $\nabla_{\mathbb{G}}$ is the gradient operator.

The above forms can already be understood as variational formulations, since the evolution is expressed by extremizing a functional or by variations or derivatives of the two functionals $\mathscr{E}$ and $\mathscr{R}$. However, for mathematical purposes it is desirable to have variational formulations for the whole solution trajectories $q:[0,T] \to \boldsymbol{X}$. One such principle can be derived on the basis of the power balance (PB) by integration in time and using the chain rule and finally employing the Young-Fenchel inequality $\Psi(w) + \Psi^*(\eta) \geq \langle \eta,w\rangle$, cf. [DMT80] or the survey [Mie14].

Theorem 5.2.2 (De Giorgi's *energy-dissipation principle*). *Under suitable technical conditions on $(\boldsymbol{X},\mathscr{E},\mathscr{R})$ a function $q:[0,T] \to \boldsymbol{X}$ satisfies (I)–(V) from above for almost all $t \in [0,T]$ if and only if the* Energy-Dissipation Principle *(EDP) holds:*

$$\textit{(EDP)} \quad \begin{cases} \mathscr{E}(T,q(T)) + \displaystyle\int_0^T \mathscr{R}(q,\dot{q}) + \mathscr{R}^*(q,-\mathrm{D}\mathscr{E}(t,q))\,\mathrm{d}t \\ \qquad\qquad\qquad \leq \mathscr{E}(0,q(0)) + \displaystyle\int_0^T \partial_t\mathscr{E}(t,q(t))\,\mathrm{d}t. \end{cases}$$

Moreover, the EDP is equivalent to the energy-dissipation balance *(EDB), where "$\leq$" in (EDP) is replaced by "$=$".*

It is obvious how to obtain (EDB) (and hence (EDP) from (I)–(V). For this one simply integrates the power balance (III) in time and uses a *abstract chain rule*

$$\mathscr{E}(t,q(t)) = \mathscr{E}(r,q(r)) + \int_r^t \langle \mathrm{D}\mathscr{E}(s,q(s)),\dot{q}(s)\rangle + \partial_s\mathscr{E}(s,q(s))\,\mathrm{d}s.$$

Starting from (EDP) and using the chain rule one easily obtains the power balance (III) as an estimate, namely $\int_0^T \mathscr{R}+\mathscr{R}^*\,\mathrm{d}t \le \int_0^T -\langle \mathrm{D}\mathscr{E},\dot{q}\rangle\,\mathrm{d}t$. However, the Young-Fenchel inequality gives $\mathscr{R}+\mathscr{R}^* \ge -\langle \mathrm{D}\mathscr{E},\dot{q}\rangle$ for almost all $t \in [0,T]$, so that the power balance (III) has to hold.

Remark 5.2.3 (Brézis-Ekeland-Nayroles principle [BE76, Nay76]). This principle exactly serves the same purpose, namely providing a minimum principle for whole trajectories. It is similar to the EDP but in some sense dual. It applies to the case that $\boldsymbol{X}$ is a Hilbert space, $\mathscr{R}(q,\dot{q}) = \frac{1}{2}\|\dot{q}\|^2_{\boldsymbol{X}}$, and an energy $\mathscr{E}(t,q) = \Phi(q) - \langle \ell(t),q)\rangle$, where Φ is convex. Then, $q : [0,T] \to \boldsymbol{X}$ solves the force balance (FB), now in the form $\dot{q}+\partial\Phi(q) \neq \ell(t)$ if and only if

$$\int_0^T \Phi(q(t)) + \Phi^*(\ell(t)-\dot{q}(t)) - \langle \ell(t),q(t)\rangle\,\mathrm{d}t + \frac{1}{2}\|q(T)\|^2_{\boldsymbol{X}} - \frac{1}{2}\|q(0)\|^2_{\boldsymbol{X}} \le 0,$$

where the estimate "≥ 0" always holds. We refer to [Ste08] for a discussion and a generalization to the doubly nonlinear case $\mathscr{R}(q,\dot{q}) = \Psi(\dot{q})$.

The importance of the EDP is that a discrete counterpart can be derived based on the incremental minimization problem (5.2) and De Giorgi's variational interpolants $\widetilde{q}_\tau$. In a classical Banach-space setting on can use the piecewise constant right and left-continuous interpolants $\underline{q}_\tau$ and $\overline{q}_\tau$ as well as the piecewise affine interpolant $\widehat{q}_\tau$ (all satisfying $q_\tau(t_k) = q_k$) and obtains the discrete version of EDP in the form

$$\mathscr{E}(t_k,\widehat{q}_\tau(t_k))+\int_{t_l}^{t_k}\mathscr{R}(\underline{q}_\tau,\dot{\widehat{q}})+\mathscr{R}^*(\underline{q}_\tau,-\mathrm{D}\mathscr{E}(t,\widetilde{q}_\tau))\,\mathrm{d}t \le \mathscr{E}(t_l,\widehat{q}_\tau(t_l)) + \int_{t_l}^{t_k}\partial_t\mathscr{E}(t,\overline{q}_\tau)\,\mathrm{d}t.$$

Under suitable assumptions it is possible to take the time-step limit $\tau \to 0$ and arrive at the notion of *weak energy-dissipation solutions*, defined by the condition that

$$\mathscr{E}(t,q(t))+\int_r^t\mathscr{R}(q,\dot{q})+\mathscr{R}^*(q,-\mathrm{D}\mathscr{E}(s,q))\,\mathrm{d}s \le \mathscr{E}(r,q(r)) + \int_r^t \partial_s\mathscr{E}(s,q)\,\mathrm{d}s$$

holds for all $t \in [0,T]$, $s = 0$, and almost all $s \in [0,T]$. An existence proof for weak energy-dissipation solutions for a model of finite-strain *viscoplasticity* using the multiplicative decomposition is given in [MRS15]. There it is not possible to derive the missing chain-rule estimate to return back to the differential inclusions (I)–(V).

Another very useful variational principle is only valid for classical gradient systems, where it is possible to define a dissipation distance $\mathscr{D}$. If the energy functionals $\mathscr{E}(t,\cdot)$ are geodesically λ-convex, then one reformulate the evolutionary problem via a so-called evolutionary variational inequality (EVI), see [AGS05, Mie14]. For an application of this theory of geodesically λ-convex gradient systems in one-dimensional viscoelasticity we refer to [MOŞ14]. This one-dimensional existence theory, where $q = y$, relies on time-incremental minimization problems

$$y^{k+1} = \operatorname{Arg\,min}\Big(\frac{1}{2(t_{k+1}-t_k)}\mathscr{D}(w,y^k)^2 + \mathscr{E}(w)\Big)$$

and establishes strong convergence of the solution even in the case of nonconvex $\mathscr{E}$.

An approximative variational characterization of whole trajectories can be obtained by the *weighted energy-dissipation functional* (WED functional), which is defined via

$$\mathscr{W}_\varepsilon(q) = \int_0^T \mathrm{e}^{-t/\varepsilon}\big(\mathscr{R}(q(t),\dot q(t)) + \frac{1}{\varepsilon}\mathscr{E}(t,q(t))\big)\,\mathrm{d}t, \quad q(0) = q_0.$$

and which was introduced in [MO08] in the context of material modeling. It was used earlier for elliptic regularization of parabolic problems, see [LM72] and [Ilm94] for the treatment of Brakke's approach to the mean-curvature flow.

Under sufficient smoothness of $\mathscr{E}$ and $\mathscr{R}$ we see that the Euler-Lagrange equation takes the form

$$\mathrm{D}_{\dot q}\mathscr{R}(q,\dot q)+\mathrm{D}_q\mathscr{E}(t,q)=\varepsilon\Big(\frac{\mathrm{d}}{\mathrm{d}t}\big(\mathrm{D}_{\dot q}\mathscr{R}(q,\dot q)\big)-\mathrm{D}_q\mathscr{R}(q,\dot q)\Big), \quad \mathrm{D}_{\dot q}\mathscr{R}(q(T);\dot q(T))=0.$$

Thus, we obtain an "elliptic regularization" of the original evolutionary problem. The advantage is that showing the existence of minimizers $\widehat q_\varepsilon : [0,T] \to \boldsymbol{X}$ for $\mathscr{W}_\varepsilon$ is usually much easier than establishing the existence of solutions for the gGS. Yet, the major problem then is to pass to the limit $\varepsilon \to 0$ to find a limit q of the approximations $\widehat q_\varepsilon$. For the rate-independent case $\mathscr{R}(q,v) = \Psi(v)$ this was done in [MO08] obtaining energetic solutions q. For classical gradient system $\mathscr{R}(q,v) = \frac{1}{2}\langle \mathbb{G}v, v\rangle$ with $\mathbb{G}$ independent of q the convergence $\widehat q_\varepsilon \to q$ was established in [MS11].

The general aim of introducing the WED functional in [MO08] was the possibility of using relaxation techniques that are invented originally only for stationary problems also in the context of evolutionary problems. First results on such relaxations are presented in [MO08, Sec. 4.4+5], mainly in the context of RIS. For a proper relaxation of a viscous PDE we refer to [CO08, Sec. 4], where the case

$$\boldsymbol{X} = \mathrm{L}^2(\Omega), \quad \mathscr{E}(q) = \int_\Omega F(\nabla q(x)) - f(t,)q\,\mathrm{d}x, \quad \mathscr{R}(\dot q) = \frac{1}{2}\int_\Omega \dot q^2\,\mathrm{d}x$$

was considered, with $\Omega \subset \mathbb{R}^2$ and $F(A) = 0$ for $A \in K := \{\pm(1,0), \pm(0,1)\}$ and ∞ else. It is proved that quasiminimizers $\widetilde q_\varepsilon$ of $\mathscr{W}_\varepsilon$ converge to solutions of the relaxed evolution defined via the differential inclusion

$$\dot q = \frac{1}{2}\operatorname{div}\sigma + \frac{1}{2}f, \quad \text{where } \sigma(t,x) \in \partial\chi_S(\nabla u(t,x)),$$

where $S = \operatorname{conv} K = \{\,(A_1,A_2) \in \mathbb{R}^2 \mid |A_1|+|A_2| \le 1\,\}$ and χ_S is indicator function of convex analysis, i.e. $\chi_S(A) = 0$ for $A \in S$ and ∞ otherwise.

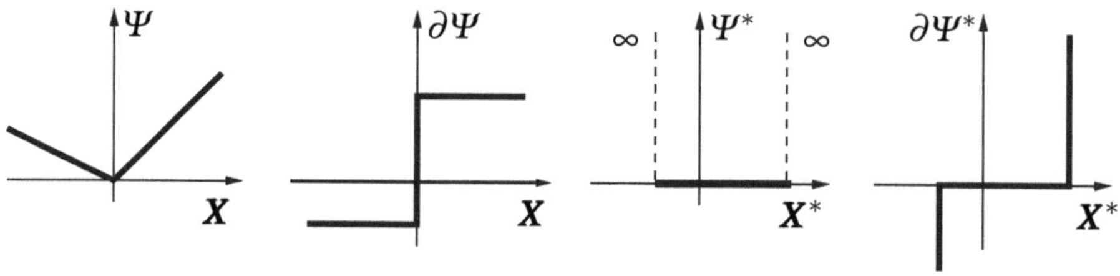

Fig. 5.1 Primal and dual dissipation potential Ψ for RIS

5.2.2 Rate-Independent Systems and Energetic Solutions

The case of purely rate-independent dissipation is distinct from the general dissipation potentials. It is characterized by the condition on $\mathscr{R}(q,\lambda v) = \lambda\mathscr{R}(q,v)$ for all $\lambda > 0$. In that case we call $(\boldsymbol{X},\mathscr{E},\mathscr{R})$ a *rate-independent system (RIS)*. Then, the force-velocity relation $v \mapsto \partial_v\mathscr{R}(q,v)$ is meant in the sense of subdifferentials of convex functions, which are set-valued:

$$\partial\Psi(v) = \{\, \eta \in \boldsymbol{X}^* \,|\, \forall w \in \boldsymbol{X} :\ \Psi(w) \geq \Psi(v) + \langle \eta, w-v\rangle \,\}.$$

For rate-independent cases we have

$$\partial_v\mathscr{R}(q,\lambda v) = \partial_v\mathscr{R}(q,v) = \{\, \eta \in K(q) \,|\, \mathscr{R}(q,v) = \langle \eta, v\rangle \,\},$$

where $K(q) := \partial_v\mathscr{R}(q,0)$ is called the elastic domain. Moreover, for the dual dissipation potential we find the simple form

$$\mathscr{R}^*(q,\xi) = \chi_{K(q)}(\xi) = \begin{cases} 0 & \text{for } \xi \in K(q), \\ \infty & \text{for } \xi \notin K(q), \end{cases}$$

see Figure 5.1.

In principle the five formulations I to V of the previous subsection are still valid for RIS. However, one can use the special structure of $\partial_v\mathscr{R}$ and $\mathscr{R}^*$ to simplify the presentation. For instance, the maximum-dissipation principle reduces to the simpler form

$$\text{rate-independent MDP:} \qquad \mathrm{D}_q\mathscr{E}(t,q) = \operatorname*{Arg\,max}_{\xi \in K(q)} \langle \xi, \dot q\rangle. \tag{5.4}$$

Second the energy-dissipation principle in the rate-independent case takes a simpler form as $\mathscr{R}^*$ is either 0 or ∞. A differentiable function $q : [0,T] \to \boldsymbol{X}$ solves I to V if and only if

$$\text{(S)}_{\text{loc}} \quad -\mathrm{D}_q\mathscr{E}(t,q) \in K(q) := \partial_v\mathscr{R}(q,0),$$

$$\text{(E)} \qquad \mathscr{E}(T;q(T)) + \int_0^T \mathscr{R}(q,\dot q)\,\mathrm{d}t = \mathscr{E}(0,q(0)) + \int_0^T \partial_t\mathscr{E}(t,q)\,\mathrm{d}t.$$

We call the first condition a *local stability condition*, since the system stays in a state $q(t)$ in which the driving force $\xi(t) = \mathrm{D}_q\mathscr{E}(t,q(t))$ is not big enough to overcome the possible dissipative forces $\eta \in K(q)$.

The major problem for RIS is that the solutions will in general develop jumps, i.e. the three values $q(t-0) := \lim_{s\nearrow t} q(s)$, $q(t)$, and $q(t+0) := \lim_{s\searrow t} q(s)$ may be different. In such a discontinuous situation the differential formulations are not really useful. Of course, if there is enough convexity in the system the solution will not develop jumps and the above formulations are optimal.

In general cases, the notion of energetic solutions can be used to characterize solutions with jump in a variational way. In that case the infinitesimal dissipation potential $\mathscr{R}$, which in mathematical terms plays the role of a infinitesimal Finsler metric, is not suitable and has to be replaced by a *dissipation distance* $\mathscr{D} : \boldsymbol{X} \times \boldsymbol{X} \to [0,\infty]$ which is assumed to satisfy the triangle inequality $\mathscr{D}(q_1,q_3) \leq \mathscr{D}(q_1,q_2)+\mathscr{D}(q_2,q_3)$, but the symmetry $\mathscr{D}(q_1,q_2) = \mathscr{D}(q_2,q_1)$ is not needed. The triple $(\boldsymbol{X},\mathscr{E},\mathscr{D})$ is called an *energetic rate-independent systems* (ERIS), and a function $q : [0,T] \to \boldsymbol{X}$ is called an *energetic solution* if for all $t \in [0,T]$ the *global stability (S)* and the *energy balance (E)* hold:

$$\begin{aligned} &\text{(S)} \quad \mathscr{E}(t,q(t)) \leq \mathscr{E}(t,\widetilde{q}) + \mathscr{D}(q(t),\widetilde{q}) \text{ for all } \widetilde{q} \in \boldsymbol{X}; \\ &\text{(E)} \quad \mathscr{E}(T,q(T)) + \mathrm{Diss}_{\mathscr{D}}(q;[0,T]) = \mathscr{E}(0,q(0)) + \int_0^T \partial_s\mathscr{E}(s,q)\,\mathrm{d}s, \end{aligned} \tag{5.5}$$

where the total dissipation along a possibly discontinuous solutions is defined via

$$\mathrm{Diss}_{\mathscr{D}}(q;[r,s]) := \{\sum_{j=1}^{N} \mathscr{D}(q(t_{j-1}),q(t_j)) \,|\, N \in \mathbb{N},\ r \leq t_0 < t_1 < \cdots < t_N \leq s\}. \tag{5.6}$$

For energetic solutions, possible jumps can be given a natural physical interpretation. First, (E) implies the exact energy conservation $\mathscr{E}(t,q(t+0)) = \mathscr{E}(t,q(t-0)) - \mathscr{D}(q(t-0),q(t+0))$. Second, (S) implies that a jump immediately occurs if it is possible, which is called the principle of realizability in [MTL02].

The notion of energetic solutions was first introduced in [MTL02], and under suitable technical assumptions it was shown that all limits of the piecewise constant interpolants of the solutions of the time-incremental minimization problems

$$q^{k+1} \in \operatorname*{Arg\,min}_{\widetilde{q} \in \boldsymbol{X}} \Big(\mathscr{D}(q^k,\widetilde{q}) + \mathscr{E}(t_{k+1},\widetilde{q}) \Big) \tag{5.7}$$

converge to energetic solutions. We refer to [Mie11b, MR15] for a detailed account of this theory.

Note that in the incremental problems (5.7) one is doing a global minimization, which is reflected in the global stability condition (S). This leads to a jump behavior which is sometimes unrealistic, since potential barriers are not seen. To define a notion of solutions that do not show the problem of too early jumps, one can treat RIS as limits of rate-dependent systems, i.e. systems with a small viscosity proportional to ε and then consider the vanishing-viscosity limit $\varepsilon \to 0$. The corresponding notion of solutions is called *Balanced-Viscosity solutions*, which will be discussed in Section 5.4.3.

The two major stimuli in the development of the theory of energetic solutions for RIS were the theory of crack evolution in brittle materials, see [DFT05] for linearized elasticity and [DL10] for finite-strain elasticity, and the theory of *finite-strain elastoplasticity*, see [MM09, Mie10]. In the former case the name *irreversible quasistatic evolution* is used for what is called energetic solutions here. In both cases, there is not a useful underlying linear structure in a function space $\boldsymbol{X}$, and the full strength of the abstract definition of energetic solutions is needed.

5.3 Evolutionary Γ-Convergence

Following the notions in the survey article [Mie14] we consider families of gGS $(\boldsymbol{X},\mathscr{E}_\varepsilon,\mathscr{R}_\varepsilon)_{\varepsilon\in]0,1[}$ and ask the questions whether the solutions q_ε for these system have a limit q for $\varepsilon \to 0$ and whether the limit q is again a solution to a gGS $(\boldsymbol{X},\mathscr{E}_0,\mathscr{R}_0)$. Ideally, one might hope that it is sufficient that $\mathscr{E}_\varepsilon$ and $\mathscr{R}_\varepsilon$ convergence in a suitable topology to $\mathscr{E}_0$ and $\mathscr{R}_0$, respectively. We will show that such results exist, but we will also discuss situations where we start with quadratic $\mathscr{R}_\varepsilon$ and end up with a limiting dissipation $\mathscr{R}_0$ that is rate independent.

We first give the general definition of *pE-convergence*, which is a short name of *evolutionary Γ-convergence* with wellprepared initial conditions. Hence, the letter"E" stands for both, 'E'volutionary convergence and 'E'nergy convergence. while the letter "p" stands for well'P'reparedness of the initial conditions, in contrast to E-convergence, where the latter is not needed.

Definition 5.3.1 (pE-convergence of $(\boldsymbol{X},\mathscr{E}_\varepsilon,\mathscr{R}_\varepsilon)$). We say that the generalized gradient systems $(\boldsymbol{X},\mathscr{E}_\varepsilon,\mathscr{R}_\varepsilon)$ pE-converge to $(\boldsymbol{X},\mathscr{E}_0,\mathscr{R}_0)$, and write $(\boldsymbol{X},\mathscr{E}_\varepsilon,\mathscr{R}_\varepsilon) \overset{\text{pE}}{\to} (\boldsymbol{X},\mathscr{E}_0,\mathscr{R}_0)$, if

$$\left.\begin{array}{c} q_\varepsilon : [0,T] \to \boldsymbol{X} \\ \text{is sol. of } (\boldsymbol{X},\mathscr{E}_\varepsilon,\mathscr{R}_\varepsilon), \\ q_\varepsilon(0) \to q^0, \text{ and} \\ \mathscr{E}_\varepsilon(0,q_\varepsilon(0)) \to \mathscr{E}_0(0,q^0)<\infty \end{array}\right\} \Longrightarrow \left\{\begin{array}{c} \exists q \text{ sol. of } (\boldsymbol{X},\mathscr{E}_0,\mathscr{R}_0) \text{ with } q(0){=}q^0 \\ \text{and a subsequence } \varepsilon_k \to 0: \\ \forall t \in]0,T]\text{: } q_{\varepsilon_k}(t) \to q(t) \text{ and} \\ \mathscr{E}_{\varepsilon_k}(q_{\varepsilon_k}(t)) \to \mathscr{E}_0(q(t)). \end{array}\right. \tag{5.8}$$

Similarly, we define the pE-convergence for ERIS $(\mathscr{Q},\mathscr{E}_\varepsilon,\mathscr{D}_\varepsilon) \overset{\text{pE}}{\to} (\mathscr{Q},\mathscr{E}_0,\mathscr{D}_0)$, if "solution" is understood in the sense of energetic solutions.

In the following subsection we discuss some abstract results for pE-convergence.

5.3.1 *pE-convergence for Generalized Gradient Systems*

The first general approach to the evolutionary Γ-convergence for classical gradient systems, where the variational structure was exploited systematically, goes back to [SS04], see also [Ser11, Mie14]. This approach is based on the energy-dissipation principle for the gGS $(\boldsymbol{X}, \mathscr{E}_\varepsilon, \mathscr{R}_\varepsilon)$ presented in Theorem 5.2.2, which transforms the evolutionary system $0 \in \partial_{\dot{q}}\mathscr{R}_\varepsilon(q_\varepsilon, \dot{q}_\varepsilon) + \mathrm{D}_q\mathscr{E}_\varepsilon(t, q_\varepsilon)$ into the upper energy-dissipation estimate

$$\mathscr{E}_\varepsilon(t, q_\varepsilon(T)) + \mathscr{J}_\varepsilon(q_\varepsilon(\cdot)) \leq \mathscr{E}_\varepsilon(0, q_\varepsilon(0)) + \int_0^T \partial_s \mathscr{E}_\varepsilon(s, q_\varepsilon(s))\,\mathrm{d}s,$$

$$\text{where } \mathscr{J}_\varepsilon(q) := \int_0^T \mathscr{R}_\varepsilon(q(t), \dot{q}(t)) + \mathscr{R}_\varepsilon^*(q(t), -\mathrm{D}_q\mathscr{E}_\varepsilon(t, q(t)))\,\mathrm{d}t$$

Having a variational principle for the whole trajectory, one can now use variational techniques to pass to the limit $\varepsilon \to 0$. First we observe that the first term on the right-hand converges to the desired limit by the assumption of the wellpreparedness of the initial conditions. For the second term on the right-hand side we may assume that it is lower order and can be handled by compactness. In fact, often one has $\mathscr{E}_\varepsilon(t,q) = \mathscr{U}_\varepsilon(q) - \langle \ell_\varepsilon(t), q\rangle$, then $\partial_t \mathscr{E}(t,q) = -\langle \dot{\ell}_\varepsilon(t), q\rangle$ is linear in q and strong convergence of $\dot{\ell}_\varepsilon(t) \to \dot{\ell}(t)$ is $\boldsymbol{X}^*$ is sufficient.

Hence, it remains to estimate the two terms on the left-hand side. Here we can take advantage that we only need an estimate from above, i.e. the liminf estimates

$$\mathscr{E}_0(T, q(T)) \leq \liminf_{\varepsilon \to 0} \mathscr{E}_\varepsilon(T, q_\varepsilon(T)) \quad \text{and} \quad \mathscr{J}_0(q(\cdot)) \leq \liminf_{\varepsilon \to 0} \mathscr{J}_\varepsilon(q_\varepsilon(\cdot))$$

are sufficient. For this, one has to derive suitable a priori estimates on the solutions q_ε such that one is able to extract a subsequence q_{ε_k} which converges in a sufficiently strong topology to establish the desired liminf estimates.

The famous Sandier-Serfaty approach [SS04, Ser11] relies on the two liminf estimates

$$\begin{aligned} &\textstyle\int_0^T \mathscr{R}_0(q_0(t), \dot{q}_0(t))\,\mathrm{d}t \leq \liminf_{\varepsilon \to 0} \int_0^T \mathscr{R}_\varepsilon(q_\varepsilon(t), \dot{q}_\varepsilon(t))\,\mathrm{d}t \quad \text{and} \\ &\mathscr{R}_0^*(q_0, -\mathrm{D}_q\mathscr{E}_0(t, q_0)) \leq \liminf_{\varepsilon \to 0} \mathscr{R}_\varepsilon^*(q_\varepsilon, -\mathrm{D}_q\mathscr{E}_\varepsilon(t, q_\varepsilon)). \end{aligned}$$

However, the energy-dissipation principle (EDP) is even more flexible, since we do not need these two separate lower bounds. In passing to the liminf for the total dissipation $\int_0^T \mathscr{R}_\varepsilon + \mathscr{R}_\varepsilon^*\,\mathrm{d}t$ we may even give up the special dual form $\mathscr{R} + \mathscr{R}^*$ of the integrand. This idea, which was applied successfully in [AMP+12, Mie12, MPR14, LMPR15], can be summarized as follows.

Defining the functional $\mathscr{J}_\varepsilon : \mathrm{W}^{1,1}([0,T]; \boldsymbol{X}) \to [0, \infty]$ via

$$\mathscr{J}_\varepsilon(u) := \int_0^T \mathscr{R}_\varepsilon(u, \dot{u}) + \mathscr{R}_\varepsilon^*(u, -\mathrm{D}\mathscr{E}_\varepsilon(u))\,\mathrm{d}t,$$

we have to find a sufficiently good lower bound for the Γ-liminf, namely

$$\text{(i)} \quad u_\varepsilon(\cdot) \overset{*}{\rightharpoonup} u(\cdot) \text{ in } \mathrm{L}^\infty([0,T]; \boldsymbol{X}) \implies \int_0^T \mathscr{M}_0(u(t), \dot{u}(t))\,\mathrm{d}t \leq \liminf_{\varepsilon \to 0} \mathscr{J}_\varepsilon(u_\varepsilon),$$

where the integrand $\mathscr{M}_0$ does not need to be of the form $\mathscr{R}_0+\mathscr{R}_0^*$. Hence, finding the best (i.e. largest) $\mathscr{M}_0$ is nothing else than finding the (static) Γ-limit of the functionals $\mathscr{J}_\varepsilon$. It suffices to find $(\boldsymbol{X},\mathscr{E}_0,\mathscr{R}_0)$ and $\mathscr{M}_0$ such that

(ii) $\mathscr{E}_\varepsilon \xrightarrow{\Gamma} \mathscr{E}_0$;
(iii) $\mathscr{M}_0(u,v) \geq -\langle \mathrm{D}\mathscr{E}_0(u),v\rangle$;
(iv) the chain rule holds for$(\boldsymbol{X},\mathscr{E}_0,\mathscr{R}_0)$;
(v) $\mathscr{M}_0(u,v) = -\langle \mathrm{D}\mathscr{E}_0(u),v\rangle \implies \mathscr{R}_0(u,v)+\mathscr{R}_0^*(u,-\mathrm{D}\mathscr{E}_0(u)) = -\langle \mathrm{D}\mathscr{E}_0(u),v\rangle.$

As before, we can start from the EDP $\mathscr{E}_\varepsilon(u_\varepsilon(T))+\mathscr{J}_\varepsilon(u_\varepsilon)=\mathscr{E}_\varepsilon(u_\varepsilon(0))$. Using the wellpreparedness of the initial datum, (i), and (ii) we pass to the limit and obtain the EDP

$$\mathscr{E}_0(u(T)) + \int_0^T \mathscr{M}_0(u(t),\dot u(t))\,\mathrm{d}t \leq \mathscr{E}_0(u(0)).$$

Now using the (iii) and the chain rule (iv) we find

$$\begin{aligned}\mathscr{E}_0(u(0)) &\overset{\text{(iv)}}{=} \mathscr{E}_0(u(T)) - \int_0^T \langle \mathrm{D}\mathscr{E}(u(t)),\dot u(t)\rangle\,\mathrm{d}t \\ &\overset{\text{(iii)}}{\leq} \mathscr{E}_0(u(T)) + \int_0^T \mathscr{M}_0(u(t),\dot u(t))\,\mathrm{d}t \leq \mathscr{E}_0(u(0)).\end{aligned}$$

Thus, we conclude that we must have equality in (iii) for almost all $t\in[0,T]$, such that we can use (v) to conclude that u is a solution for $(\boldsymbol{X},\mathscr{E}_0,\mathscr{R}_0)$. Hence, the pE-convergence $(\boldsymbol{X},\mathscr{E}_\varepsilon,\mathscr{R}_\varepsilon) \overset{\text{pE}}{\to} (\boldsymbol{X},\mathscr{E}_0,\mathscr{R}_0)$ is established.

Section 5.4.1 summarizes the results of [Mie12, MT12], which show that the above strategy can even be applied to justify the passage from small viscous dissipation (i.e. $\mathscr{R}_\varepsilon(u,\cdot)$ is quadratic) to a limit problem with large rate-independent dissipation (i.e. $\mathscr{R}_0(u,\cdot)$ is positively homogeneous of degree 1, see Section 5.2.2).

In fact, under a slight and natural strengthening of the conditions (i) to (v), it is possible to construct $\mathscr{R}_0$ directly from $\mathscr{M}_0$. Indeed, assume that $\mathscr{M}_0(u,\cdot)$ is additionally even, convex, $\mathbb{R}$-valued, and lower semicontinuous, then $\mathscr{R}_{\mathscr{M}}$ defined via

$$\mathscr{R}_{\mathscr{M}}(u,v) := \mathscr{M}_0(u,v) - \mathscr{M}_0(u,0)$$

is a dissipation potential. Moreover, using property (iii) we find the estimate

$$\mathscr{R}^*_{\mathscr{M}}(u,-\mathrm{D}\mathscr{E}_0(u)) = \sup_{v\in\boldsymbol{X}}\Big(\langle -\mathrm{D}\mathscr{E}_0(u),v\rangle - \mathscr{M}_0(u,v) + \mathscr{M}_0(u,0)\Big) \leq \mathscr{M}_0(u,0).$$

Thus, we find the desired EDP $\mathscr{E}_0(u(T)) + \int_0^T \mathscr{R}_{\mathscr{M}} + \mathscr{R}^*_{\mathscr{M}}\,\mathrm{d}t \leq \mathscr{E}_0(u(0))$. We emphasize that the choice $\mathscr{R}_0 = \mathscr{R}_{\mathscr{M}}$ in (iv) and (v) is admissible, but not unique. In particular, it may be possible to find simpler $\mathscr{R}_0$ as is the case in the application discussed in Section 5.4.1.

5.3.2 pE-convergence for Rate-Independent Systems

A quite general theory of evolutionary Γ-convergence for ERIS $(\boldsymbol{X},\mathscr{E}_\varepsilon,\mathscr{D}_\varepsilon)$ was already developed in [MRS08], see also [MR15] for more details and applications. For simplicity, here we restrict to the case that the energies have the form

$$\mathscr{E}_\varepsilon(t,q)=\mathscr{F}_\varepsilon(q)-\langle \ell_\varepsilon(t),q\rangle, \tag{5.9a}$$

where $\boldsymbol{X}$ is a reflexive Banach space. We allow for the case that $\mathscr{F}$ is not convex and that the dissipation distances $\mathscr{D}_\varepsilon$ are not translation invariant. A typical set of assumptions reads as follows:

$$\exists c,C>0\,\forall\varepsilon\in[0,1],\ q\in\boldsymbol{X}:\ \mathscr{F}_\varepsilon(q)\geq c\|q\|^2-C; \tag{5.9b}$$

$$\forall\varepsilon\in[0,1]:\quad \mathscr{F}_\varepsilon:\boldsymbol{X}\to\mathbb{R}_\infty \text{ is weakly lower semicontinuous;} \tag{5.9c}$$

$$\exists C>0\,\forall\varepsilon\in[0,1]:\ \|\ell_\varepsilon\|_{\mathrm{C}^1([0,T])}\leq C; \tag{5.9d}$$

$$\forall t\in[0,T]:\quad \dot\ell_\varepsilon(t)\to\dot\ell_0(t) \text{ in } \boldsymbol{X}^* \text{ as } \varepsilon\to 0; \tag{5.9e}$$

$$\forall\varepsilon\in[0,1]\,\forall q_j\in\boldsymbol{X}:\ \begin{cases}\mathscr{D}_\varepsilon(q_1,q_3)\leq\mathscr{D}_\varepsilon(q_1,q_2)+\mathscr{D}_\varepsilon(q_2,q_3),\\ \quad\mathscr{D}_\varepsilon(q_1,q_2)=0\implies q_1=q_2.\end{cases} \tag{5.9f}$$

In general, these conditions together with Γ convergence of the energies and the dissipation are not strong enough to show pE-convergence. Even for existence for a fixed ε we need additional conditions, e.g. weak continuity of $\mathscr{D}_\varepsilon$ is sufficient.

Our first result on pE-convergence for ERIS assumes that the dissipation distances $\mathscr{D}_\varepsilon$ weakly continuously converge to $\mathscr{D}_0$, viz.

$$\mathscr{D}_\varepsilon\stackrel{\mathrm{C}}{\rightharpoonup}\mathscr{D}_0,\quad\text{which means that } q_\varepsilon\rightharpoonup q_0,\ \widehat q_\varepsilon\rightharpoonup\widehat q_0\implies\mathscr{D}_\varepsilon(q_\varepsilon,\widehat q_\varepsilon)\to\mathscr{D}_0(q_0,\widehat q_0).$$

Theorem 5.3.2 (pE-convergence for ERIS). *Assume that the ERIS $(\boldsymbol{X},\mathscr{E}_\varepsilon,\mathscr{D}_\varepsilon)$ satisfy (5.9), $\mathscr{E}_\varepsilon\stackrel{\Gamma}{\rightharpoonup}\mathscr{E}_0$, and $\mathscr{D}_\varepsilon\stackrel{\mathrm{C}}{\rightharpoonup}\mathscr{D}_0$ in $\boldsymbol{X}$; then $(\boldsymbol{X},\mathscr{E}_\varepsilon,\mathscr{D}_\varepsilon)\stackrel{\mathrm{pE}}{\rightharpoonup}(\boldsymbol{X},\mathscr{E}_0,\mathscr{D}_0)$.*

We refer to [MRS08] for the first proof and to [Mie14, Thm. 5.4] for a shorter proof. In fact, it is rather straightforward to establish the EDP, i.e. (E) in (5.5) where "$=$" is replaced by "$\leq$". The major difficulty lies in showing that the global stability condition (S) holds for the limit $\varepsilon=0$. This stability then implies a "chain-rule estimate", which show that (E) holds even with equality "$=$".

The major tool for passing to the limit in the stability condition is the existence of so-called *mutual recovery sequences*. (A very similar condition is already very useful in showing existence of energetic solutions.) Given a family $(q_\varepsilon)_{\varepsilon\in[0,1]}$ with $q_\varepsilon\rightharpoonup q_0$ and a test state $\widehat q_0$, we say that the family $(\widehat q_\varepsilon)_{\varepsilon\in]0,1[}$ is a mutual recovery sequences at time t, if

$$\limsup_{\varepsilon\to 0}\big(\mathscr{E}_\varepsilon(t,\widehat q_\varepsilon)-\mathscr{E}_\varepsilon(t,q_\varepsilon)+\mathscr{D}_\varepsilon(q_\varepsilon,\widehat q_\varepsilon)\big)\ \leq\ \mathscr{E}_0(t,\widehat q)-\mathscr{E}_0(t,q_0)+\mathscr{D}_0(q_0,\widehat q_0). \tag{5.10}$$

Clearly, if all q_ε satisfy the stability condition at time t, then all term in the limsup are nonnegative; hence we conclude that the right-hand side is nonnegative, which is the stability of q_0 if the test state $\widehat{q}_0$ can be chosen arbitrary. Under the conditions of the above Theorem 5.3.2 we see that the existence of mutual recovery sequence easily holds, since it suffices to choose recovery sequences for the energy $\mathscr{F}_\varepsilon$ and use the weak continuity of $\mathscr{D}_\varepsilon$ and $\langle \ell_\varepsilon(t), \cdot \rangle$.

In the case that $\boldsymbol{X}$ is a Hilbert space $\boldsymbol{H}$, the energies are quadratic, and the dissipation distances are translationally invariant, viz.

$$\mathscr{F}_\varepsilon(q) = \frac{1}{2}\langle A_\varepsilon q, q\rangle \geq c\|q\|_{\boldsymbol{H}}^2 \quad \text{and} \quad \mathscr{D}_\varepsilon(q_1, q_2) = \Psi_\varepsilon(q_2 - q_1), \tag{5.11}$$

one can construct mutual recovery sequences in the form $\widehat{q}_\varepsilon = q_\varepsilon + w_\varepsilon$ with $w_\varepsilon \to \widehat{q}_0 - q_0$ and exploit the better convergence $\widehat{q}_\varepsilon - q_\varepsilon = w_\varepsilon \to \widehat{q}_0 - q_0$ (strong convergence in $\boldsymbol{H}$!) in the following terms:

$$\mathscr{F}_\varepsilon(\widehat{q}_\varepsilon) - \mathscr{F}_\varepsilon(q_\varepsilon) = \frac{1}{2}\langle A_\varepsilon w_\varepsilon, \widehat{q}_\varepsilon + q_\varepsilon\rangle \quad \text{and} \quad \mathscr{D}_\varepsilon(q_\varepsilon, \widehat{q}_\varepsilon) = \Psi_\varepsilon(w_\varepsilon). \tag{5.12}$$

Using this, the following result was derived in [LM11] and [MR15, Ch. 3.5.4]. Here the Mosco convergence $\mathscr{E}_\varepsilon \overset{\mathrm{M}}{\to} \mathscr{E}_0$ means $\mathscr{E}_\varepsilon(t,\cdot) \overset{\Gamma}{\to} \mathscr{E}_0(t,\cdot)$ and $\mathscr{E}_\varepsilon(t,\cdot) \overset{\Gamma}{\rightharpoonup} \mathscr{E}_0(t,\cdot)$ for all $t \in [0,T]$.

Theorem 5.3.3 (pE-convergence for quadratic ERIS). *Let* $(\boldsymbol{H}, \mathscr{E}_\varepsilon, \Psi_\varepsilon)_{\varepsilon\in[0,1]}$ *satisfy* (5.9) *and* (5.11). *If* $\mathscr{E}_\varepsilon \overset{\mathrm{M}}{\to} \mathscr{E}_0$, $\Psi_\varepsilon \overset{\mathrm{C}}{\to} \Psi_0$, *and* $\Psi_\varepsilon \overset{\Gamma}{\rightharpoonup} \Psi_0$, *then* $(\boldsymbol{H}, \mathscr{E}_\varepsilon, \Psi_\varepsilon) \overset{\mathrm{pE}}{\rightharpoonup} (\boldsymbol{H}, \mathscr{E}_0, \Psi_0)$.

In contrast to Theorem 5.3.2 we need the continuous convergence $\Psi_\varepsilon \overset{\mathrm{C}}{\to} \Psi_0$ here only in the strong topology of $\boldsymbol{H}$. Applications of this theory occur in linearized elastoplasticity in the context of homogenization in [MT07, GM11, Han11] and in the derivation of elastoplastic plate models.

A highly non-trivial application of pE-convergence is treated in [MS13], where the ERIS $(\boldsymbol{X}, \mathscr{E}_\varepsilon, \mathscr{D}_\varepsilon)$ for $\varepsilon > 0$ describe models for *finite-strain elastoplasticity* for which existence of energetic solutions was established in [MM09, Mie10]. In [MS13], the energy, the dissipation distance, and the loadings are scaled by $\varepsilon > 0$ in such a way that the system converges to *linearized elastoplasticity* in the sense of pE-convergence. The major assumption is that the yield stress (contained in $\mathscr{D}_\varepsilon$) scales in the same way as the displacement. Thus, *linearized elastoplasticity* is a justifiable model only under the condition that the yield stress is so small that even small strains can generate plastic effects.

5.4 Justification of Rate-Independent Models

In this section we discuss two distinct cases in which RIS arise as limits of rate-dependent systems. The typical situation we are interested in is a system with slow loading, where we always assume that the loading time $t \in [0,T]$ is our relevant time scale. In fact, in mechanics this time scale is often called process time, since it may be significantly larger than the intrinsic time scales inside the material.

In Section 5.4.1 we consider purely viscous systems, i.e. with a quadratic dissipation potential $\mathcal{R}_\varepsilon(q,v) = \frac{\varepsilon^\alpha}{2}\langle\mathbb{G}(q)v,v\rangle$, where the small parameter ε indicates that the relaxation times due to viscous effects are much smaller, namely of order $O(\varepsilon^\alpha)$. However, to prevent the system to relax into a global minimum for each macroscopic time we consider an energy that has microscopic wiggles that keeps the system outside macroscopic minimizers.

In Section 5.4.3 we consider gGS with a dissipation potential consisting of a fixed rate-independent and a small rate-dependent part, e.g. $\mathcal{R}_\varepsilon(q,v) = \mathcal{R}_{\mathrm{ri}}(q,v) + \frac{\varepsilon}{2}\langle\mathbb{G}(q)v,v\rangle$. For $\varepsilon > 0$ the solutions q_ε will be absolutely continuous with respect to $t \in [0,T]$ and the task is to characterize the jumps that develop in the vanishing-viscosity limit $\varepsilon \to 0$.

We also refer to [LOR07] for a derivation of macroscopic rate-independent behavior in the case of crack propagation.

5.4.1 Wiggly Energies Give Rise to Rate-Independent Friction

This section deals with the question how macroscopic RIS can arise from purely viscous systems in the limit of vanishing viscosity $\varepsilon \to 0$. We refer to [PT02, MT12, Mie12] for the full details. We stay in the framework of evolutionary Γ-convergence of gGS $(\boldsymbol{X},\mathcal{E}_\varepsilon,\mathcal{R}_\varepsilon)$. In particular, we will start with the cases $\mathcal{R}_\varepsilon(q,v) = \frac{\varepsilon^\alpha}{2}\langle\mathbb{G}v,v\rangle$, where obviously $\mathcal{R}_\varepsilon \to 0$, and end up with a limit system $(\boldsymbol{X},\mathcal{E}_0,\mathcal{R}_0)$, where $\mathcal{R}_0$ is rate-independent. The first example will show very clearly that $\mathcal{R}_0$ is determined not by $\mathcal{R}_\varepsilon$, but by microscopic variations in the energies $\mathcal{E}_\varepsilon$, hence one uses the name *wiggly energies*.

In [Mie12] the following slight generalization of the wiggly-energy model of [Jam96] was studied. The latter was analyzed already in [PT02, PT05], but the gradient structure was first exploited in [Mie12]. As viscous gradient system $(\boldsymbol{X},\mathcal{E}_\varepsilon,\mathcal{R}_\varepsilon)$ it takes the form

$$\boldsymbol{X} = \mathbb{R}, \qquad \mathcal{E}_\varepsilon(t,q) = \mathcal{F}(q) + \varepsilon W(q,\tfrac{1}{\varepsilon}q) - \ell(t)q, \qquad \mathcal{R}_\varepsilon(v) = \frac{\varepsilon^\alpha}{2}v^2.$$

Here $\mathcal{F} \in \mathrm{C}^2(\mathbb{R})$ denotes the macroscopic part of the energy, $W \in \mathrm{C}^2(\mathbb{R}\times\mathbb{S}^1)$ denotes the wiggly part, and $\ell \in \mathrm{C}^1([0,T])$ is the given time-dependent loading. Here $\mathbb{S}^1 = \mathbb{R}/\mathbb{Z}$ indicated that W is nontrivially periodic with period 1 in the second variable. In particular, writing $W = W(q,p)$, we assume

$$\rho_+(q) := \max\{\,\mathrm{D}_pW(q,p)\,|\,p\in\mathbb{S}^1\,\} > 0 \ \text{ and} \tag{5.13a}$$

$$\rho_-(q) := \min\{\,\mathrm{D}_pW(q,p)\,|\,p\in\mathbb{S}^1\,\} < 0. \tag{5.13b}$$

Defining $\mathcal{E}_0(t,q) = \mathcal{F}(q) - \ell(t)q$, we see that the energies $\mathcal{E}_\varepsilon$ uniformly converge to the macroscopic limit $\mathcal{E}_0$ via $|\mathcal{E}_\varepsilon(t,q) - \mathcal{E}_0(t,q)| \le C\varepsilon$, i.e. the wiggles are not seen on the energetic level. However, for the restoring force $\mathrm{D}_q\mathcal{E}_\varepsilon(t,q)$ we see a strong deviation from $\mathrm{D}_q\mathscr{E}_0(t,q)$. In particular, the functions $q \mapsto \mathrm{D}_q\mathscr{E}_\varepsilon(t,q)$ has many zeros (local equilibria of $\mathscr{E}_\varepsilon$).

The ODE $0 = \mathrm{D}_{\dot q}\mathscr{R}_\varepsilon(\dot q) + \mathrm{D}_q\mathscr{E}_\varepsilon(t,q)$ generated by $(\mathbb{R}, \mathscr{E}_\varepsilon, \mathscr{R}_\varepsilon)$ reads

$$0 = \varepsilon^\alpha \dot q + \mathscr{F}'(q) + \mathrm{D}_p W(q, \tfrac{1}{\varepsilon}q) - \varepsilon \mathrm{D}_q W(q, \tfrac{1}{\varepsilon}q) - \ell(t). \tag{5.14}$$

The aim of evolutionary Γ-convergence is to show that the solutions q_ε of the viscous gradient system $(\mathbb{R}, \mathscr{E}_\varepsilon, \mathscr{R}_\varepsilon)$ converge to a solutions of the RIS $(\mathbb{R}, \mathscr{E}_0, \mathscr{R}_0)$, where the macroscopic energy $\mathscr{E}_0$ is given above and the rate-independent dissipation potential $\mathscr{R}_0$ is defined via

$$\mathscr{R}_0(z,v) := \begin{cases} \rho_+(z)v & \text{for } v \geq 0, \\ \rho_-(z)v & \text{for } v \leq 0. \end{cases} \tag{5.15}$$

Hence the solutions q of the limiting RIS $(\mathbb{R}, \mathscr{E}_0, \mathscr{R}_0)$ are given by the differential inclusion

$$0 \in \partial_{\dot q}\mathscr{R}_0(q, \dot q) + \mathrm{D}_q\mathscr{E}_0(t,q). \tag{5.16}$$

We emphasize that the definition of $\mathscr{R}_0$ does only involve characteristics of the wiggly microscopic energy landscape of $\mathscr{E}_\varepsilon$, namely the p-derivate of the wiggle function $W(q,p)$.

The main convergence result states that the solutions q_ε of (5.14) converge to solutions of the RIS $(\mathbb{R}, \mathscr{E}_0, \mathscr{R}_0)$.

Theorem 5.4.1 ([PT02, Mie12]). *Let $\mathscr{F}$, W, ℓ, $\mathscr{E}_\varepsilon$, and $\mathscr{R}_\varepsilon$ be as described above, $\alpha > 0$, and assume that the mutual-convexity condition*

$$\inf\{\, \mathscr{F}''(q) \mid q \in \mathbb{R} \,\} > \sup\{\, |\mathrm{D}_q \mathrm{D}_p W(q,p)| \mid q \in \mathbb{R},\ p \in \mathbb{S}^1 \,\} \tag{5.17}$$

holds. Then $(\mathbb{R}, \mathscr{E}_\varepsilon, \mathscr{R}_\varepsilon) \xrightarrow{\mathrm{E}} (\mathbb{R}, \mathscr{E}_0, \mathscr{R}_0)$.

The proof in [Mie12] relies on three major pillars, namely (a) suitable a priori estimates, (b) a liminf-estimate for the energy-dissipation principle, and (c) uniqueness of the limiting systems. For (a) and (c) the standard energy estimates and the mutual-convexity condition (5.17) are used. The major difficulty lies in the limit passage (b) for the energy-dissipation principle as described in Section 5.3.1. For this we define the total dissipation functional

$$\mathscr{J}_\varepsilon(q) = \int_0^T \mathscr{M}_\varepsilon(t, q_\varepsilon(t), \dot q_\varepsilon(t))\,\mathrm{d}t \text{ with } \mathscr{M}_\varepsilon(t,q,v) = \mathscr{R}_\varepsilon(q,v) + \mathscr{R}_\varepsilon^*(q, -\mathrm{D}_q\mathscr{E}_\varepsilon(t,q)).$$

Inserting the specific forms of $\mathscr{R}_\varepsilon$, $\mathscr{R}_\varepsilon^*$, and $\mathscr{E}_\varepsilon$ we find

$$\mathscr{M}_\varepsilon(t,q,v) = \frac{\varepsilon^\alpha}{2}v^2 + \frac{1}{2\varepsilon^\alpha}\big|\mathscr{F}'(q) - \ell(t) + \mathrm{D}_p W(q, q/\varepsilon) + \varepsilon \mathrm{D}_q W(q, q/\varepsilon)\big|^2.$$

Homogenization arguments from [Bra02, Sect. 3] yield the liminf estimate

$$\liminf_{\varepsilon\to 0} \mathscr{J}_\varepsilon(q_\varepsilon) \geq \mathscr{J}_0(q) := \int_0^T \mathscr{M}_0(t,q,\dot{q})\,\mathrm{d}t \text{ with } \mathscr{M}_0(t,q,v) = \mathfrak{P}(v,\mathscr{F}'(q)-\ell(t)),$$

$$\mathfrak{P}(q,\xi) := |v|K(q,\xi) + \chi_{[\rho_-(q),\rho_+(q)]}(\xi), \text{ and } K(q,\xi) = \int_{\mathbb{S}^1} |\xi + \mathrm{D}_p W(q,p)|\,\mathrm{d}p.$$

It is easy to check the conditions (ii)–(v) in Section 5.3.1 for $\mathscr{E}_0$ and $\mathscr{R}_0$ given above. First note that (ii) and (iv) are trivial. Next observe $K(q,\xi) \geq |\xi|$, which implies (iii). For the crucial condition (v) we use that $\mathscr{M}_0(t,q,v) = -v\mathrm{D}_q\mathscr{E}_0(t,q)$ means $\xi = \mathrm{D}_q\mathscr{E}_0(t,q) \in [\rho_-(q),\rho_+(q)]$ and $|v|K(q,\xi) = -v\xi$. However, $K(q,\xi) = |\xi|$ holds if and only if $\xi \notin]\rho_-(q),\rho_+(q)[$. Thus, the equivalence to $0 \in \partial_v\mathscr{R}_0(q,v) + \xi$ (or any other of the five equivalent formulations in Proposition 5.2.1) follows easily.

5.4.2 1D Elastoplasticity as Limit of a Chain of Bistable Springs

A second evolutionary Γ-limit with wiggly energies is established in [MT12]. The system models a chain of N *bistable springs* with small viscous damping. Denoting by e_j the strain in the jth spring, the system reads

$$\left.\begin{aligned} &\nu\dot{e}_j = -\mathscr{F}'_{\mathrm{biq}}(e_j) + \mu_j^N + G(t,j/N) + \sigma(t) \quad \text{for } j=1,\ldots,N; \\ &\mathscr{C}_N((e_j)) := \tfrac{1}{N}\textstyle\sum_{j=1}^N e_j = \ell(t), \end{aligned}\right\} \tag{5.18}$$

where the biquadratic double-well potential $\mathscr{F}_{\mathrm{biq}}(e) := \frac{k}{2}\min\{(e+a)^2,(e-a)^2\}$ generates the bistability. The coefficients μ_j^N are biases that act as quenched disorder (time-independent) and are chosen randomly, namely independently and identically distributed according to a probability density $f \in \mathrm{L}^1([-\mu_*,\mu_*])$ with average 0.

The system is driven by the volume loading $G \in \mathrm{C}^1([0,T]\times[0,1])$ and the constraint $\mathscr{C}_N$ corresponding to a Dirichlet loading $\ell \in \mathrm{C}^1([0,T])$ prescribing the total length of the chain, where σ is the Lagrange parameter for this constraint.

Using $\boldsymbol{e} = (e_1,\ldots,e_N)$ as a state vector, the system has the energy functional $\mathscr{E}_N$ and the viscous dissipation potential $\mathscr{R}_N$:

$$\mathscr{E}_N(t,\boldsymbol{e}) = \frac{1}{N}\sum_{j=1}^N \Big(\mathscr{F}_{\mathrm{biq}}(e_j) - \mu_j^N e_j + G(t,j/N)e_j\Big) \text{ and } \mathscr{R}_N(\boldsymbol{e},\dot{\boldsymbol{e}}) = \frac{\nu}{2N}\sum_{j=1}^N \dot{e}_j^2.$$

The total system can now be written abstractly as a viscous gradient flow via

$$0 = \mathrm{D}_{\dot{\boldsymbol{e}}}\mathscr{R}_N(\boldsymbol{e},\dot{\boldsymbol{e}}) + \mathrm{D}_{\boldsymbol{e}}\mathscr{E}_N(t,\boldsymbol{e}) + \sigma(t)\mathrm{D}\mathscr{C}_N(\boldsymbol{e}) \quad \text{with } \mathscr{C}_N(\boldsymbol{e}) = \ell(t).$$

Our small parameter is now $\varepsilon = 1/N$, which is the ratio between the length of the springs and the total length. Clearly, the energy $\mathscr{E}_N$ is wiggly in the sense that there are many local minimizers for a given constraint $\mathscr{C}_N(\boldsymbol{e}) = \ell$, namely up to 2^N.

The limit of particle number $N \to \infty$ and viscosity $\nu \to 0$ can be studied by embedding the system into a spatially continuous setting on the physical domain

$\Omega =]0,1[$. The potential $\mathscr{F}_{\text{biq}}$ has two wells and hence two phases for each spring, which we characterize by the phase indicators $z_j = \text{sign}(e_j) \in \{-1,0,1\}$. With the indicator functions

$$\varphi_j^N(x) := \begin{cases} 1 \text{ for } x \in \big((j-1)/N, j/N\big), \\ 0 \text{ otherwise.} \end{cases} \tag{5.19}$$

we define elastic and plastic strains via $(\overline{e}^N(t), \overline{p}^N(t)) := \mathscr{P}_N(e^N(t))$, where

$$\mathscr{P}_N : \begin{cases} \mathbb{R}^N \quad \to \quad \mathrm{L}^2(\Omega) \times \mathrm{L}^2(\Omega), \\ \boldsymbol{e} = (e_j)_{j=1,\dots,N} \mapsto \Big(\sum_{j=1}^N e_j^N \varphi_j^N, a\sum_{j=1}^N z_j^N \varphi_j^N \Big) \end{cases} \tag{5.20}$$

The definition of $(\overline{e}^N, \overline{p}^N)$ is such that we obtain a linear stress-strain relation

$$\mathscr{F}'_{\text{biq}}(\overline{e}^N(t,x)) = k\big(\overline{e}^N(t,x) - \overline{p}^N(t,x)\big),$$

since the nonlinearity is moved into the definition of $\overline{p}$ via $z_j = \text{sign}(e_j)$.

The limiting gGS $(\boldsymbol{H}, \mathscr{E}_0, \mathscr{R}_0)$ describes *linearized elastoplasticity* with hardening and is defined via

$$\boldsymbol{H} = \mathrm{L}^2(\Omega) \times \mathrm{L}^2(\Omega), \qquad \mathscr{R}_0(\dot{\overline{p}}) = \int_\Omega ka\big|\dot{\overline{p}}(x)\big|\,\mathrm{d}x,$$
$$\mathscr{E}_0(\overline{e},\overline{p}) = \int_\Omega \frac{k}{2}(\overline{e}(x) - \overline{p}(x))^2 + H_f(\overline{p}(x)) + G(t,x)\overline{e}(x)\,\mathrm{d}x,$$

where the hardening potential H_f is a convex function that is uniquely determined by the distribution function f for the random biases μ_j^N. Indeed, defining L_f such that $L_f'' = f$ one obtains H_f as Legendre transform of L_f, see [MT12].

Together with the constraint $\mathscr{C}_0(\overline{e}) := \int_\Omega \overline{e}(x)\,\mathrm{d}x = \ell(t)$, we obtain the RIS $(\boldsymbol{H}, \mathscr{E}_0, \mathscr{R}_0, \mathscr{C}_0)$ with a 1-homogeneous dissipation potential $\mathscr{R}_0$ given in terms of the "yield stress ka". The associated differential inclusion

$$\begin{aligned} &0 = \mathrm{D}_{\overline{e}}\mathscr{E}(\overline{e},\overline{p}) + \sigma(t)\mathrm{D}\mathscr{C}(\overline{e}) = k(\overline{e} - \overline{p}) + \sigma, \qquad \mathscr{C}(\overline{e}) = \ell(t), \\ &0 \in \partial\mathscr{R}(\dot{\overline{p}}) + \mathrm{D}_{\overline{p}}\mathscr{E}(\overline{e},\overline{p}) = ka\mathrm{Sign}(\dot{\overline{p}}) + k(\overline{p} - \overline{e}) + \partial H_f(\overline{p}). \end{aligned} \tag{5.21}$$

describes one-dimensional elastoplasticity with Dirichlet loading $u(t,0) = 0$ and $u(t,1) = \ell(t)$, if the displacement is defined by $u(t,x) = \int_0^x \overline{e}(t,y)\,\mathrm{d}y$.

The following convergence result shows that the rate-independent evolution (5.21) is indeed the evolutionary Γ-limit of the finite-dimensional viscous systems (5.18).

Theorem 5.4.2 ([MT12, Thm. 5.2]). *Assume $\nu_N = 1/N^\alpha$ for a fixed $\alpha > 1$. Consider the solutions $\boldsymbol{e}^N : [0,T] \to \mathbb{R}^N$ of the gradient system $(\mathbb{R}^N, \mathscr{E}_N, \mathscr{R}_N)$, where the biases μ_j^N are chosen randomly (and independently and identically distributed) according to the distribution f. Then, with probability 1 with respect to the random biases μ_j^N we have $(\mathbb{R}^N, \mathscr{E}_N, \mathscr{R}_N) \xrightarrow{\text{pE}} (\boldsymbol{H}, \mathscr{E}_0, \mathscr{R}_0)$ in the sense of the embedding $\mathscr{P}_N$: If the initial conditions $\boldsymbol{e}^N(0)$ satisfy $e_j^N(0) < 0$ for all j,*

$$\mathscr{P}_N(\boldsymbol{e}^N(0)) \rightharpoonup (\overline{e}_0, \overline{p}_0) \text{ in } \boldsymbol{H}, \quad \text{and} \quad \mathscr{E}^N(0, \boldsymbol{e}^N(0)) \to \mathscr{E}(0, \overline{e}_0, \overline{p}_0);$$

then, for all $t \in [0,T]$ *we have*

$$\mathscr{P}_N(\boldsymbol{e}^N(t)) \rightharpoonup (\overline{e}(t), \overline{p}(t)) \text{ in } \boldsymbol{H} \quad \text{and} \quad \mathscr{E}^N(t, \boldsymbol{e}^N(t)) \to \mathscr{E}(t, \overline{e}(t), \overline{p}(t)),$$

where $(\overline{e}, \overline{p})$ *is the unique solution of* (5.21).

We again emphasize that the limiting dissipation potential $\mathscr{R}_0$ is not related to the original quadratic potentials $\mathscr{R}_N$. In the definition of $\mathscr{R}_0$ the constants k and a appear, which are part of the definition of the double-well potential $\mathscr{F}_{\text{biq}}$.

5.4.3 Balanced-Viscosity Solutions as Vanishing-Viscosity Limits

Assuming rate independence for an evolutionary system is always an approximation: the loading time-scale is taken to be much slower than all the internal relaxation processes. Moreover, in most material models there are two kinds of variables, i.e. we write the state variable q as a couple $q = (y, z)$, where y denotes the elastic or fast variables, usually containing the elastic deformation $\phi : \Omega \to \mathbb{R}^d$ or the small displacement $u : \Omega \to \mathbb{R}^d$. The variable z are taken to be internal variables which are slower and may be modeled by rate-independent friction such as plastic yields or activated phase transformation. Hence, a typical quasistatic material model (where we still neglect inertial terms) will have the form of a coupled system

$$0 = \varepsilon^\alpha \mathbb{G}_1(y,z)\dot{y} + \mathrm{D}_y\mathscr{E}(t,y,z), \qquad 0 \in \partial\Psi(y,z,\dot{z}) + \varepsilon\mathbb{G}_2(y,z)\dot{z} + \mathrm{D}_z\mathscr{E}(t,y,z),$$

where we again assume that the loading rate is scaled to be of order one, such that the viscous relaxation times for the variable y are $O(\varepsilon^\alpha)$ while the variable z has rate-independent terms (instantaneous relaxation is possible) as well as additional viscous relaxation on the time scale $O(\varepsilon)$. Clearly, we have a generalized gradient system $(\boldsymbol{X}, \mathscr{E}, \mathscr{R}_\varepsilon)$ with

$$\boldsymbol{X} = \boldsymbol{Y} \times \boldsymbol{Z} \text{ and } \mathscr{R}_\varepsilon(y,z,\dot{y},\dot{z}) = \Psi(y,z,\dot{z}) + \frac{\varepsilon^\alpha}{2}\langle \mathbb{G}_1(y,z)\dot{y},\dot{y}\rangle_{\boldsymbol{Y}} + \frac{\varepsilon}{2}\langle \mathbb{G}_2(y,z)\dot{z},\dot{z}\rangle_{\boldsymbol{Z}}.$$

Again, we can ask the question of evolutionary Γ-convergence of $(\boldsymbol{X}, \mathscr{E}, \mathscr{R}_\varepsilon)$ towards a limit system $(\boldsymbol{X}, \mathscr{E}, \Psi, \Xi)$, in the sense that solutions q_ε of the former converge to the solutions q_0 of the latter system. Here the additional structure "Ξ" indicates that the simple RIS $(\boldsymbol{X}, \mathscr{E}, \Psi)$ needs to be enhanced by some information characterizing the jumps.

To obtain a rate-independent limit, one is again interested in the case $\varepsilon \to 0$, which is called the *vanishing-viscosity limit*. Formally, it is expected that the limits $q_0 = (y_0, z_0)$ of solutions $q_\varepsilon = (y_\varepsilon, z_\varepsilon)$ will satisfy the different inclusion

$$0 = \mathrm{D}_y\mathscr{E}(t, q_0(t)) \quad \text{and} \quad 0 \in \partial_{\dot{z}}\Psi(q_0(t), \dot{z}_0(t)) + \mathrm{D}_z\mathscr{E}(t, q_0(t)) \tag{5.22}$$

for almost all $t \in [0,T]$. However, in general the limits $q_0 : [0,T] \to \boldsymbol{X}$ will develop jumps with $q_0(t-0) \neq q_0(t+0)$ and (5.22) will not be enough to characterize these jumps. Moreover, the jumps arising in the vanishing-viscosity limit will depend on the different viscosity choices $\varepsilon^\alpha \mathbb{G}_1(q)$ and $\varepsilon \mathbb{G}_2(q)$.

Indeed, in [MRS14b] the dependence of the exponent $\alpha > 0$ was investigated in a situation where $q = (y,z) \in \mathbb{R}^n \times \mathbb{R}^m$ and where $\mathcal{E}(t,\cdot,z)$ is strictly convex. It turns out that the jump behavior is quite different for the three cases $\alpha \in]0,1[$, $\alpha = 1$, and $\alpha > 1$. For $\alpha > 1$ the component y can relax into the unique minimizer of $\mathcal{E}(t,\cdot,z(t))$ much faster than any changes in z. Hence, it is possible to reduce the situation by eliminating the variable y by defining $y = Y(t,z) = \operatorname{Arg\,min}_{\widetilde{y} \in \boldsymbol{Y}} \mathcal{E}(t,\widetilde{y},z)$ and $\widehat{\mathcal{E}}(t,z) = \mathcal{E}(t,Y(t,z),z)$.

For $\alpha \leq 1$ the situation is much more difficult and new jump phenomena occur, which are not yet understood, see [MRS14b] for some first results.

In light of the above discussion for $\alpha > 1$ we restrict ourself to the case $\boldsymbol{X} = \boldsymbol{Z}$ and consider gGS $(\boldsymbol{Z},\mathcal{E},\mathcal{R}_\varepsilon)$ with the simplest "vanishing-viscosity dissipation potential"

$$\mathcal{R}_\varepsilon(v) = \Psi(v) + \frac{\varepsilon}{2}\langle \mathbb{G}v, v\rangle, \tag{5.23}$$

where Ψ is positively homogeneous of degree 1 and $\mathbb{G} = \mathbb{G}^* > 0$. The important observation is that $\mathbb{G}$ generates a Hilbert-space norm $\|v\|_{\boldsymbol{V}} := \big(\langle \mathbb{G}v, v\rangle\big)^{1/2}$, which is defines the Hilbert space $\boldsymbol{V}$. Throughout, we assume that $\boldsymbol{Z}$ is continuously embedded into $\boldsymbol{V}$, which is certainly the case for the model system studied in [Miel11b, MZ14]:

$$\text{(MS)} \quad \begin{cases} \boldsymbol{Z} = \mathrm{L}^1(\Omega), \quad \boldsymbol{V} = \mathrm{L}^2(\Omega), \quad \mathcal{R}_\varepsilon(v) = \int_\Omega |v| + \frac{\varepsilon}{2}|v|^2 \,\mathrm{d}x, \\ \text{and } \mathcal{E}(t,z) = \int_\Omega \frac{\kappa}{2}|\nabla z|^2 + W(z) - \ell(t) z \,\mathrm{d}x \text{ for } z \in \mathrm{H}^1_0(\Omega), \end{cases}$$

where $\Omega \subset \mathbb{R}^d$ is a smooth bounded domain, W is the double-well potential $W(z) = (z^2-1)^2/4$, and ℓ is a smooth loading. The evolutionary equation is

$$\begin{aligned} 0 \in \operatorname{Sign}(\dot z) + \varepsilon \dot z - \kappa \Delta z + W'(z) - \ell(t) \ \text{ for } (t,x) \in [0,T] \times \Omega, \\ z(t,x) = 0 \ \text{ for } (t,x) \in [0,T] \times \partial\Omega, \end{aligned} \tag{5.24}$$

which is extensively studied in [MZ14] by direct PDE methods.

For passing to the limit $\varepsilon \to 0$ and still controlling the jump behavior it is useful to reparametrize the solutions $t \mapsto (t, z_\varepsilon(t)) \in [0,T] \times \boldsymbol{Z}$ in the extended state space and study the convergence there. This idea was introduced in for RIS in [EM06] and turned into an energetic framework in the series of papers [MRS09, MRS12, MRS14a, MRS14b].

For the reparametrization we let $t = \mathsf{t}(s)$ and $z(t) = \mathfrak{z}(s)$, where $s \in [0,S]$ is now an arclength-like parameter. We write $\mathfrak{z}'(s) = \frac{\mathrm{d}}{\mathrm{d}s}\mathfrak{z}(s)$ and note $\dot z(\mathsf{t}(s))\mathsf{t}'(s) = \mathfrak{z}'(s)$.

Definition 5.4.3 (Parametrized solutions). Let the RIS $(\boldsymbol{Z},\mathcal{E},\Psi,\mathbb{G})$ and $\boldsymbol{V}$ be given as above. Then, a pair $(\mathsf{t},\mathfrak{z}) : [0,S] \to [0,T] \times \boldsymbol{Z}$ is called a $\mathbb{G}$-*parametrized solution*, if $(\mathsf{t},\mathfrak{z}) \in \mathrm{W}^{1,1}(0,T;\mathbb{R} \times \boldsymbol{V})$ and there exists $\lambda : [0,S] \to [0,\infty[$ such that

$$\left.\begin{array}{l}\mathfrak{t}(0)=0,\ \mathfrak{t}(S)=T,\ \mathfrak{t}'(s)\geq 0,\ \lambda(s)\geq 0,\ \lambda(s)\mathfrak{t}'(s)=0,\\ 0\in\partial\Psi(\mathfrak{z}'(s))+\lambda(s)\mathbb{G}\mathfrak{z}'(s)+\mathrm{D}_z\mathscr{E}(\mathfrak{t}(s),\mathfrak{z}(s)),\end{array}\right\}\text{ a.e. on }[0,S]. \quad (5.25)$$

The definition clearly displays the rate independence of the notion of $\mathbb{G}$-parametrized solutions, since $\mathfrak{z}'$ only occurs in the rate-independent term $\partial\Psi$ or together with λ which can be scaled freely.

For a variational approach we transform the EDP, cf. Theorem 5.2.2, by time rescaling and obtain for $(\mathfrak{t},\mathfrak{z})$ the following identity:

$$\begin{aligned}\mathscr{E}(\mathfrak{t}(S),\mathfrak{z}(S))+\int_{s=0}^{S}\mathfrak{P}_\varepsilon\big(\mathfrak{t}'(s),\mathfrak{z}'(s),-\mathrm{D}_z\mathscr{E}(\mathfrak{t}(s),\mathfrak{z}(s))\big)\,\mathrm{d}s&\\ =\mathscr{E}(mft(0),\mathfrak{z}(0))+\int_0^S\partial_t\mathscr{E}(\mathfrak{t}(s),\mathfrak{z}(s))\mathfrak{t}'(s)\,\mathrm{d}s,&\quad (5.26)\\ \text{where }\mathfrak{P}_\varepsilon(\tau,V,\xi)=\tau\mathscr{R}_\varepsilon(\tfrac{1}{\tau}V)+\tau\mathscr{R}_\varepsilon^*(\xi).&\quad (5.27)\end{aligned}$$

Using the special form of $\mathscr{R}_\varepsilon$ we obtain a quite explicit form for $\mathfrak{P}_\varepsilon$, namely

$$\mathfrak{P}_\varepsilon(\tau,V,\xi)=\Psi(V)+\frac{\varepsilon}{2\tau}\langle\mathbb{G}V,V\rangle+\frac{\tau}{2\varepsilon}M_{\boldsymbol{V}}(\xi)^2\ \text{ with } M_{\boldsymbol{V}}(\xi):=\inf_{\eta\in\partial\Psi(0)}\|\xi-\eta\|_{\boldsymbol{V}^*}.$$

It is now easy to see that the Γ-limit of $\mathfrak{P}_\varepsilon:[0,\infty[\times\boldsymbol{Z}\times\boldsymbol{V}^*\to[0,\infty]$ for $\varepsilon\to 0$ takes the form

$$\mathfrak{P}_0(\tau,V,\xi):=\begin{cases}\Psi(V)+\Psi^*(\xi) & \text{for }\tau>0,\\ \Psi(V)+\|V\|_{\boldsymbol{V}}M_{\boldsymbol{V}}(\xi) & \text{for }\tau=0.\end{cases}$$

Clearly, $\mathfrak{P}_0(\tau,V,\xi)\geq-\langle\xi,V\rangle$ for all (τ,V,ξ). Moreover, equality holds if and only if $0\in\partial\Psi(V)+\xi$ in the case $\tau>0$ and $0\in\partial\Psi(V)+\lambda\mathbb{G}V+\xi$ in the case $\tau=0$ see [MRS12, Sec. 3.2]. Thus, all parametrized solutions satisfy the limiting EDP

$$\begin{aligned}\mathscr{E}(\mathfrak{t}(S),\mathfrak{z}(S))+\int_{s=0}^{S}\mathfrak{P}_0\big(\mathfrak{t}'(s),\mathfrak{z}'(s),-\mathrm{D}_z\mathscr{E}(\mathfrak{t}(s),\mathfrak{z}(s))\big)\,\mathrm{d}s&\quad (5.28)\\ =\mathscr{E}(\mathfrak{t}(0),\mathfrak{z}(0))+\int_0^S\partial_t\mathscr{E}(\mathfrak{t}(s),\mathfrak{z}(s))\mathfrak{t}'(s)\,\mathrm{d}s,&\quad (5.29)\end{aligned}$$

and vice versa, sufficiently smooth solutions of the EDP are parametrized solutions. The advantage of (5.29) is that we do not need to assume $\mathfrak{z}\in\mathrm{W}^{1,1}([0,T];\boldsymbol{V})$. All solutions $(\mathfrak{t},\mathfrak{z})$ with $\mathfrak{t}\in\mathrm{W}^{1,1}([0,T])$ and $\mathfrak{z}\in\mathrm{BV}([0,T];\boldsymbol{Z})\cap\mathrm{C}^0([0,T];\boldsymbol{V})$ of (5.29) are called *parametrized balanced-viscosity* solutions of $(\boldsymbol{Z},\mathscr{E},\Psi,\mathbb{G})$. Here the term "balanced viscosity" relates to the subtle balance of rate-independent and viscous dissipations along jumps, that is seen in $\mathfrak{P}_0$ for $\tau=0$ in the term $\Psi(V)+\|V\|_{\boldsymbol{V}}M_{\boldsymbol{V}}(\xi)$.

The advantage of reformulating subdifferential equations like (5.24) and (5.25) in terms of the reparametrized EDP (5.27) is that we can control the limit $\varepsilon\to 0$ easily. In particular, if the define the solutions of $(\boldsymbol{Z},\mathscr{E},\Psi,\mathbb{G})$ to be parametrized balanced-viscosity solution, then we have evolutionary Γ-convergence of $(\boldsymbol{Z},\mathscr{E},\mathscr{R}_\varepsilon)$ (with $\mathscr{R}_\varepsilon$ from (5.23)) to $(\boldsymbol{Z},\mathscr{E},\Psi,\mathbb{G})$.

However, the introduction of the parametrization may appear ad hoc and disturbing. So one can define the notion of *Balanced-Viscosity solutions* as follows: $z:[0,T]\to\mathbf{Z}$ is called a BV solutions for $(\mathbf{Z},\mathcal{E},\Psi,\mathbb{G})$ if there exists a parametrized balanced-viscosity solutions $(\mathsf{t},\mathfrak{z}):[0,S]\to[0,T]\times\mathbf{Z}$ such that for all $t\in[0,T]$ there exists an $s\in[0,S]$ with $t=\mathsf{t}(s)$ and $z(t)=\mathfrak{z}(s)$. This simply means that the image of $(\mathsf{t},\mathfrak{z})$ in $[0,T]\times\boldsymbol{X}$ contains the graph of $z:[0,T]\to\mathbf{Z}$.

One major achievement in [MRS12, MRS14a] is a proper intrinsic definition of BV solutions without referring to parametrizations. For this one defines a new (time-dependent) dissipation distance $\boldsymbol{\Delta}(t,\cdot,\cdot)$ that measures the minimal dissipation according to $\mathfrak{P}_0$ along all curves connecting to states z_0 and z_1:

$$\boldsymbol{\Delta}(t,z_1,z_2):=\inf\Big\{\int_0^1\mathfrak{P}_0(0,\dot{\mathfrak{y}}(r),-\mathrm{D}_z\mathcal{E}(t,\mathfrak{y}(r)))\,\mathrm{d}r\;\Big|\; \mathfrak{y}\in\mathrm{C}^1([0,1];\boldsymbol{V}),\ \mathfrak{y}(0)=z_1,\ \mathfrak{y}(1)=z_2\Big\}. \tag{5.30}$$

Note that $\boldsymbol{\Delta}$ is defined with time t as a frozen parameter, i.e. $\mathsf{t}'(r)=\tau=0$. Clearly, we have the triangle inequality $\boldsymbol{\Delta}(t,z_0,z_2)\le\boldsymbol{\Delta}(t,z_0,z_1)+\boldsymbol{\Delta}(t,z_1,z_2)$ and the lower estimate $\boldsymbol{\Delta}(t,z_1,z_2)\ge\Psi(z_2-z_1)$. For the definition of BV solutions we use a supplemented dissipation functional $\mathrm{Diss}_{\mathfrak{p},\mathcal{E}}$ defined on functions $z\in\mathrm{BV}([0,T];\boldsymbol{X})$. Here $J(z)\subset[0,T]$ is the jump set of z, i.e. all the times t where the three values $z(t-0)$, $z(t)$, and $z(t+0)$ are not equal. The new dissipation functional $\mathrm{Diss}_{\mathfrak{M},\mathcal{E}}(z;[t_1,t_2])$ is bigger than the purely rate-independent functional Diss_Ψ defined in (5.6), because it properly accounts for the additional dissipation through the viscous terms during jumps:

$$\begin{aligned}\mathrm{Diss}_{\mathfrak{p},\mathcal{E}}(z;[t_1,t_2]):=\;&\mathrm{Diss}_\Psi(z;[t_1,t_2])+\widehat{\boldsymbol{\Delta}}(t_1,z(t_1),z(t_1^+))+\widehat{\boldsymbol{\Delta}}(t_2,z(t_2^-),z(t_2))\\&+\textstyle\sum_{t\in J(z)}\big(\widehat{\boldsymbol{\Delta}}(t,z(t^-),z(t))+\widehat{\boldsymbol{\Delta}}(t,z(t),z(t^+))\big),\end{aligned}$$

where $\widehat{\boldsymbol{\Delta}}(t,z_0,z_1):=\boldsymbol{\Delta}(t,z_0,z_1)-\Psi(z_1-z_0)\ge0$.

Definition 5.4.4 (Balanced-Viscosity solutions). A function $z\in\mathrm{BV}([0,T];\mathbf{Z})$ is called a Balanced-Viscosity solution, in short *BV solution*, for $(\mathbf{Z},\mathcal{E},\Psi,\mathbb{G})$, if

$$\forall t\in[0,T]\setminus J(z):\ \ z(t)\in\mathcal{S}_{\mathrm{loc}}(t):=\{z\in\mathbf{Z}\,|\,0\in\partial\Psi(0)+\mathrm{D}_z\mathcal{E}(t,z)\}\ \text{and} \tag{5.31a}$$

$$\forall t\in[0,T]:\ \ \mathcal{E}(t,z(t))+\mathrm{Diss}_{\mathfrak{M},\mathcal{E}}(z;[0,t])=\mathcal{E}(0,z(0))+\int_0^t\partial_t\mathcal{E}(t,z(t))\,\mathrm{d}t. \tag{5.31b}$$

It is interesting to see that the definition of BV solutions again consists of a static stability condition and an energy balance as in the case of energetic solutions, see (5.5). However, now the stability is local instead of global and it is only valid at continuity points of the solution. To compensate for this the dissipation is enhanced at jumps deriving from the additional dissipation through balanced viscosity.

We now use the advantage that BV solutions are defined as functions from the time interval $[0,T]$ into the state space $\mathbf{Z}$ like the viscous approximations. Thus, the natural question is how the solutions z_ε converge to BV solutions. This question was first answered in [MRS12] for the finite-dimensional setting and in [MRS14a, Thm. 3.9] for a general infinite-dimensional setting.

Theorem 5.4.5 (Vanishing-viscosity limit gives BV solutions). *Under suitable technical conditions on $(\mathbf{Z},\mathscr{E},\Psi,\mathbb{G})$ and the initial condition $z^0 \in \mathbf{Z}$, the solutions $z_\varepsilon : [0,T] \to \mathbf{Z}$ of $(\boldsymbol{X},\mathscr{E},\mathscr{R}_\varepsilon)$ with $z_\varepsilon(0) = z^0$ and $\mathscr{R}_\varepsilon$ from (5.23) exist and there exist a subsequence $\varepsilon_k \to 0$ and a BV solution $z : [0,T] \to \mathbf{Z}$ for $(\mathbf{Z},\mathscr{E},\Psi,\mathbb{G})$ such that*

$$\forall t \in [0,T]: \quad z_{\varepsilon_k}(t) \rightharpoonup z(t) \ \text{ in } \mathbf{Z} \ \text{ and } \ \mathscr{E}(t,z_{\varepsilon_k}(t)) \to \mathscr{E}(t,z(t)) \ \text{ for } k \to \infty.$$

Moreover, any pointwise limit z of a subsequence of $(z_\varepsilon)_{\varepsilon>0}$ is a BV solution.

Our final result concerns the vanishing-viscosity limit jointly with time discretizations, which provides an easy way of numerically calculating BV solutions. We discretize the time interval by partitions $\Pi = (t_0,t_1,\dots,t_{N_\Pi})$ with fineness $\phi(\Pi) = \max\{\, t_k - t_{k-1} \mid k = 1,\dots,N_\Pi \,\}$. The incremental minimization problem for the viscous problem reads

$$z_k^\varepsilon \in \operatorname{Arg\,min}_{z\in\mathbf{Z}} \mathscr{E}(t_k,z) + \Psi(z - z_{k-1}^\varepsilon) + \frac{\varepsilon}{2(t_k - t_{k-1})} \|z - z_{k-1}^\varepsilon\|_{\mathbf{V}}^2, \quad z_0^\varepsilon = z^0.$$

We denote by $z^{\Pi,\varepsilon} : [0,T] \to \mathbf{Z}$ the piecewise constant interpolant. The following result was first proved in [EM06, MRS12] for the finite-dimensional setting. For a quite general infinite-dimensional version we refer to [MRS14a, Thm. 3.10].

Theorem 5.4.6 (Convergence of viscous time discretizations). *Assume suitable technical conditions on $(\mathbf{Z},\mathscr{E},\Psi,\mathbb{G})$ and $z_0 \in \mathbf{Z}$ (see [Mie11b, MRS14a]) and consider a sequences $(\Pi_n)_{n\in\mathbb{N}}$ and $(\varepsilon_n)_{n\in\mathbb{N}}$ such that*

$$\varepsilon_n \to 0 \ \text{ and } \ \phi(\Pi_n)/\varepsilon_n \to 0. \tag{5.32}$$

Then, there exists a subsequence $n_l \to \infty$ and a BV solution z for $(\mathbf{Z},\mathscr{E},\Psi,\mathbb{G})$ such that the piecewise constant interpolants z^{Π_n,ε_n} satisfy

$$\forall t \in [0,T]: \ z^{\Pi_{n_l},\varepsilon_{n_l}}(t) \rightharpoonup z(t) \text{ in } \mathbf{Z} \text{ and } \mathscr{E}(t,z^{\Pi_{n_l},\varepsilon_{n_l}}(t)) \to \mathscr{E}(t,z(t)) \ \text{ for } l \to \infty.$$

Moreover, any such pointwise limit of a subsequence of $(z^{\Pi_n,\varepsilon_n})_{n\in\mathbb{N}}$ is a BV solution.

5.5 Rate-Independent Evolution of Microstructures

The theory of RIS provides an ideal framework for studying microstructures in the sense of the calculus of variations, namely those given by laminates or more general Young measures. The starting point of most of these works was the seminal paper [OR99] on microstructures in finite-strain plasticity. In the sequel a lot of work was done for the relaxation of a single elastoplastic time step, see [CHM02, CDK13b, CDK13a]. We also refer to [HK14, Hei15, Hei14] for the characterization and numerical calculation of *quasiconvex hulls*.

In contrast, the evolution of microstructures in plasticity is mathematically much less developed, see e.g. [Mie04, CT05]. However, the same theory was soon transferred to easier dissipative material models such as damage (cf. e.g. [FG06, GL09, Mie11a]) and phase transformations in elastomers (cf. e.g. [DD02]) or shape-memory materials (cf. e.g. [BCHH04, BH09, KH11, CLR15]).

In the following we discuss two applications of the evolutionary theory, both based on energetic solutions for RIS, see Section 5.2.2. The first application is treated in [HHM12] and deals with the evolution of microstructure in the form of laminates, where laminates are explicitly takes as an allowed microstructure with an appropriate dissipation distance as proposed in [KH11]. The second application reconsiders the evolutionary model from [MTL02], where the microstructure is captured by a macroscopic phase fraction $z(t,x) \in [0,1]$.

5.5.1 Laminate Evolution in Finite-Strain Plasticity

We summarize the results in [HHM12], which analyze a rate-independent model for finite-strain elastoplasticity with microstructure. The state of the system is described by the deformation $\phi : \Omega \to \mathbb{R}^d$ and by a Young measure $\Lambda : \Omega \to \mathscr{L} \subset \mathrm{Prob}(K)$, where $K := \mathbb{R}^{d\times d} \times \mathrm{SL}(\mathbb{R}^d))$, and $\mathrm{SL}(\mathbb{R}^d) = \{ P \in \mathbb{R}^{d\times d} \mid \det P = 1 \}$ is the special linear group containing the plastic strains, whereas $\mathbb{R}^{d\times d}$ will contain microfluctuations of the deformation gradient.

The main idea is to specify a physically relevant subset $\mathscr{L}$ of admissible Young measures, like laminates of a fixed order as in [OR99], to define a suitable dissipation distance between these measures, and to prevent formation of different microstructures by a suitable regularization. Following [KH11] the simplest set of admissible probability measures are laminates of first order:

$$\mathscr{L} := \{ \alpha\delta_{((1-\alpha)b\otimes n,Q)} + (1-\alpha)\delta_{(-\alpha b\otimes n,R)} \mid \alpha \in [0,1],\ b,n \in \mathbb{R}^d,\ R,Q \in \mathrm{SL}(\mathbb{R}^d) \}.$$

Of course, more complicated lamination trees on the sense of [OR99] would be possible. The point is now to define a dissipation distance $D_{\mathrm{lam}} : \mathscr{L} \times \mathscr{L} \to [0,\infty]$ between such laminates, which properly accounts for changes in the microstructure. In particular, one wants to model the fact that it is very difficult to rotate the normal vector n in such microstructures. When keeping n fixed, then the deformation fluctuation $b \in \mathbb{R}^d$ may change without dissipation, while changes of the volume fraction α dissipate according to the distance $D_{\mathrm{SL}}(Q_0,Q_1)$ or $D_{\mathrm{SL}}(R_0,R_1)$.

The ERIS is now constructed via the state space $\mathscr{Q} = \boldsymbol{Y} \times \mathscr{Z}$ with $\boldsymbol{Y} = \mathrm{W}^{1,p}(\Omega;\mathbb{R}^d)$ and $\mathscr{Z} = \{ \Lambda \in \mathrm{YM}(\Omega;K) \mid \Lambda(x) \in \mathscr{L} \text{ a.e. } \}$ and the energy functional

$$\mathscr{E}(t,\phi,\Lambda) = \int_\Omega \int_{\mathscr{L}} \big(W(\nabla\phi(I_d{+}A)P^{-1}) + H(P)\big)\Lambda(\mathrm{d}A,\mathrm{d}P)\,\mathrm{d}x$$
$$\qquad + \sigma\mathscr{G}(\Lambda) - \langle \ell(t),\phi\rangle \ \text{ with } \mathscr{G}(\Lambda) := \int_\Omega\int_\Omega \frac{d_{\mathrm{W}}(\Lambda(x),\Lambda(y))^p}{|x-y|^{d+\theta p}}\,\mathrm{d}x\,\mathrm{d}y,$$

where d_{W} defines a 1-Wasserstein like norm on $\mathscr{L}$, namely

$$d_{\mathrm{W}}(\Lambda_0,\Lambda_1) := \sup\{ \int_K g(A,P)\Lambda_1(\mathrm{d}A,\mathrm{d}P) - \int_K g(B,Q)\Lambda_0(\mathrm{d}B,\mathrm{d}Q) \mid \mathrm{Lip}_K(g) \leq 1 \}.$$

Thus, $\mathscr{G}(\Lambda)$ serves as a spatial regularization for the laminate field $\Lambda : \Omega \to \mathscr{L}$ which prevents the formation of further more complicated microstructures.

The dissipation distance $\mathscr{D} : \mathscr{Z} \times \mathscr{Z} \to [0,\infty]$ is defined as

$$\mathscr{D}(\Lambda_0,\Lambda_1) = \int_\Omega D_{\mathrm{lam}}(\Lambda_0(x),\Lambda_1(x))\,\mathrm{d}x$$

Under suitable assumptions on the polyconvex energy density W and the hardening energy H it is shown in [HHM12, Thm. 2.4] that the ERIS $\mathscr{Q},\mathscr{E},\mathscr{D})$ has for each stable initial condition (ϕ^0,Λ^0) an *energetic solution* describing the *laminate evolution*. Indeed, using the regularizing term $\mathscr{G}$ one has a compactness for the laminate fields, which allows to establish suitable lower semicontinuity results for $\mathscr{E}$ and $\mathscr{D}$ as well as mutual recovery sequences in the sense of (5.10).

5.5.2 A Two-Phase Shape-Memory Model for Small Strains

Finally we present some new results for the *two-phase model* for introduced in [MTL02]. In fact, this model was the origin for the development of energetic solutions.

The two elastic phases are described by linearized elasticity with the same elastic tensor $\mathbb{C}$, but have different transformation strains A_j. On the microscopic level one may use the stored energy density

$$\widehat{W}(e) = \min\{\frac{1}{2}(e{-}A_1){:}\mathbb{C}(e{-}A_1)+c_1, \frac{1}{2}(e{-}A_2){:}\mathbb{C}(e{-}A_2)+c_2\},$$

where $e = e(u) := \frac{1}{2}(\nabla u{+}\nabla u^\top)$ is the infinitesimal strain tensor. The *relaxation* of $\widehat{W}$ with given volume fraction $z \in [0,1]$ for phase 2 was derived in [Koh91]:

$$W(e,z) = (1{-}z)\big(\tfrac{1}{2}(e{-}A_1){:}\mathbb{C}(e{-}A_1)+c_1\big) + z\big(\tfrac{1}{2}(e{-}A_2){:}\mathbb{C}(e{-}A_2)+c_2\big) - \rho z(1{-}z),$$

where the relaxation coefficient $\rho > 0$ can be calculated explicitly.

The ERIS studied in [MTL02] is given by $\boldsymbol{Q} = \mathrm{H}^1_{\mathrm{D}}(\Omega;\mathbb{R}^d) \times \mathrm{L}^1(\Omega;[0,1])$,

$$\mathscr{E}(t,u,z) = \int_\Omega W(e(u),z) - \ell(t)\cdot u\,\mathrm{d}x, \ \text{ and } \mathscr{D}(z_0,z_1) = \delta\|z_1{-}z_0\|_{\mathrm{L}^1} \tag{5.33}$$

for some smooth loading and some dissipation coefficient $\delta > 0$. A first existence result for energetic solutions was obtained in [MTL02, Thm. 5.1] under the unnatural assumption that the energy $\mathscr{E}(t,\cdot)$ is convex. A corresponding numerical algorithm using space-time discretization and incremental minimizations (cf. (5.7)) was developed in [CP01]. Using the abstract theory for ERIS in [Mie11b, MR15], the existence theory was recently improved, see [HM15], by a new construction of mutual recovery sequences, see (5.10).

Theorem 5.5.1 ([HM15]). *The ERIS* (5.33) *with* $\ell \in \mathrm{W}^{1,1}([0,T];\mathrm{H}^1_D(\Omega)^*)$ *has, for each stable initial state* $q_0 = (u_0, z_0)$, *an* energetic solution $(u,z) : [0,T] \to \boldsymbol{Q}$.

The proof relies in reducing the system to a problem in z alone. For this note that the equation $\mathrm{D}_u\mathscr{E}(t,u,z) = 0$ is a linear elliptic PDE for u with a right-hand side that is linear in z and ℓ. Hence, the unique solution $u = U(z,\ell) \in \mathrm{H}^1_D(\Omega;\mathbb{R}^d)$ can be inserted into $\mathscr{E}$ to obtain the reduced ERIS $(\mathscr{Z},\mathscr{I},\mathscr{D})$ with

$$\mathscr{Z} := \mathrm{L}^1(\Omega;[0,1]) \quad \text{and} \quad \mathscr{I}(t,z) = \mathscr{E}(t,U(z,\ell(t)),z) = \frac{1}{2}\langle Lz+\gamma(t),z\rangle + \alpha(t).$$

Here L is a pseudo-differential operator of order 0, and the symbol, which can be calculated explicitly, is non-negative by the explicit formula for ρ from [Koh91]. The symbol attains the value 0 along the optimal laminates and ρ is the largest number such that the symbol remains non-negative.

Because of the constraint $z \in [0,1]$ the quadratic trick indicated in (5.12) cannot be used for showing the closedness of the set of stable states. Indeed, from the incremental minimization problem (5.7) we obtain piecewise constant interpolants $z^\tau : [0,T] \to \mathscr{Z}$ that are globally stable, i.e. (S) in (5.5) holds at $t = k\tau$ for $k \in \mathbb{N}_0$. For a subsequence $\tau_k \to 0$ we have $z^{\tau_n}(t) \rightharpoonup z(t)$ and we have to show that $z(t)$ is stable as well.

Since stability is a static concept we can fix t and drop it for notational convenience. To establish stability of z we start from the stability of z_n in the form

$$\mathscr{I}(t,\widehat{z}_n) + \mathscr{D}(z_n,\widehat{z}_n) - \mathscr{I}(t,z_n) \geq 0 \text{ for all } \widehat{z}_n \in \mathscr{Z}.$$

To pass to the limit we can only use $z_n \rightharpoonup z$, but may choose a suitable mutual recovery sequence $\widehat{z}_n \rightharpoonup \widehat{z}$ for a given test state $\widehat{z}$. In [HM15] the following choice was introduced:

$$\widehat{z}_n(x) = \widehat{z}(x) + g(x)\,(z_n(x){-}z(x)), \text{ where } g(x) = \begin{cases} \frac{\widehat{z}(x)}{z(x)} & \text{for } \widehat{z}(x) < z, \\ 1 & \text{for } \widehat{z}(x) = z, \\ \frac{1-\widehat{z}(x)}{1-z(x)} & \text{for } \widehat{z}(x) > z. \end{cases}$$

Clearly we have $\widehat{z}_n \in \mathscr{Z}$, $\widehat{z}_n \rightharpoonup \widehat{z}$ and $\mathrm{sign}(\widehat{z}_n{-}z_n) = \mathrm{sign}(\widehat{z}{-}z)$. Decomposing Ω into Ω_+ and Ω_- such that $\widehat{z} \geq z$ and $\widehat{z} < z$, respectively, we obtain

$$\begin{aligned} \tfrac{1}{r}\mathscr{D}(z_n,\widehat{z}_n) &= \|z_n{-}\widehat{z}_n\|_{\mathrm{L}^1} = \int_{\Omega_+} \widehat{z}_n{-}z_n \,\mathrm{d}x + \int_{\Omega_-} z_n{-}\widehat{z}_n \,\mathrm{d}x \\ &= \int_{\Omega_+} \tfrac{\widehat{z}-z}{1-z}(1{-}z_n)\,\mathrm{d}x + \int_{\Omega_-} \tfrac{z-\widehat{z}}{z} z_n \,\mathrm{d}x \;\to\; \int_{\Omega_+} \widehat{z}{-}z \,\mathrm{d}x + \int_{\Omega_-} z{-}\widehat{z}\mathrm{d}x = \tfrac{1}{r}\mathscr{D}(z,\widehat{z}). \end{aligned}$$

To control the energy differences $\mathscr{I}(t,\widehat{z}_n) - \mathscr{I}(t,z_n)$ we exploit the quadratic form of the energy. In fact, the sequence $v_n := z_n{-}z \rightharpoonup 0$ generates an *H-measure* $\mu \geq 0$ which exactly characterizes the limit of the quadratic energy, namely

$$\lim_{n\to\infty} \mathscr{I}(t,z_n) = \mathscr{I}(t,z) + \int_\Omega \int_{\omega\in\mathbb{S}^{d-1}} \Sigma_L(\omega)\mu(x,\mathrm{d}\omega)\,\mathrm{d}x,$$

where $\Sigma_L(\omega) \geq 0$ is the symbol of L. The construction of $\widehat{z}_n$ gives $\widehat{v}_n := \widehat{z}_n - \widehat{z} = gv_n \rightharpoonup 0$, such that $\widehat{v}_n$ generates the H-measure $g^2\mu$. Thus, we obtain

$$\begin{aligned}&\lim_{n\to\infty} \big(\mathscr{I}(t,\widehat{z}_n) - \mathscr{I}(t,z_n)\big)\\ &= \mathscr{I}(t,\widehat{z}) - \mathscr{I}(t,z) + \int_\Omega \int_{\omega\in\mathbb{S}^{d-1}} (g(x)^2-1)\Sigma_L(\omega)\mu(x,\mathrm{d}\omega)\,\mathrm{d}x.\end{aligned}$$

Now, using $g^2 \leq 1$ we conclude the desired limsup estimate

$$0 \leq \limsup_{n\to\infty} \Big(\mathscr{I}(t,\widehat{z}_n) + \mathscr{D}(z_n,\widehat{z}_n) - \mathscr{I}(t,z_n)\Big) \leq \mathscr{I}(t,\widehat{z}) + \mathscr{D}(z,\widehat{z}) - \mathscr{I}(t,z).$$

Since $\widehat{z}$ was arbitrary, the global stability (S) of z is established.

We refer to [HM15] for a detailed analysis, which includes the convergence of space-time discretizations in suitable finite-element spaces as well as the strong convergence of certain Riesz projections related to the directions of the microstructures between the two phases.

Acknowledgements. This paper surveys the research done in the subproject P5 *Microstructures in evolutionary material models* in the years 2007 to 2014 which was largely supported by Deutsche Forschungsgemeinschaft within the Research Unit FOR 797 *Analysis and computation of microstructure in finite plasticity*. The author is grateful to his collaborators Sebastian Heinz, Klaus Hackl, Michael Ortiz, Christoph Ortner, Riccarda Rossi, Tomáš Roubíček, Guiseppe Savaré, Yasemin Şengül, Ulisse Stefanelli, Marita Thomas, and Sergey Zelik.

References

[AGS05] Ambrosio, L., Gigli, N., Savaré, G.: Gradient flows in metric spaces and in the space of probability measures. Lectures in Mathematics ETH Zürich. Birkhäuser Verlag, Basel (2005)

[AMP$^+$12] Arnrich, S., Mielke, A., Peletier, M.A., Savaré, G., Veneroni, M.: Passing to the limit in a Wasserstein gradient flow: from diffusion to reaction. Calc. Var. Part. Diff. Eqns. 44, 419–454 (2012)

[BCHH04] Bartels, S., Carstensen, C., Hackl, K., Hoppe, U.: Effective relaxation for microstructure simulations: algorithms and applications. Comput. Methods Appl. Mech. Engrg. 193, 5143–5175 (2004)

[BE76] Brézis, H., Ekeland, I.: Un prinicpe varationnel associeé à certaines équations paraboliques. Compt. Rendues Acad. Sci. Paris 282, 971–974, 1197–1198 (1976)

[BH09] Bartel, T., Hackl, K.: A micromechanical model for martensitic phase-transformations in shape-memory alloys based on energy-relaxation. Z. angew. Math. Mech (ZAMM) 89(10), 792–809 (2009)

[Bio55] Biot, M.A.: Variational principles in irreversible thermodynamics with applications to viscoelasticity. Phys. Review 97(6), 1463–1469 (1955)

[Bra02] Braides, A.: Γ-Convergence for Beginners. Oxford University Press (2002)

[CDK13a] Conti, S., Dolzmann, G., Kreisbeck, C.: Relaxation and microstructure in a model for finite crystal plasticity with one slip system in three dimensions. Discr. Cont. Dynam. Systems Ser. S 6, 1–16 (2013)

[CDK13b] Conti, S., Dolzmann, G., Kreisbeck, C.: Relaxation of a model in finite plasticity with two slip systems. Math. Models Meth. Appl. Sci. (M^3AS) 23, 2111–2128 (2013)

[CHM02] Carstensen, C., Hackl, K., Mielke, A.: Non–convex potentials and microstructures in finite–strain plasticity. Proc. Royal Soc. London Ser. A 458(2018), 299–317 (2002)

[CLR15] Conti, S., Lenz, M., Rumpf, M.: Hysteresis in magnetic shape memory composites: modeling and limulation. J. Mech. Physics Solids (submitted, 2015)

[CO08] Conti, S., Ortiz, M.: Minimum principles for the trajectories of systems governed by rate problems. J. Mech. Physics Solids 56(5), 1885–1904 (2008)

[CP01] Carstensen, C., Plecháč, P.: Numerical analysis of a relaxed variational model of hysteresis in two-phase solids. M2AN Math. Model. Numer. Anal. 35(5), 865–878 (2001)

[CT05] Conti, S., Theil, F.: Single-slip elastoplastic microstructures. Arch. Rational Mech. Anal. 178, 125–148 (2005)

[DD02] DeSimone, A., Dolzmann, G.: Macroscopic response of nematic elastomers via relaxation of a class of SO(3)-invariant energies. Arch. Ration. Mech. Anal. 161(3), 181–204 (2002)

[DFT05] Dal Maso, G., Francfort, G., Toader, R.: Quasistatic crack growth in nonlinear elasticity. Arch. Rational Mech. Anal. 176, 165–225 (2005)

[DL10] Dal Maso, G., Lazzaroni, G.: Quasistatic crack growth in finite elasticity with non-interpenetration. Ann. Inst. H. Poinc. Anal. Non Lin. 27(1), 257–290 (2010)

[DMT80] De Giorgi, E., Marino, A., Tosques, M.: Problems of evolution in metric spaces and maximal decreasing curve. Atti Accad. Naz. Lincei Rend. Cl. Sci. Fis. Mat. Natur (8) 68(3), 180–187 (1980)

[EM06] Efendiev, M., Mielke, A.: On the rate–independent limit of systems with dry friction and small viscosity. J. Convex Anal. 13(1), 151–167 (2006)

[Fen49] Fenchel, W.: On conjugate convex functions. Canadian J. Math. 1, 73–77 (1949)

[FG06] Francfort, G., Garroni, A.: A variational view of partial brittle damage evolution. Arch. Rational Mech. Anal. 182, 125–152 (2006)

[GL09] Garroni, A., Larsen, C.J.: Threshold-based quasi-static brittle damage evolution. Arch. Rational Mech. Anal. 194(2), 585–609 (2009)

[GM11] Giacomini, A., Musesti, A.: Two-scale homogenization for a model in strain gradient plasticity. ESAIM Control Optim. Calc. Var. 17(4), 1035–1065 (2011), doi:10.1051/cocv/2010036

[Han11] Hanke, H.: Homogenization in gradient plasticity. Math. Models Meth. Appl. Sci. (M^3AS) 21(8), 1651–1684 (2011)

[Hei14] Heinz, S.: On the structure of the quasiconvex hull in planar elasticity. Calc. Var. Part. Diff. Eqns. 50, 481–489 (2014)

[Hei15] Heinz, S.: Quasiconvexity equals lamination convexity for isotropic sets of 2x2 matrices. Adv. Calc. Var. 8, 43–53 (2015)

[HF08] Hackl, K., Fischer, F.D.: On the relation between the principle of maximum dissipation and inelastic evolution given by dissipation potentials. Proc. R. Soc. A 464, 117–132 (2008)

[HHM12] Hackl, K., Heinz, S., Mielke, A.: A model for the evolution of laminates in finite-strain elastoplasticity. Z. Angew. Math. Mech (ZAMM) 92(11-12), 888–909 (2012)

[HK14] Heinz, S., Kruzik, M.: Computations of quasiconvex hulls of isotropic sets. WIAS preprint no. 2049 (2014)

[HM15] Heinz, S., Mielke, A.: Analysis and numerics of a phase-transformation model (in preparation, 2015)

[Ilm94] Ilmanen, T.: Elliptic regularization and partial regularity for motion by mean curvature. Mem. Amer. Math. Soc. 108(520), x+90 (1994)

[Jam96] James, R.D.: Hysteresis in phase transformations. In: ICIAM 1995, Hamburg. Math. Res., vol. 87, pp. 135–154. Akademie Verlag, Berlin (1996)

[KH11] Kochmann, D.M., Hackl, K.: The evolution of laminates in finite crystal plasticity: a variational approach. Contin. Mech. Thermodyn. 23(1), 63–85 (2011)

[Koh91] Kohn, R.V.: The relaxation of a double-well energy. Contin. Mech. Thermodyn. 3(3), 193–236 (1991)

[LM72] Lions, J.-L., Magenes, E.: Non-homogeneous boundary value problems and applications, vol. I. Springer, New York (1972)

[LM11] Liero, M., Mielke, A.: An evolutionary elastoplastic plate model derived via Γ-convergence. Math. Models Meth. Appl. Sci (M^3AS) 21(9), 1961–1986 (2011)

[LMPR15] Liero, M., Mielke, A., Peletier, M.A., Renger, D.R.M.: On the microscopic origin of generalized gradient structures (in preparation, 2015)

[LOR07] Larsen, C., Ortiz, M., Richardson, C.: Fracture paths from front kinetics: relaxation and rate-independence. Arch. Rational Mech. Anal.

[Mie04] Mielke, A.: Deriving new evolution equations for microstructures via relaxation of variational incremental problems. Comput. Methods Appl. Mech. Engrg. 193(48-51), 5095–5127 (2004)

[Mie10] Mielke, A.: Existence theory for finite-strain crystal plasticity with gradient regularization. In: Hackl, K. (ed.) IUTAM Symposium on Variational Concepts with Applications to the Mechanics of Materials, Proceedings of the IUTAM Symposium on Variational Concepts, Bochum, Germany, September 22-26. Springer (2008)

[Mie11a] Mielke, A.: Complete-damage evolution based on energies and stresses. Discr. Cont. Dynam. Systems Ser. S 4(2), 423–439 (2011)

[Mie11b] Mielke, A.: Differential, energetic, and metric formulations for rate-independent processes. In: Ambrosio, L., Savaré, G. (eds.) Nonlinear PDE's and Applications, Cetraro, Italy. C.I.M.E. Summer School, Lect. Notes Math., vol. 2028, pp. 87–170. Springer (2028)

[Mie12] Mielke, A.: Emergence of rate-independent dissipation from viscous systems with wiggly energies. Contin. Mech. Thermodyn. 24(4), 591–606 (2012)

[Mie14] Mielke, A.: On evolutionary Γ-convergence for gradient systems. WIAS Preprint 1915, 2014. To appear in Proc. Summer School in Twente University (June 2012)

[MM09] Mainik, A., Mielke, A.: Global existence for rate-independent gradient plasticity at finite strain. J. Nonlinear Sci. 19(3), 221–248 (2009)

[MO08] Mielke, A., Ortiz, M.: A class of minimum principles for characterizing the trajectories of dissipative systems. ESAIM Control Optim. Calc. Var. 14, 494–516 (2008)

[MOŞ14] Mielke, A., Ortner, C., Şengül, Y.: An approach to nonlinear viscoelasticity via metric gradient flows. SIAM J. Math. Analysis 46(2), 1317–1347 (2014)

[MPR14] Mielke, A., Peletier, M., Renger, M.: On the relation between gradient flows and the large-deviation principle, with applications to Markov chains and diffusion. Potential Analysis 41(4), 1293–1327 (2014)

[MR15] Mielke, A., Roubíček, T.: Rate-Independent Systems: Theory and Application. Springer (2015)

[MRS08] Mielke, A., Roubíček, T., Stefanelli, U.: Γ-limits and relaxations for rate-independent evolutionary problems. Calc. Var. Part. Diff. Eqns. 31, 387–416 (2008)

[MRS09] Mielke, A., Rossi, R., Savaré, G.: Modeling solutions with jumps for rate-independent systems on metric spaces. Discr. Cont. Dynam. Systems Ser. A 25(2), 585–615 (2009)

[MRS12] Mielke, A., Rossi, R., Savaré, G.: BV solutions and viscosity approximations of rate-independent systems. ESAIM Control Optim. Calc. Var. 18(1), 36–80 (2012)

[MRS14a] Mielke, A., Rossi, R., Savaré, G.: Balanced-viscosity (BV) solutions to infinite-dimensional rate-independent systems. J. Europ. Math. Soc. (to appear, 2014), `http://arxiv.org/abs/1309.6291` WIAS preprint (1845)

[MRS14b] Mielke, A., Rossi, R., Savaré, G.: Balanced-Viscosity solutions for multi-rate systems. J. Physics, Conf. Series (2014); WIAS Preprint (2001) (submitted)

[MRS15] Mielke, A., Rossi, R., Savaré, G.: Existence results for viscoplasticity at finite strain (in preparation, 2015)

[MS11] Mielke, A., Stefanelli, U.: Weighted energy-dissipation functionals for gradient flows. ESAIM Control Optim. Calc. Var. 17, 52–85 (2011)

[MS13] Mielke, A., Stefanelli, U.: Linearized plasticity is the evolutionary Γ-limit of finite plasticity. J. Europ. Math. Soc. 15(3), 923–948 (2013)

[MT07] Mielke, A., Timofte, A.M.: Two-scale homogenization for evolutionary variational inequalities via the energetic formulation. SIAM J. Math. Analysis 39(2), 642–668 (2007)

[MT12] Mielke, A., Truskinovsky, L.: From discrete visco-elasticity to continuum rate-independent plasticity: rigorous results. Arch. Rational Mech. Anal. 203(2), 577–619 (2012)

[MTL02] Mielke, A., Theil, F., Levitas, V.I.: A variational formulation of rate–independent phase transformations using an extremum principle. Arch. Rational Mech. Anal. 162, 137–177 (2002)

[MZ14] Mielke, A., Zelik, S.: On the vanishing viscosity limit in parabolic systems with rate-independent dissipation terms. Ann. Sc. Norm. Sup. Pisa Cl. Sci (5) XIII, 67–135 (2014)

[Nay76] Nayroles, B.: Deux théorèmes de minimum pour certains systèmes dissipatifs. Compt. Rendues Acad. Sci. Paris 282, A1035–A1038 (1976)

[Ons31] Onsager, L.: Reciprocal relations in irreversible processes, I+II. Physical Review 37, 405–426, Part II, 38, 2265–2279 (1931)

[OR99] Ortiz, M., Repetto, E.: Nonconvex energy minimization and dislocation structures in ductile single crystals. J. Mech. Phys. Solids 47(2), 397–462 (1999)

[PT02] Puglisi, G., Truskinovsky, L.: Rate independent hysteresis in a bi-stable chain. J. Mech. Phys. Solids 50(2), 165–187 (2002)

[PT05] Puglisi, G., Truskinovsky, L.: Thermodynamics of rate-independent plasticity. J. Mech. Phys. Solids 53, 655–679 (2005)

[Ray71] Rayleigh, L., Strutt, H.J.W.: Some general theorems relating to vibrations. Proc. London Math. Soc. s1-4, 357–368 (1871)

[Ser11] Serfaty, S.: Gamma-convergence of gradient flows on Hilbert spaces and metric spaces and applications. Discr. Cont. Dynam. Systems Ser. A 31(4), 1427–1451 (2011)

[SS04] Sandier, E., Serfaty, S.: Gamma-convergence of gradient flows with applications to Ginzburg-Landau. Comm. Pure Appl. Math. LVII, 1627–1672 (2004)

[Ste08] Stefanelli, U.: The Brezis-Ekeland principle for doubly nonlinear equations. SIAM J. Control Optim. 47(3), 1615–1642 (2008)

Chapter 6
Energy Estimates, Relaxation, and Existence for Strain-Gradient Plasticity with Cross-Hardening

Keith Anguige and Patrick W. Dondl

Abstract. We consider a variational formulation of gradient elasto-plasticity subject to a class of single-slip side conditions, and show how the non-convexity effects induced by such conditions can be not only resolved mathematically, but also tested physically. We first show that, for a large class of plastic deformations, a given single-slip condition (specification of Burgers' vectors and slip planes) can be relaxed by introducing a microstructure through a two-stage process of mollification and lamination. This yields a relaxed side condition which only prescribes certain slip planes, and allows for arbitrary slip directions in these planes. The relaxed model should be a useful tool for simulating macroscopic plastic behavior without the need to resolve arbitrarily fine spatial scales. After deriving the relaxed model, we discuss a partial result on the existence of minimizers. Finally, we apply the relaxed model to a specific physical system, in order to be able to compare the analytical results with experiments. In particular, a rectangular shear sample in which only two slip planes are active is clamped at each end, and is subjected to a prescribed horizontal shear, which requires a certain amount of energy. We show that above some critical aspect ratio the energy is strictly positive and below that aspect ratio it is zero. Moreover, in the respective regimes determined by the aspect ratio, we prove energy scaling bounds, expressed in terms of the amount of prescribed shear, and we show that the scalings as well as the critical aspect ratio change radically if the single-slip condition or the strain gradient penalization is neglected.

Keith Anguige
Fakultät fur Mathematik, TU-Dortmund, Dortmund, Germany
e-mail: `keith.anguige@math.tu-dortmund.de`

Patrick W. Dondl
Department of Mathematical Sciences, Durham University, Durham, UK
e-mail: `patrick.dondl@durham.ac.uk`

S. Conti and K. Hackl (eds.), *Analysis and Computation of Microstructure in Finite Plasticity*, Lecture Notes in Applied and Computational Mechanics 78, DOI: 10.1007/978-3-319-18242-1_6

6.1 Introduction

The formation of complex subgrain dislocation patterns in plastically deformed crystals has been observed by many authors [HBO10, HLCH97, HCLH98, NGH85] over several decades. Such dislocation microstructures are known to have a substantial effect on a number of aspects of plastic behaviour, such as work hardening and the Bauschinger and Hall-Petch effects. At present, these effects are commonly modeled in a phenomenological manner. However, the resulting models often fail to capture correctly the behavior of specimens on the micron scale [UDFN04, DRR10], and the appropriate modelling approach in this case is still under debate.

In the framework of continuum plasticity, Ortiz and Repetto proposed a mechanism for subgrain pattern formation in their seminal work [OR99]. The basic idea is to model plastic evolution by an incremental time-stepping procedure, where in each step the sum of the potential energy and the dissipation potential is minimized. Under certain conditions, such discretizations can be shown to converge to a suitable gradient flow in the limit of small time steps [MM09], and furthermore the framework of incremental energy minimization makes the models amenable to variational treatment. The dissipation term in such models of crystal plasticity is naturally non-convex: if dislocations from different slip systems meet, then they form energetically favorable (but sessile) atomistic reaction products, such as the so-called Lomer-Cottrell locks. This increases the dissipation potential for any plastic deformation that does not occur in single slip and is the origin of microstructure formation in those models. We will introduce this non-convexity in a simplified manner, precluding material deformation in more than one slip system, i.e., enforcing a single-slip condition which states that at each material point, the plastic deformation has to occur in single slip. For a detailed discussion on non-convex potentials and their relation to microstructure formation in plasticity, see [CHM02, Mie06].

A strain-gradient term, such as $\int |\operatorname{curl} F_{\mathrm{pl}}|$, modeling the potential-energy contribution of geometrically necessary dislocations, is commonly added to the energetic formulation, and in fact such a term may be necessary to recover a physically realistic description of hardening in some models [FCO14]. Historically, the introduction of a curl-type strain-gradient term goes back to Nye [Nye53] and, independently, to Kondo [Kon52].

In the following we will, similarly as in [CT05, CDK13, ACD09], adopt the framework of incremental energy minimization proposed by Ortiz and Repetto, applied in particular to a strain-gradient plasticity model with cross hardening. Section 6.2 will give a term-by-term introduction to the modeling ingredients. The model introduced there, in both its geometrically linear and nonlinear forms, will serve as the basis of our further investigations. In particular, we introduce a special curl term which accounts for non-cancellation of dislocations in collinear slip in Section 6.2.3.

In Section 6.3, following closely the procedure in [AD14b], we discuss an inherent problem with our single-slip condition, showing that, for the purposes of efficient numerical simulation and (relatedly) obtaining existence of minimizers, this condition needs to be relaxed in a suitable way. In particular, the strain-gradient term does *not* regularise the problem fully, since we show that the resulting energy is not lower-semicontinuous. Hildebrand *et al.* [HM12] already observed this issue in numerical simulations. They used exactly a model of single-slip plasticity with penalization of geometrically necessary dislocations, and discovered a mesh-dependent microstructure when the sample was placed under a shear load within a slip plane, but not in a Burgers' vector direction. There, the problem was solved by introducing an ultra-fine full-gradient-penalty regularisation, thus rendering the minimization problem well-posed. Such a penalty term, however, introduces an even finer length scale which needs to be resolved in simulations. In order to remedy this, we thus propose a relaxation of the single-slip condition to a single-plane condition (where coplanar slip-systems are allowed to be active at the same point), and show that the resulting relaxed energy is an upper bound for the lower-semicontinuous envelope of our model.

Under exactly what assumptions our relaxed model does in fact admit a minimizer is currently an open question. We are, however, able to show that the single-plane side-condition is preserved along converging minimizing sequences, which is a necessary condition for lower-semicontinuity. The main ideas underlying the proof of this result are presented in Section 6.4.

It is of course natural to ask whether the strain-gradient term and the cross-hardening side condition are in fact relevant for modeling, and whether their implications are visible in experiments. Therefore, in Section 6.5, we propose an experiment that can be used to measure the effect of the combination of these two modeling ingredients. In fact, it turns out that their implications are quite drastic, since they yield a change in the optimal scaling of the energy of a specimen undergoing shear deformation. This section closely follows the corresponding discussion in [AD14a].

Finally, some conclusions and a look ahead to some further possible avenues of investigation are presented in Section 6.6. In particular, we continue our discussion about the existence and regularity of minimizers of our relaxed model, which still poses a number of unresolved questions.

6.2 A Continuum Model for Strain-Gradient Plasticity with Cross Hardening

In the following, we will introduce the individual components of our continuum model for crystal plasticity—except for the specific form of the penalty term for geometrically necessary dislocations, this model is commonly used when considering the mathematical foundations for continuum plasticity.

6.2.1 Plastic Shear

We consider an elasto-plastic body in its reference configuration $\Omega \subset \mathbb{R}^3$, together with a mapping

$$y\colon \Omega \to \mathbb{R}^3$$

which satisfies suitable boundary conditions. For sufficiently smooth deformations y, we define the deformation gradient $F = \nabla y$, such that the row-wise curl of this tensor vanishes: $\operatorname{Curl} F = 0$.

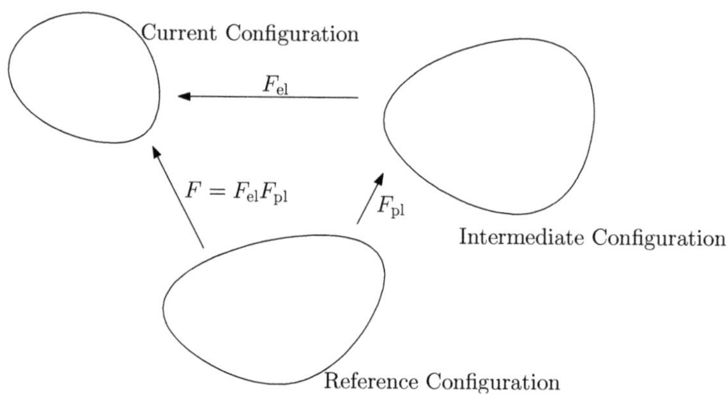

Fig. 6.1 The deformation described by the deformation gradient F is decomposed into a plastic part F_{pl} and and elastic part F_{el}. The plastic deformation generates an intermediate configuration.

Now, following [LL67], we make the fundamental assumption that this deformation gradient can be decomposed into, first, a purely plastic strain, consisting only of a rearrangement of atoms through plastic slip, generating an intermediate configuration, and, second, a purely elastic deformation. We denote the plastic strain by F_{pl} and the elastic strain by F_{el}, and refer to Figure 6.1 for an illustration. We note that the validity of this multiplicative decomposition is still under debate, and refer to [RC14] for a recent discussion of the issue.

The decomposition thus allows us to identify an elastic energy for the crystalline specimen depending only on the elastic strain $F_{\text{el}} = FF_{\text{pl}}^{-1}$. We furthermore postulate that the dissipation potential, in the sense of the implicit time discretization, can be written as a function of the increment in plastic strain. In the following, we will restrict ourselves to a treatment of the first such time-step, and thus arrive at an energy for the elastic strain and the dissipation of the form

$$\int_\Omega W_{\text{el}}(FF_{\text{pl}}^{-1}) + \int_\Omega W_{\text{pl}}(F_{\text{pl}}),$$

with a suitable frame-indifferent elastic energy density W_{el} and a dissipation potential W_{pl}.

It is well known that plastic deformation in crystals is ultimately mediated by the motion of dislocations through the body. At moderate temperature, each such dislocation travels predominantly on a given crystallographic plane, a so-called slip-plane - after the dislocation has passed a point on the plane, the atoms above the plane have been moved by one Burgers' vector relative to those below it. Each type of dislocation is associated with a fixed slip-plane and a fixed Burgers' vector lying in the slip plane, and each such pair constitutes a *slip system*. Kinematically, therefore, the plastic strain can be written as the product of a number of simple shear deformations, each with a given crystallographically determined shear normal and shear direction, and arbitrary shear angle. In the common face-centered-cubic crystalline metals, for example, there are twelve such slip systems, given by four slip-plane normals and three different Burgers' vectors in each slip-plane.

6.2.2 Locks and Cross-Hardening

Cross-hardening [WBL91, Koc64, FBZ80] describes the phenomenon whereby activity in one slip system suppresses activity in all other slip systems at the same point, and this immediately leads to a loss of convexity in the dissipation potential W_{pl} introduced above [OR99]. Corresponding to this, $W_{\mathrm{pl}}(F_{\mathrm{pl}})$ will (roughly speaking) be minimal if F_{pl} is a pure simple shear in one of the given crystallographic slip systems.

Mostly in fatigue experiments, but also in experiments involving only a single-pass deformation, lamination-type microstructures with alternating slip-system activity are observed [RP80, JW84, BHHK92, DDMR09]—this effect is believed to stem from cross-hardening [OR99, DDMR09]. In the following, we make the simplifying assumption of infinite cross hardening, such that F_{pl} must be in single slip at each point. Similar assumptions have been made for example in [CDK13, CDK09], among others. In [DDMR09] it was shown that the predicted laminate microstructure arising from the assumption of infinite cross hardening does indeed match experimental results, while evolution models of such laminate structures have been analyzed in [KH11]. Our main result, namely that the single-slip condition should be relaxed in a specific manner, carries over to models where multiple slip is penalized, for example by an off-diagonally-dominant hardening matrix.

The effect of cross hardening arises from the formation of energetically favorable, and *sessile* dislocation products, when two dislocations from different slip-planes meet. These so-called Lomer-Cottrell locks have been obeserved in experiments [FZ82] and studied in detail in atomistic simulations [RP99]. In order to continue the plastic deformation with activity in more than one slip-plane, these locks have to either be broken or new dislocation loops have to be formed, thus leading to an increased dissipation in directions involving more than one slip-system in F_{pl}.

In line with the above discussion, we will thus assume that the crystal structure admits a set of slip-plane normals $\mathcal{M} = \{m_j\}_{j=1}^{N}$, each with a given set of Burgers' vectors $\mathcal{B}_j = \{b_{ij}\}_{i=1}^{K(j)}$, and that F_{pl} takes the form

$$F_{\mathrm{pl}}(x) = \mathrm{Id} + \sum_{j=1}^{N} \sum_{i=1}^{K(j)} s_{ij}(x) b_{ij} \otimes m_j, \tag{6.1}$$

with the single slip condition $s_{ij}s_{kl} = 0$ if $i \neq k$ or $j \neq l$ for almost every point $x \in \Omega$. Note that, under this condition, the product of simple shears assumed above simplifies immediately, such that there is only one non-zero summand at each point, and that the plastic deformation proposed above therefore satisfies $\det F_{\mathrm{pl}} = 1$ everywhere (despite its simplified form as a sum, not a product of simple shears).

6.2.3 Geometrically Necessary Dislocations

As noted in the introduction, a strain-gradient penalty term is often included in models of crystal plasticity. Such a length-scale-introducing term corresponds to a penalization of geometrically necessary dislocations in the crystal. The basic idea is the following: while the deformation gradient F is required to be kinematically compatible, i.e., there exists a sufficiently regular function $y\colon \Omega \to \mathbb{R}^3$ such that $F = \nabla y$, neither the plastic strain F_{pl} nor the elastic strain F_{el} need admit this property. Consider, for example, a cubic specimen with a slip plane aligned with one of the axes, and then shear two parts of the crystal in opposing directions (forming a small-angle grain boundary). For an illustration see for example [Lej10], Fig 2.2. Clearly, F_{pl}, and therefore also F_{el}, are not gradients. The surface where the two differently sheared subdomains meet admits a density of geometrically necessary dislocations. In [CG01] an argument is made that the correct term for the density of geometrically necessary dislocations must be

$$\left| \frac{1}{\det F_{\mathrm{pl}}} (\mathrm{Curl}\, F_{\mathrm{pl}}) F_{\mathrm{pl}}^{\mathrm{T}} \right|, \tag{6.2}$$

(see also [MM06] for a brief discussion of this matter, as well as a generalization to arbitrary dimension). Here and below, Curl denotes the row-wise curl of a 3×3-matrix, and $|A| = \sqrt{\mathrm{Tr}\, A^T A}$ for a matrix A.

Considering the term in (6.2), together with the fact that our single-slip side condition yields a very specific form of F_{pl}, it is easy to see that both the volumetric prefactor as well as the multiplication with F_{pl}^T can be disregarded. We thus arrive at the simpler term

$$|\mathrm{Curl}\, F_{\mathrm{pl}}| .$$

This term does, however, have certain undesirable properties. To see this, consider a specimen with two abutting subdomains deformed in collinear slip (i.e., slip in different slip planes, but in the same Burgers' vector direction). Such a plastic strain is completely curl free, since the subdomains are rank-one connected, and thus the geometric dislocation density tensor would vanish due to cancellations of dislocations with opposite sign. However, it has been shown in atomistic simulations by Devincre *et. al.* [DHK05, DKH07], that these cancellations are not complete, and a density of dislocations remains on the surface between the subdomains. We thus introduce a different measure for the dislocation density in order to avoid such cancellations of dislocations from different slip planes. For a discussion of this assumption in a simplified scalar model, see [CO05, Chapter 4]. Concretely, our dislocation density is taken to be

$$\mathcal{G}(F_{\mathrm{pl}}) = \sum_{j=1}^{N} \sum_{i=1}^{K(j)} \left| \nabla_{m_j^\perp} s_{ij} \right| , \tag{6.3}$$

i.e., we take the sum of the lengths of the vectors given by the planar gradients (gradients only in the directions perpendicular to m_j) of the respective plastic slips s_{ij}. We note that this term, as well as the curl-terms above, are in general non-negative measures.

Regarding the physical validity of our modelling assumptions, we have shown analytically in [AD14a] that in rectangular shear samples of small size, the strain-gradient energy and single-slip condition, taken together, play a dramatic role. Specifically, for a particular crystallographic configuration, there are three qualitatively different energy-scaling regimes, which are determined by the aspect ratio of the specimen—moreover, if either the curl penalty or the single-slip condition are dropped then the intermediate scaling regime vanishes. Corresponding to the high-aspect-ratio case, one consequence of this is that a micron-sized sample consisting of only a few grains will easily shear off if a slip system under stress connects free surfaces [UDFN04]. This is laid out in more detail in Section 6.5.

6.2.4 The Model

The geometrically nonlinear elasto-plastic energy, which encodes the elements described above, is taken to be

$$E(u, F_{\mathrm{pl}}) = \begin{cases} \int_\Omega W_{\mathrm{el}}(F_{\mathrm{el}})\, \mathrm{d}x + \int_\Omega \mathcal{G}(F_{\mathrm{pl}}) + \int_\Omega |F_{\mathrm{pl}}|\, \mathrm{d}x & : \text{if } \textbf{(SSC)} \text{ holds,} \\ +\infty & : \text{otherwise,} \end{cases} \tag{6.4}$$

Note that we have taken $W_{\mathrm{pl}}(F_{\mathrm{pl}})$ to be a simple L^1-rate-independent dissipation potential. As mentioned above, cross-hardening is taken account of in the single-slip condition **(SSC)**:

$$F_{\mathrm{pl}}(x) = \mathrm{Id} + s(x) \otimes m(x) \quad \text{almost everywhere,}$$

with

$$m(x) = m_{j(x)} \in \mathcal{M} \text{ and } s(x) \in \bigcup_{i=1}^{K(j)} \mathrm{Span}\{b_{ij(x)}\} \subset \mathbb{R}^3.$$

6.3 Relaxation of the Single-Slip Condition

The goal of this section is to show that instead of the (SSC) introduced above, one should only enforce the **relaxed slip condition (RSC)**

$$m(x) = m_{j(x)} \in \mathcal{M} \text{ and } s(x) \in \mathrm{Span} \bigcup_{i=1}^{K(j)} \{b_{ij(x)}\} \subset \mathbb{R}^3.$$

We claim that this **(RSC)** is the natural relaxation, preserving the exclusivity of slip in different slip planes, but allowing the mixing of coplanar slip.

Suppose thus that we have a displacement $u \in H^1$ on $\Omega \subset \mathbb{R}^3$ satisfying some Dirichlet condition, and a C^1, relaxed plastic strain F_{pl} of the form (6.1). Furthermore, for each slip-plane normal, m_j, we make a fixed choice of two admissible Burgers' vectors, b_{1j}, b_{2j} such that (with coefficients c_{ij}),

$$F_{\mathrm{pl}} = \mathrm{Id} + \sum_{j=1}^{N} s_j \otimes m_j \qquad \text{and} \qquad s_j = \sum_{i=1}^{2} c_{ij} b_{ij}, \tag{6.5}$$

thus exactly satisfying **(RSC)**.

Then, for any such selection of Burgers' vectors, we claim that the following functional is a good candidate for the relaxation of the energy E:

$$E_{\mathrm{rel}}(u, F_{\mathrm{pl}}) = \begin{cases} \int_\Omega W_{\mathrm{el}}(F_{\mathrm{el}}) + \int_\Omega \mathcal{G}_{\mathrm{lam}}(F_{\mathrm{pl}}) + \int_\Omega |F_{\mathrm{pl}}|_{\mathrm{lam}} & : \text{if } \mathbf{(RSC)} \text{ holds,} \\ +\infty & : \text{otherwise,} \end{cases} \tag{6.6}$$

where the *laminated curl* is given by

$$\mathcal{G}_{\mathrm{lam}}(F_{\mathrm{pl}}) = \sum_{j=1}^{N} \sum_{i=1}^{2} \left| \nabla_{m_j^\perp} c_{ij} \right| . \tag{6.7}$$

Note that we explicitly take the curl of the (unique, as per our assumption) coefficients of the decomposition of the relaxed slip in its Burgers' vector components. Moreover, the *laminated dissipation* is given by

$$|F_{\rm pl}|_{\rm lam} = \sum_{j=1}^{N}\sum_{i=1}^{2} |c_{ij}| \, \mathrm{d}x, \tag{6.8}$$

again, taking the norm not of the slip, but of its decomposition.

The reason why we take (6.6) as our expression for the relaxed energy is that a smooth, relaxed strain $F_{\rm pl}$ can be approximated by a sequence of laminated single-slip strains $(F_{\rm pl}^n)_{n=1}^{\infty}$ in which the slip alternates between the two chosen Burgers' vectors in each slip patch, and such that

$$\int_{\Omega} \mathcal{G}(F_{\rm pl}^n) \to \int_{\Omega} \mathcal{G}_{\rm lam}(F_{\rm pl}), \tag{6.9}$$

and

$$\int_{\Omega} |F_{\rm pl}^n| \to \int_{\Omega} |F_{\rm pl}|_{\rm lam}, \tag{6.10}$$

as $n \to \infty$.

Our main theorem is then

Theorem 1. *Suppose that $W_{el} : \mathbb{M}^{3\times 3} \mapsto [0,\infty)$ is continuous and satisfies a p-growth condition*

$$-c_1 + c_2|F|^p \le W_{el}(F) \le C_1 + C_2|F|^p \tag{6.11}$$

*for $1 < p < \infty$. Suppose furthermore that we have a Lipschitz domain $\Omega \subset \mathbb{R}^3$ and that we have $(u, F_{\rm pl})$ on Ω, such that $u \in W^{1,p}$ satisfies a Dirichlet condition on a Lipschitz subset of $\partial\Omega$, $F_{\rm pl}$ satisfies the relaxed-slip condition **(RSC)** with the j-th slip normal active only on $\Omega_j \subset \Omega$, and the relaxed energy* (6.6) *is finite. Assume that the sets $\{\Omega_j\}_{j=1}^N$ where $F_{\rm pl} = Id + s \otimes m_j$ satisfy the regularity condition $\mathcal{H}^2(\partial\Omega_j \setminus \mathcal{F}\Omega_j) = 0$.*

*Then, for each $\varepsilon > 0$, there exists a pair of test functions $(u_\varepsilon, F_{{\rm pl},\varepsilon})$ satisfying the same Dirichlet condition and the single-slip condition **(SSC)**, such that $u_\varepsilon \in W^{1,p}$, $F_{{\rm pl},\varepsilon} \in L^\infty$ and*

$$E(u_\varepsilon, F_{{\rm pl},\varepsilon}) \le E_{\rm rel}(u, F_{\rm pl}) + \varepsilon. \tag{6.12}$$

This theorem provides an upper bound for the relaxation of the original functional at points with a certain regularity. The question of whether it is in fact the relaxation is open at present, in particular, it is not clear that the $E_{\rm rel}$ is necessarily lower-semicontinuous. For a partial result see Section 6.4.

A major technical obstacle to proving this result is to show that a non-smooth $F_{\rm pl}$ satisfying the relaxed slip condition can be approximated by a smooth $F_{{\rm pl},\varepsilon}$ without violating **(RSC)** or increasing the curl. This, however, can be done by a double-mollification-and-cut-off procedure, provided the slip patches Ω_j satisfy the mild (but rather irksome) regularity condition $\mathcal{H}^2(\partial\Omega_j \setminus \mathcal{F}\Omega_j) = 0$—i.e., the reduced boundary of each Ω_j should be $\mathcal{H}^2$-almost-all of $\partial\Omega_j$, which very roughly means that Ω_j must have no cuts.

The necessary single-slip lamination of an $F_{\rm pl}$ which satisfies **(RSC)** is done as follows. We fill each Ω_j with a stack of bi-layers, each parallel to $m_j^{\perp}$ and having thickness $\frac{1}{2^n}$, $n \in \mathbb{N}$, and then define on each successive bi-layer an alternating (in the m_j-direction), single-slip plastic strain, $F_{\rm pl}^n$, by

$$F_{\rm pl}^n = \begin{cases} 2c_{1j}b_{1j} \otimes m_j : \text{top slice}, \\ 2c_{2j}b_{2j} \otimes m_j : \text{bottom slice}, \end{cases} \tag{6.13}$$

where the s_{ij} (coming from (6.1)) are evaluated on the centre-plane of the bi-layer in (6.13), all slices have the same thickness, and the $(n+1)$-th laminate is obtained from the n-th by bisecting each of the bi-layers along a slip plane.

The laminated curl and hardening which result from this construction has the following property which is useful for obtaining the energy scalings in our proposed shear experiment (Section 6.5).

Proposition 1. *If we have a fixed* $m_j \in \mathcal{M}$, *then for a finite-energy* $F_{\rm pl} = Id + s \otimes m_j$ *with* $s = \sum_{i=1}^{2} c_{ij}b_{ij}$, *we have*

$$\int_\Omega \sum_{j=1}^{N} \left|\nabla_{m_j^{\perp}} s\right| \le (|b_{2j}.b_{1j}| + |b_{2j}.b_{1j}^{\perp}|) \int_\Omega \mathcal{G}_{\rm lam}(F_{\rm pl}) \le \sqrt{2} \int_\Omega \mathcal{G}_{\rm lam}(F_{\rm pl}) \tag{6.14}$$

and

$$\int_\Omega \mathcal{G}_{\rm lam}(F_{\rm pl}) \le \sqrt{2} \left(\frac{1 + |b_{1j}.b_{2j}|}{|b_{1j}^{\perp}.b_{2j}|}\right) \int_\Omega \sum_{j=1}^{N} \left|\nabla_{m_j^{\perp}} s\right| \tag{6.15}$$

and an analogous statement holds for the laminated dissipation.

Furthermore, a convexity property regarding the laminated curl and dissipation holds [AD14b, Proposition 2.3]. This ensures that the laminated curl is lower-semicontinuous with respect to weak convergence of the gradient of s, despite its somewhat cumbersome definition involving the gradients of the decomposition in Burgers' vectors.

A final ingredient in the proof of Theorem 1 is to note that, in order to accommodate the laminated plastic strain without excessive expenditure of elastic energy, a small zig-zag perturbation must be added to the displacement of the relaxed test function, u. This perturbation, denoted by $\hat{u}_n$, is constructed as follows. On a given bi-layer of the $F_{\rm pl}^n$-laminate, we set $\hat{u}_n = 0$ on the bottom boundary, then on the b_{ij}-slice we set

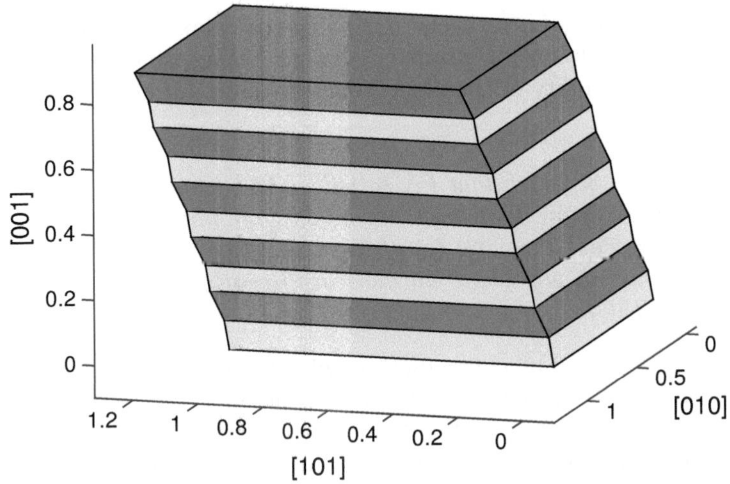

Fig. 6.2 Laminate construction of the plastic displacement. The figure illustrate the decomposition of a deformation with gradient in the direction $[011] \otimes (100)$, i.e., $s = [011]$ and $m = (100)$, into alternating components in Burgers' vector directions $[010] \otimes (100)$ and $[001] \otimes (100)$. The curl of such a construction vanishes if s is constant.

$$\frac{\partial \hat{u}_n}{\partial t_j} = s - 2c_{ij}b_{ij}, \tag{6.16}$$

where the right-hand side of (6.16) is evaluated on the centre-plane of the bi-layer. By construction, this implies that $\hat{u}_n = 0$ on top of the bi-layer, and so this procedure can be carried out consistently on the whole slip patch Ω_j. Also, by the differentiability of F_{pl}, both $\hat{u}_n$ and $\nabla_{m_j^\perp} \hat{u}_n$ are $\mathcal{O}\left(\frac{1}{2^n}\right)$, uniformly on Ω_j. Thus, by defining $u_n = u + \hat{u}_n$ and $\widehat{F}^n_{\text{pl}} = F^n_{\text{pl}} - F_{\text{pl}}$, we get

$$\nabla \hat{u}_n = \widehat{F}^n_{\text{pl}} + \mathcal{O}\left(\frac{1}{2^n}\right), \tag{6.17}$$

and then an easy calculation shows that the elastic energy of (u_n, F^n_{pl}) converges to that of (u, F_{pl}) as $n \to \infty$: in other words, the lamination commutes sufficiently well with the multiplicative decomposition of the strain.

Strictly speaking, the above perturbation applies to u away from $\partial\Omega_j$. However, since we only need to assume that the support of our relaxed plastic strain restricted to any Ω_j is Lipschitz and compactly included in Ω_j, a standard argument can be made to taper the displacement perturbation down to zero near each $\partial\Omega_j$ with negligible cost in elastic energy. An illustration of the laminate construction is given in Figure 6.2.

6.4 Some Remarks about Existence of Minimizers

While a complete existence theory regarding the relaxed energy has not yet been obtained, the following result is very promising, since it demonstrates the lower semicontinuity of the plastic energy along minimizing sequences.

Theorem 2. *In a domain $\Omega \subset \mathbb{R}^n$, for $n = 2$ or $n = 3$, consider $F_{\mathrm{pl}}^j \rightharpoonup F_{\mathrm{pl}}$ in L^q, $q \geq 1$, as $j \to \infty$, such that F_{pl}^j satisfies (RSC) and that $\int_\Omega \mathcal{G}_{\mathrm{lam}}(F_{\mathrm{pl}}^j) < \infty$. Assume furthermore that there are either four non-degenerate slip-plane normals, in the sense that any set of three is a basis (as in fcc crystals, for example), or that all slip plane normals are linearly independent. Then F_{pl} also satisfies (RSC).*

This result is proved by contradiction, with the aid of a very useful lemma which encodes the relaxed slip condition, due to Conti & Ortiz (Lemma 4.3,[CO05]). The lemma basically allows us to show that arbitrarily fine mixtures of slip patches are inconsistent with finite curl. In some sense the proof is reminiscent of that of a div-curl-lemma.

Thus, if the limit F_{pl} does not satisfy (RSC), then 'phase-mixing' occurs on a set of positive measure, S: assume the slip normals m_1 and m_2 are active on S. The set S can be covered by small parallelepipeds whose edges are aligned with m_1, m_2 and $m_1 \times m_2$. Then, since F_{pl}^n satisfies (RSC), one can show that Lemma 4.3 of Conti & Ortiz, together with the Poincaré inequality, implies that the modified curl is large if the parallelepipeds are small and n is large enough. Finally, letting the size of the parallelepipeds tend to zero, while $n \to \infty$, we see that the curl becomes infinite, which is a contradiction.

6.5 Energy Estimates for a Shear Experiment

In this section we address the following question: can one devise an experiment to determine whether the single-slip condition (introduced above), taken together with a surface energy which penalises geometrically necessary dislocations, is a relevant constraint which needs to be factored into macroscopic models of crystal plasticity? The model considered will, for analytical simplicity, be a geometrically linear version of the energy E introduced earlier.

Kinematically, we restrict ourselves to the case of a cubic crystal structure with $\langle 011 \rangle \{100\}$ slip systems. This means that plastic deformation occurs only on planes with normal parallel to one of the three cube axes, and in the direction of one of two Burgers' vectors lying diagonally in these planes. The reference configuration is a cuboid, given by $\Omega_L = (0,1) \times (0,L) \times (0,1) \subset \mathbb{R}^3$, and we consider the geometrically linear problem of minimizing the energy

$$E_L(y, F_{\mathrm{pl}}) = \int_{\Omega_L} |(F - F_{\mathrm{pl}})_{\mathrm{sym}}|^2 \,\mathrm{d}x + \sigma \int_{\Omega_L} \mathcal{G}_{\mathrm{lam}}(F_{\mathrm{pl}}) + \tau \int_{\Omega_L} |F_{\mathrm{pl}}|_{\mathrm{lam}} \,\mathrm{d}x, \tag{6.18}$$

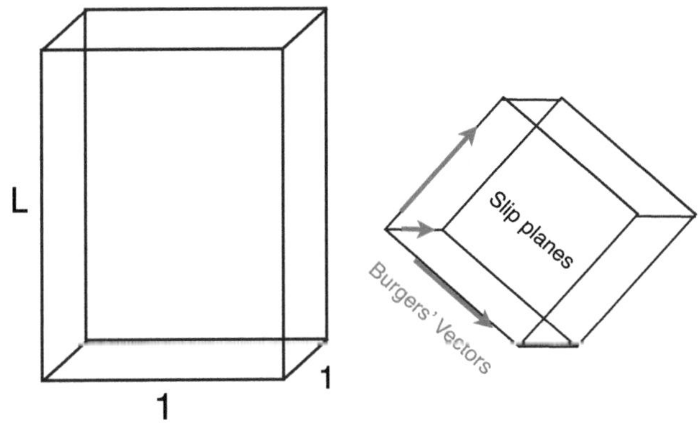

Fig. 6.3 Geometry of the sample and arrangement of slip systems. There are three slip planes corresponding to the faces of the 45 degree rotated cube, and two Burgers' vectors in each slip plane (on the edges of the rotated cube).

among vector-valued displacements $y\colon \Omega_L \to \mathbb{R}^3$ and matrix-valued plastic strains $F_{\mathrm{pl}}\colon \Omega_L \to \mathbb{R}^{3\times 3}$, for non-negative coefficients σ and τ, such that $F = \nabla y$. Here, the subscript 'sym' denotes taking the symmetric part of the matrix in parenthesis. The first term in (6.18) is the linearised elastic energy of the specimen, and the second and third term are as in the geometrically nonlinear E: the third term will mostly be neglected by taking $\tau = 0$.

The setup we describe here is the same as in [AD14a]. For convenience, we repeat the main points in this review. The admissible displacements for the minimisation problem are all functions $y(x) = x + u(x)$ such that $u(\cdot, 0, \cdot) = (0, 0, 0)$, $u(\cdot, L, \cdot) = \gamma(1, 0, 0)$ for some parameter $\gamma \geq 0$, and such that (y, F_{pl}) has finite energy for some F_{pl}. This results in a clamped specimen undergoing a shear of magnitude γ with free boundary conditions on the sides (since other conditions here would not be realizable in an experiment). The plastic strain F_{pl} is not constrained by any boundary conditions, however its row-wise curl must be a finite measure.

The single slip condition is as follow. For $\beta = F_{\mathrm{pl}} - \mathrm{Id}$, we have

$$\beta(x) \in \left\{ s(x)\begin{pmatrix} 1/\sqrt{2} \\ -1/\sqrt{2} \\ 1 \end{pmatrix} \otimes \begin{pmatrix} 1 \\ 1 \\ 0 \end{pmatrix}, s(x)\begin{pmatrix} 1/\sqrt{2} \\ -1/\sqrt{2} \\ -1 \end{pmatrix} \otimes \begin{pmatrix} 1 \\ 1 \\ 0 \end{pmatrix}, \right. \tag{6.19}$$
$$s(x)\begin{pmatrix} 1/\sqrt{2} \\ 1/\sqrt{2} \\ 1 \end{pmatrix} \otimes \begin{pmatrix} 1 \\ -1 \\ 0 \end{pmatrix}, s(x)\begin{pmatrix} 1/\sqrt{2} \\ 1/\sqrt{2} \\ -1 \end{pmatrix} \otimes \begin{pmatrix} 1 \\ -1 \\ 0 \end{pmatrix},$$
$$\left. s(x)\begin{pmatrix} 1 \\ 1 \\ 0 \end{pmatrix} \otimes \begin{pmatrix} 0 \\ 0 \\ 1 \end{pmatrix}, s(x)\begin{pmatrix} 1 \\ -1 \\ 0 \end{pmatrix} \otimes \begin{pmatrix} 0 \\ 0 \\ 1 \end{pmatrix} \right\},$$

a.e. in Ω_L for some $s\colon \Omega_L \to \mathbb{R}$. Materials with B2 (caesium-chloride) structure, such as the intermetallic compounds Yttrium-Zinc [CBW+10] or Nickel-Aluminium, indeed exhibit such a slip-system structure. The arrangement of slip systems and the geometry of the specimen are illustrated in Figure 6.3.

Given these assumptions, it is then possible to prove the following statement.

Theorem 3. *Let* $\Omega_L = (0,1) \times (0,L) \times (0,1)$, *and consider* $E_L(u, F_{\text{pl}}) = \int_{\Omega_L} ||(\nabla u - \beta)_{\mathsf{sym}}||^2 + \sigma \int_{\Omega_L} \mathcal{G}_{\text{lam}}(F_{\text{pl}})$, *i.e.,* (6.18) *with* $\tau = 0$, *subject to the boundary conditions* $u(\cdot, 0, \cdot) = 0, u(\cdot, L, \cdot) = (\gamma, 0, 0)$. *Then*

$$\begin{array}{ll} \inf E_L = 0 & : \; L \geq 2, \\ \frac{c_L \sigma \gamma^2}{\sigma + \sqrt{\sigma^2 + 2c_L\gamma^2}} \leq \inf E_L \leq \min\left\{\frac{\gamma^2}{2L}, 2\sqrt{2}\gamma\sigma\right\} & : \; 1 \leq L < 2, \\ \frac{\gamma^2}{2L}(1-L) \leq \inf E_L \leq \min\left\{\frac{\gamma^2}{2L}, \frac{\gamma^2}{2L}(1-L) + c_L\sigma\gamma\right\} & : \; L < 1, \end{array}$$

if $\sigma > 0$ *and the single slip side condition* (6.19) *is enforced, and*

$$\begin{array}{ll} \inf E_L = 0 & : \; L \geq 1, \\ \inf E_L = \frac{\gamma^2}{2L}(1-L) & : \; L < 1, \end{array}$$

otherwise.

As usual, the upper bounds are proved by construction of explicit test functions, while the lower bounds require a little more analysis. In particular, for the intermediate case when $\sigma > 0$ and (6.19) holds, we have to use a new version of Korn's inequality which is specially adapted to the L^1 curl penalty in order to get the required energy scaling.

These results can be described heuristically as follows. We have shown that the inclusion of both cross-hardening and surface energy significantly affects the energetic scaling in a very specific simple-shear experiment which is realisable for single crystals with B2 structure. Without either cross hardening or surface energy, the energy infimum in our system is zero (or equal to a certain amount of plastic work if a strain-hardening energy is included) when the aspect ratio of the crystal (base-square side-length/height) is below one, while the energy increases quadratically with the imposed shear magnitude when the aspect ratio is above one. The behaviour is qualitatively different when multiple slip is forbidden and geometrically necessary dislocations penalised: now the energy only vanishes for aspect ratios smaller than one half, while, again, for aspect ratios above one, the energy grows quadratically with the strain imposed by the boundary conditions. For intermediate aspect ratios, a new regime of linearly growing energy arises. Thus, we conclude that the experiment proposed here can be used to discriminate between those models with surface energy and cross hardening, and those without, and hence to determine whether the inclusion of these effects in macroscopic models for single-pass plastic deformation is physically reasonable.

6.6 Conclusions

We have shown that, in crystal plasticity, instead of the single-slip side condition, one should use a relaxed condition such that mixing coplanar slip is allowed, but slip in two or more different slip planes is still forbidden at every point of the crystal domain. Such a relaxation should improve numerical simulations by removing the need to resolve an artificial small length scale, which would otherwise be required to prevent mesh-dependent microstructures from forming.

There are two main open questions relating to the relaxed model. First, in our existence result, we assume weak convergence of the plastic strain. This is of course perfectly fine if one introduces an L^q, $q > 1$-type dissipation into the hardening. However, there are some indications that such a hardening is not present in single crystals undergoing single slip [FCO14]. Thus, if one only relies on a more natural L^1-bound on F_{pl}, the limits of minimising sequences will in general be measures, and then lower-semicontinuity results are much more difficult to obtain.

The second open question is whether minimizers of our energy, assuming for example L^q, $q > 1$ hardening, are regular enough to allow the lamination necessary for relaxation. As indicated in Theorem 1, up until now, we have only managed to derive an upper bound for the relaxed energy evaluated at test functions with sufficient regularity.

Acknowledgements. This work was supported by the DFG Research Unit FOR 797 *Analysis and computation of microstructures in finite plasticity*, subproject P6. The authors are grateful to Alexander Mielke, Sergio Conti, Georg Dolzmann and Stefan Mller for many inspiring discussions.

References

[ACD09] Albin, N., Conti, S., Dolzmann, G.: Infinite-order laminates in a model in crystal plasticity. Proc. Roy. Soc. Edinburgh Sect. A 139(4), 685–708 (2009)

[AD14a] Anguige, K., Dondl, P.W.: Optimal energy scaling for a shear experiment in single-crystal plasticity with cross-hardening. Z. Angew. Math. Phys. 65(5), 1011–1030 (2014)

[AD14b] Anguige, K., Dondl, P.W.: Relaxation of the single-slip condition in strain-gradient plasticity. R. Soc. Lond. Proc. Ser. A Math. Phys. Eng. Sci. 470(2169) (2014)

[BHHK92] Bay, B., Hansen, N., Hughes, D.A., Kuhlmann-Wilsdorf, D.: Overview no-96 - evolution of FCC deformation structures in polyslip. Acta Metall. Mater. 40(2), 205–219 (1992)

[CBW+10] Cao, G.H., Becker, A.T., Wu, D., Chumbley, L.S., Lograsso, T.A., Russell, A.M., Gschneidner, K.A.: Mechanical properties and determination of slip systems of the B2 YZn intermetallic compound. Acta Mater. 58(12), 4298–4304 (2010)

[CDK09] Conti, S., Dolzmann, G., Klust, C.: Relaxation of a class of variational models in crystal plasticity. R. Soc. Lond. Proc. Ser. A Math. Phys. Eng. Sci. 465(2106), 1735–1742 (2009)

[CDK13] Conti, S., Dolzmann, G., Kreisbeck, C.: Relaxation of a model in finite plasticity with two slip systems. Math. Models Methods Appl. Sci. 23(11), 2111–2128 (2013)

[CG01] Cermelli, P., Gurtin, M.E.: On the characterization of geometrically necessary dislocations in finite plasticity. J. Mech. Phys. Solids 49(7), 1539–1568 (2001)

[CHM02] Carstensen, C., Hackl, K., Mielke, A.: Non-convex potentials and microstructures in finite-strain plasticity. R. Soc. Lond. Proc. Ser. A Math. Phys. Eng. Sci. 458(2018), 299–317 (2002)

[CO05] Conti, S., Ortiz, M.: Dislocation microstructures and the effective behavior of single crystals. Arch. Ration. Mech. An. 176(1), 103–147 (2005)

[CT05] Conti, S., Theil, F.: Single-slip elastoplastic microstructures. Arch. Ration. Mech. An. 178(1), 125–148 (2005)

[DDMR09] Dmitrieva, O., Dondl, P.W., Mueller, S., Raabe, D.: Lamination microstructure in shear deformed copper single crystals. Acta Mater. 57(12), 3439–3449 (2009)

[DHK05] Devincre, B., Hoc, T., Kubin, L.P.: Collinear interactions of dislocations and slip systems. Mater. Sci. Engrng. A 400-401, 182–185 (2005)

[DKH07] Devincre, B., Kubin, L., Hoc, T.: Collinear superjogs and the low-stress response of fcc crystals. Scripta Mater. 57(10), 905–908 (2007)

[DRR10] Demir, E., Raabe, D., Roters, F.: The mechanical size effect as a mean-field breakdown phenomenon: Example of microscale single crystal beam bending. Acta Mater. 58(5), 1876–1886 (2010)

[FBZ80] Franciosi, P., Berveiller, M., Zaoui, A.: Latent hardening in copper and aluminum single-crystals. Acta Metall. 28(3), 273–283 (1980)

[FCO14] Fokoua, L., Conti, S., Ortiz, M.: Optimal scaling in solids undergoing ductile fracture by void sheet formation. Arch. Ration. Mech. An. 212(1), 331–357 (2014)

[FZ82] Franciosi, P., Zaoui, A.: Multislip Tests on Copper-Crystals - a Junctions Hardening Effect. Acta Metall. 30(12), 2141–2151 (1982)

[HBO10] Hansen, B.L., Bronkhorst, C.A., Ortiz, M.: Dislocation subgrain structures and modeling the plastic hardening of metallic single crystals. Model. Simul. Mater. Sc. 18(5), 055001 (2010)

[HCLH98] Hughes, D., Chrzan, D., Liu, Q., Hansen, N.: Scaling of misorientation angle distributions. Phys. Rev. Lett. 81(21), 4664–4667 (1998)

[HLCH97] Hughes, D.A., Liu, Q., Chrzan, D.C., Hansen, N.: Scaling of microstructural parameters: Misorientations of deformation induced boundaries. Acta Mater. 45(1), 105–112 (1997)

[HM12] Hildebrand, F., Miehe, C.: Variational phase field modeling of laminate deformation microstructure in finite gradient crystal plasticity. Proc. Appl. Math. Mech. 12(1), 37–40 (2012)

[JW84] Jin, N.Y., Winter, A.T.: Dislocation-structures in cyclically deformed [001] copper-crystals. Acta Metall. 32(8), 1173–1176 (1984)

[KH11] Kochmann, D.M., Hackl, K.: The evolution of laminates in finite crystal plasticity: a variational approach. Continuum Mech. Thermodyn. 23(1), 63–85 (2011)

[Koc64] Kocks, U.F.: Latent hardening and secondary slip in aluminum and silver. Trans. Metall. Soc. AIME 230(5), 1160–1167 (1964)

[Kon52] Kondo, K.: On the geometrical and physical foundations of the theory of yielding. In: Proc. 2nd Japan Nat. Congr. Applied Mechanics, pp. 41–47 (1952)

[Lej10] Lejcek, P.: Grain Boundary Segregation in Metals. Springer Series in Materials Science, vol. 136. Springer Science & Business Media, Heidelberg (2010)

[LL67] Lee, E.H., Liu, D.T.: Finite-Strain Elastic-Plastic Theory with Application to Plane-Wave Analysis. J. Appl. Phys. 38(1), 19–27 (1967)

[Mie06] Miehe, C.: Microstructure development in standard dissipative solids based on energy minimization. GAMM-Mitt. 29(2), 247–272 (2006)

[MM06] Mielke, A., Müller, S.: Lower semicontinuity and existence of minimizers in incremental finite-strain elastoplasticity. Z. Angew. Math. Mech. 86(3), 233–250 (2006)

[MM09] Mainik, A., Mielke, A.: Global existence for rate-independent gradient plasticity at finite strain. J. Nonlinear Sci. 19(3), 221–248 (2009)

[NGH85] Nix, W.D., Gibeling, J.C., Hughes, D.A.: Time-dependent deformation of metals. Metall. Trans. A 16(12), 2215–2226 (1985)

[Nye53] Nye, J.F.: Some geometrical relations in dislocated crystals. Acta Metall. 1(2), 153–162 (1953)

[OR99] Ortiz, M., Repetto, E.A.: Nonconvex energy minimization and dislocation structures in ductile single crystals. J. Mech. Phys. Solids 47(2), 397–462 (1999)

[RC14] Reina, C., Conti, S.: Kinematic description of crystal plasticity in the finite kinematic framework: a micromechanical understanding of $F = F^{e}F^{p}$. J. Mech. Phys. Solids 67, 40–61 (2014)

[RP80] Rasmussen, K.V., Pedersen, O.B.: Fatigue of copper polycrystals at low plastic strain amplitudes. Acta Metall. 28(11), 1467–1478 (1980)

[RP99] Rodney, D., Phillips, R.: Structure and strength of dislocation junctions: An atomic level analysis. Phys. Rev. Lett. 82(8), 1704–1707 (1999)

[UDFN04] Uchic, M., Dimiduk, D., Florando, J., Nix, W.: Sample dimensions influence strength and crystal plasticity. Science 305(5686), 986–989 (2004)

[WBL91] Wu, T.Y., Bassani, J.L., Laird, C.: Latent hardening in single-crystals.1. theory and experiments. R. Soc. Lond. Proc. Ser. A Math. Phys. Eng. Sci. 435, 1–19 (1893)

Chapter 7
Gradient Theory for Geometrically Nonlinear Plasticity via the Homogenization of Dislocations

Stefan Müller, Lucia Scardia, and Caterina Ida Zeppieri

Abstract. This article gives a short description and a slight refinement of recent work [MSZ15], [SZ12] on the derivation of gradient plasticity models from discrete dislocations models. We focus on an array of parallel edge dislocations. This reduces the problem to a two-dimensional setting. As in the work Garroni, Leoni & Ponsiglione [GLP10] we show that in the regime where the number of dislocation N_ε is of the order $\log \frac{1}{\varepsilon}$ (where ε is the ratio of the lattice spacing and the macroscopic dimensions of the body) the contributions of the self-energy of the dislocations and their interaction energy balance. Upon suitable rescaling one obtains a continuum limit which contains an elastic energy term and a term which depends on the homogenized dislocation density. The main novelty is that our model allows for microscopic energies which are not quadratic and reflect the invariance under rotations. A key mathematical ingredient is a rigidity estimate in the presence of dislocations which combines the nonlinear Korn inequality of Friesecke, James & Müller [FJM02] and the linear Bourgain & Brezis estimate [BB07] for vector fields with controlled divergence. The main technical improvement of this article compared to [MSZ15] is the removal of the upper bound $W(F) \leq C\mathrm{dist}^2(F, \mathrm{SO}(2))$ on the stored energy function.

7.1 Introduction

Classical theories of plasticity are scale independent. Nonetheless experiments show a notable size dependence of plastic behaviour in the micron and submicron range.

Stefan Müller
Hausdorff Center for Mathematics and Institute for Applied Mathematics, University of Bonn, Bonn, Germany
e-mail: sm@hcm.uni-bonn.de

Lucia Scardia
Department for Mathematical Sciences, University of Bath, Bath, UK
e-mail: l.scardia@bath.ac.uk

Caterina Ida Zeppieri
Applied Mathematics, University of Münster, Münster, Germany
e-mail: caterina.zeppieri@uni-muenster.de

S. Conti and K. Hackl (eds.), *Analysis and Computation of Microstructure in Finite Plasticity*, Lecture Notes in Applied and Computational Mechanics 78, DOI: 10.1007/978-3-319-18242-1_7

A number of phenomenological strain gradient theories have been developed to capture this behaviour, see, e.g., Fleck & Hutchinson [FH93] or Hutchinson [Hut00] for a discussion and further references to the literature.

Here our goal is to understand how strain gradient theories can arise from a mathematical limiting process if one starts with dislocation based models.

One option is to start with fully discrete dislocation models. Recent models include the model of Ariza & Ortiz [AO05], which is based on the concept of discrete eigenstrains and harmonic interactions and employs ideas of algebraic topology. Another recent model is the general nonlinear model of Luckhaus & Mugnai [LM10] which is set purely in the actual configuration and avoids the recourse to any global reference configuration (see also Luckhaus & Wohlgemuth [LW14] for further development of this model).

Another option is to start from a continuum model of dislocations (see the classical works by Kröner [Krö58] and Kondo [Kon64] and further developments in the physics literature by Kleman & Toulouse [TK76, KT77], Mermin [Mer77] and Michel [Mic80]; in the mathematics literature Davini & Parry [DP91a, DP91b] have classified all invariants and Cermelli [Cer99] has put the idea of patching together local deformations in the natural setting of flat torsion-free connections and uses DeRham currents supported on the defect manifold to describe the singularities). One then uses a cut-off on the lattice scale to avoid diverging strains. Such a cut-off seems reasonable if one focusses on the total energy of the system since the energy in the lattice-size core of the dislocation is by a factor of $|\log \varepsilon|$ smaller than the self-energy of the dislocation (here and in the following ε denotes the lattice size or, more precisely, the ratio between the lattice size and the macroscopic size of the body under consideration). This heuristic argument has been made precise, e.g., in the case of an array for screw dislocations, see Ponsiglione [Pon07]. In the following we will use the continuum model of dislocations with a lattice-scale cut-off as our starting point.

To simplify the geometry and the analysis we focus on an ensemble of straight and parallel edge dislocations. This reduces the problem to a two-dimensional domain $\Omega \subset \mathbb{R}^2$. So far most rigorous results have been restricted to this setting, but in recent work of Conti, Garroni & Ortiz [CGO15] the line tension limit has been derived in a full three-dimensional setting for linear elasticity and in the dilute limit (see also [CGM14], [SvG14]). Thus an extension of the results below to a more general three-dimensional setting might be possible, but is far from obvious. In the following we will always assume that $\Omega \subset \mathbb{R}^2$ is *simply connected*, bounded and has a Lipschitz boundary. We will also focus solely on the stored energy and will not discuss dissipation or dynamics. To set the stage we first recall the conventional setting of linearized elasticity.

In this case we consider the elastic strain $\gamma : \Omega \to \mathbb{R}^{2\times 2}$ and the quadratic elastic energy

$$\int_\Omega W(\gamma)\,dx \quad \text{with} \quad W(\xi) = \frac{1}{2}\mathbb{C}\xi \cdot \xi,$$

where $\mathbb{C}$ is a symmetric linear operator from $\mathbb{R}^{2\times 2}$ to itself which satisfies

$$c_1|\operatorname{sym}\xi|^2 \leq \mathbb{C}\xi\cdot\xi \leq c_2|\operatorname{sym}\xi|^2 \quad \forall \xi \in \mathbb{R}^{2\times 2} \quad \text{with} \quad 0 < c_1 \leq c_2. \tag{7.1}$$

Note that the elastic energy is independent of the skew-symmetric part of γ, reflecting invariance under infinitesimal rotations. If no dislocations are present then the elastic strain γ is the gradient of the displacement field: $\gamma = \nabla u$ where $u : \Omega \to \mathbb{R}^2$. This in particular implies that along any closed curve c in Ω one has $\int_c \gamma t\, ds = 0$, where t is the unit tangent vector of the curve and ds denotes the length measure. If a dislocation with Burgers vector b is present at a point $x_0 \in \Omega$ then we have instead

$$\int_c \gamma t\, ds = b$$

for any curve which does not contain x_0 and encircles x_0 once (and in the positive sense). In particular we have for sufficiently small ε

$$\int_{\partial B_\varepsilon(x_0)} \gamma t\, ds = b. \tag{7.2}$$

Local minimization of the elastic energy subject to the constraint $\operatorname{curl}\gamma = b\delta_{x_0}$ gives the strain field of a single dislocation. This field diverges like $|x-x_0|^{-1}$ near x_0. Hence the integral $\int_{\Omega\setminus\overline{B_\delta(x_0)}} \mathbb{C}\gamma\cdot\gamma\, dx$ diverges logarithmically in δ. Let ε denote the lattice spacing. One introduces a core cut-off and looks at the regularized energy

$$\int_{\Omega\setminus\overline{B_\varepsilon(x_0)}} W(\gamma)\, dx.$$

As mentioned above this is motivated by the observation that in fully discrete models one expects the energy in a lattice size core to be a factor of $\log\frac{1}{\varepsilon}$ smaller than the energy in a region $B_\rho(x_0)\setminus\overline{B_\varepsilon(x_0)}$ (see also Section 7.3).

We are interested in the cumulative effect of many dislocations. Assume that there are N dislocations at positions x_i with Burgers vectors b_i. We assume that the dislocations are well separated, i.e., there exists $\rho_\varepsilon \gg \varepsilon$ such that

$$|x_i - x_j| \geq 2\rho_\varepsilon \quad \text{if } i \neq j.$$

We consider the total dislocation density

$$\mu = \sum_{i=1}^N b_i \delta_{x_i}$$

and set

$$\Omega_\varepsilon = \Omega \setminus \bigcup_{i=1}^N \overline{B_\varepsilon(x_i)}.$$

Then we seek to minimize

$$\int_{\Omega_\varepsilon} W(\gamma)\, dx \qquad \text{subject to} \qquad \int_{\partial B_\varepsilon(x_i)} \gamma t\, ds = b_i \quad \forall i = 1,\dots,N. \tag{7.3}$$

In order to make connection with the continuum theory of strain gradient plasticity we want to take a limit as $\varepsilon \to 0$ and $N = N_\varepsilon \to \infty$. From the point of view of physics it is more natural to fix the lattice spacing and to consider domains $\frac{1}{\varepsilon}\Omega$ of increasing size. Upon elasticity scaling both points of view are equivalent and fixing Ω rather than the lattice spacing is more convenient for the analysis. Thus ε really is a dimensionless parameter of the order of lattice spacing divided by the macroscopic dimension of the body. For brevity we will nonetheless often refer to ε as the lattice spacing.

The main result of [GLP10] is the following (see Section 7.4 for a more precise statement in the context of the geometrically nonlinear theory). Assume that

$$N_\varepsilon \sim \log \frac{1}{\varepsilon}$$

and that we are given dislocations at (well separated) points x_i^ε ($i = 1, \dots, N_\varepsilon$) with Burgers vectors $\varepsilon\xi_i^\varepsilon$, where ξ_i^ε are of order one. Define the ($\mathbb{R}^2$-valued) measure

$$\mu_\varepsilon := \sum_{i=1}^{N_\varepsilon} \varepsilon\, \xi_i^\varepsilon\, \delta_{x_i^\varepsilon}$$

and set $\Omega_\varepsilon = \Omega \setminus \bigcup_i \overline{B_\varepsilon(x_i^\varepsilon)}$. Then the functionals

$$\frac{1}{\varepsilon^2|\log\varepsilon|^2}\int_{\Omega_\varepsilon} \frac{1}{2}\mathbb{C}\gamma\cdot\gamma\, dx \qquad \text{subject to} \qquad \int_{\partial B_\varepsilon(x_i)} \gamma t\, ds = \varepsilon\xi_i^\varepsilon \quad \forall i = 1,\dots,N_\varepsilon \tag{7.4}$$

Γ-converge as $\varepsilon \to 0$ to the functional

$$\int_\Omega \frac{1}{2}\mathbb{C}\gamma\cdot\gamma\, dx + \int_\Omega \varphi\left(\frac{d\mu}{d|\mu|}\right) d|\mu| \qquad \text{subject to} \qquad \operatorname{curl}\gamma = \mu. \tag{7.5}$$

Here $\varphi : \mathbb{R}^2 \to \mathbb{R}$ is a function which can be determined explicitly by solving an auxiliary problem for a single dislocation in $\mathbb{R}^2$ and $\frac{d\mu}{d|\mu|}$ denotes the Radon-Nikodym derivative of the $\mathbb{R}^2$-valued Radon measure μ (see Section 7.4 for a more detailed description).

Moreover for any family $(\gamma_\varepsilon, \mu_\varepsilon)$ which satisfies the constraint in (7.4) and for which the energy in (7.4) is uniformly bounded there exists subsequence $\varepsilon_j \to 0$ such that $\frac{1}{\varepsilon_j|\log\varepsilon_j|}\gamma_{\varepsilon_j}$ converges weakly to γ in L^2 (after subtraction of constant skew-symmetric matrices) and $\frac{1}{\varepsilon_j|\log\varepsilon_j|}\mu_{\varepsilon_j}$ converges weak* in the sense of Radon measures to μ. Together with the Γ-convergence this implies that a subsequence of minimizers of the problem (7.4) converges to a minimizer of the problem (7.5). Of course the minimization of the limit functional in μ *and* γ yields the trivial solution $\mu = 0$ and $\gamma = 0$, but since Γ-convergence is a very robust notion it is not difficult to show that convergence of (a subsequence of) minimizers still holds if one adds a Dirichlet type boundary condition of the form

$$\gamma_\varepsilon t = \varepsilon|\log\varepsilon| g \quad \text{on } \partial\Omega \qquad \text{where } t \text{ denotes the unit tangent of } \partial\Omega.$$

Note that the relation between γ_ε and μ_ε implies the compatibility conditions

$$\int_{\partial\Omega} g = \frac{1}{\varepsilon|\log\varepsilon|}\mu_\varepsilon(\Omega) = \frac{1}{|\log\varepsilon|}\sum_i \xi_i,$$

where $\partial\Omega$ is positively oriented. Then in the limit problem one obtains the boundary condition $\gamma t = g$ on $\partial\Omega$ and this in general leads to nontrivial minimizers.

The first term in the limit functional (7.5) is the usual elastic energy of the continuum theory and the second term reflects the strain gradient energy. In this setting the strain gradient energy depends only on the dislocation density $\mu = \operatorname{curl}\gamma$ and is one-homogeneous in μ. The first term arises from the interaction energy of the dislocations while the second term arises from the self-energy of the dislocations. The scaling $N_\varepsilon \sim \log\frac{1}{\varepsilon}$ is distinguished because only in this scaling both terms are of equal strength. If $N_\varepsilon \ll |\log\varepsilon|$ then the self-energy of the dislocations dominates, while for $N_\varepsilon \gg |\log\varepsilon|$ the interaction energy dominates, see Section 7.3 for a heuristic discussion of the scaling regimes and [GLP10], Theorems 12, 15 and 18 for precise statements.

To go beyond the linearized theory and in particular to incorporate invariance under rigid motions (and not just infinitesimal rigid motions) one has to replace the quadratic elastic energy density by a frame indifferent energy density $W(\beta)$ where β now denotes the local elastic deformation gradient (i.e., $\gamma = \beta - \mathrm{Id}$ in the linearized theory). Natural assumptions on W are

$$W(RF) = W(F) \quad \forall R \in SO(2), F \in \mathbb{R}^{2\times 2} \qquad \text{(frame indifference)}, \tag{7.6}$$

$$W(F) \geq W(\mathrm{Id}) = 0 \quad \forall F \in \mathbb{R}^{2\times 2} \qquad \text{(normalization)} \tag{7.7}$$

$$W \text{ is } C^2 \text{ in a neighbourhood of } SO(2) \qquad \text{(local smoothness)}, \tag{7.8}$$

$$\exists c > 0 \text{ such that} \quad W(F) \geq c\,\operatorname{dist}^2(F, SO(2)) \quad \forall F \in \mathbb{R}^{2\times 2}. \tag{7.9}$$

Assumption (7.9) is a natural counterpart of (7.1). Indeed if we assume (7.8) and set $\mathbb{C} := D^2W(\mathrm{Id})$, then (7.1) holds if and only if (7.9) holds for all F close to SO(2). Requiring the inequality $W(F) \geq \operatorname{dist}^2(F, SO(2))$ for all F amounts to the additional assumptions that $W(F) > 0$ if $F \notin SO(2)$ and that $W(F)$ grows quadratically as $F \to \infty$.

Since

$$\int_{\partial B_\varepsilon(x_0)} \mathrm{Id}\, t\, ds = 0$$

the constraint (7.2) corresponds to

$$\int_{\partial B_\varepsilon(x_0)} \beta\, t\, ds = b.$$

In [MSZ15] it is shown that in the same scaling regime as above and under an additional technical growth condition one obtains the same limiting problem (7.5), after the elimination of a global constant rotation. We give a precise statement and a slightly improved result in Section 7.4.

In the nonlinear setting it is actually more natural to use a map from the actual (or 'deformed' or 'spatial') configuration back to a (locally defined) reference lattice to define the Burgers vector. Thus let $\Omega^{sp} \subset \mathbb{R}^2$ now denote the actual configuration and let $G : \Omega^{sp} \to \mathbb{R}^{2\times 2}$ denote a field of linear maps such that $G(x)$ infinitesimally maps the actual configuration back to a (local) reference lattice (in a setting without dislocations and a global reference configuration $G(x)$ is just the inverse of the deformation gradient). If c is a curve in Ω^{sp} which encircles a single dislocation at x_0 then the Burgers vector b is defined as

$$\int_{\partial B_\varepsilon(x_0)} Gt\, ds = -b \tag{7.10}$$

(the change of sign is natural if one notes that $G = \beta^{-1}$ and considers the linearization $\beta \approx Id + \gamma$ which implies $G \approx \mathrm{Id} - \gamma$). If $W(A)$ denotes as before the energy (per unit reference volume) needed to perform an affine deformation A of the reference lattice then the elastic energy associated to the field G is given by

$$\int_{\Omega^{sp}} W(G^{-1}) \det G\, dx.$$

Thus if we define

$$\tilde{W}(G) = W(G^{-1}) \det G$$

then we are back to the previous formulation with W replaced by $\tilde{W}$, β replaced by G, Ω replaced by Ω^{sp} and b replaced by $-b$. Note that assumption (7.6) is equivalent to

$$\tilde{W}(GR) = \tilde{W}(G) \quad \forall R \in \mathrm{SO}(2),\, \forall G \in \mathbb{R}^{2\times 2}. \tag{7.11}$$

The assumptions (7.7) and (7.8) hold for $\tilde{W}$ if and only if they hold for W. The inequality $\tilde{W}(G) \geq c\,\mathrm{dist}^2(G, \mathrm{SO}(2))$ holds for G close to SO(2) if and only if it holds for W and G close to SO(2). Moreover it is easy to give reasonable conditions on W such that this inequality holds for all G.

To put the results discussed so far in perspective let us briefly mention some related work in scaling limits of nonlinear elasticity and elastoplasticity. First, immediately after the appearance of the nonlinear counterpart of Korn's inequality (see (7.14) below) Dal Maso, Negri & Percivale [DMNP02] realized that it could be used to show rigorously that nonlinear elasticity converges to linearized elasticity in the low-energy limit. Our scaling also corresponds to a low-energy limit, so it should not be surprising that the limit problem is essentially linear (up to the freedom of a single rotation). Very recently Mielke & Stefanelli [MS13] solved the much more difficult problem to rigorously show that the evolutionary problem of finite strain elastoplasticity converges to linearized plasticity. A key idea is to extend variational concepts such as Γ-convergence to evolution equations, see Mielke's survey [Mie15] in this book. An existence theory for a single time step in elasto-plasticity theory with multiplicative decomposition was developed in [MM06]. For an existence theory for energies $\int_\Omega W(\beta)\,dx$ with the constraint that $\operatorname{curl}\beta$ is given by a single dislocation loop see [MP08]. In the static setting a hierarchy of scaling limits of a discrete dislocation model is discussed in [GPPS13] and [SPPG14] in the context of dislocation pile-ups.

It is also interesting to compare the setting discussed above to the setting of minimization problems in crystal plasticity which is discussed in the articles by Anguige & Dondl [AD15] and by Dmitrieva el al. [DRMD15] in this book (see also [AD14b, AD14a]). For this comparison consider first a single dislocation in the geometrically linear setting. For definiteness assume that the dislocation is at 0 with Burgers vector εe_1. In the geometrically linear setting we have been looking for $\gamma : \Omega \setminus \overline{B_\varepsilon(0)} \to \mathbb{R}^{2\times 2}$ with

$$\int_{\partial B_\varepsilon(0)} \gamma t\,ds = b = \varepsilon e_1, \qquad \operatorname{curl}\gamma = 0 \quad \text{in } \Omega \setminus \overline{B_\varepsilon(0)}.$$

No such γ can be the gradient ∇u of a C^1 map $u : \Omega \setminus \overline{B_\varepsilon(0)} \to \mathbb{R}^2$. There exist, however, maps u which are smooth away from a half-line Σ, satisfy

$$\int_{\partial B_r(0)\setminus\Sigma} \nabla u t\,ds = \varepsilon e_1 \quad \text{for all } r \geq \varepsilon \text{ with } B_r(0) \subset \Omega$$

and jump across Σ. Indeed one can take, e.g.,

$$u(x) = \varepsilon e_1 \frac{1}{2\pi}\theta, \quad \text{where } x = r(\cos\theta, \sin\theta), \text{ with } r > 0,\ 0 \leq \theta < 2\pi.$$

Then u jumps across the half-line $\Sigma := \{(x_1, 0) : x_1 > 0\}$ and is smooth otherwise. The distributional derivative Du of u consists of a smooth part (written as $\nabla u \mathscr{L}^2$, where $\mathscr{L}^2$ denotes the two-dimensional Lebesgue measure) and a singular part which is given by a one-dimensional measure concentrated on the half-line Σ:

$$Du = \nabla u - \varepsilon e_1 \otimes e_2 \mathscr{H}^1 \angle \Sigma.$$

Note that e_2 is the normal to Σ. The space of L^1 functions for which the distributional derivative is the sum of an L^1 function and a Radon measure concentrated on a codimension 1 set is known as SBV (special functions of bounded variation).

One usually interprets the regular part as the elastic strain γ and the singular part as the plastic strain

$$\gamma_p := -\varepsilon e_1 \otimes e_2 \mathscr{H}^1 \angle \Sigma.$$

The plastic strain then corresponds to a tangential slip with Burgers vector εe_1 along the slip line Σ with normal e_2. Note that $\operatorname{curl} Du = 0$ and

$$\operatorname{curl}\gamma_p = -\operatorname{curl}\nabla u = -\varepsilon e_1 \delta_{x_0}$$

in the sense of distributions in Ω.

One is thus led to consider the minimization of

$$\frac{1}{2}\int_{\Omega\setminus\overline{B_\varepsilon(0)}} \mathbb{C}(Du-\gamma_p)\cdot(Du-\gamma_p)\,dx$$

over SBV functions u and plastic strains γ_p. If one admits all plastic strains γ_p which satisfy

$$\operatorname{curl}\gamma_p = 0 \quad \text{in } \Omega\setminus\overline{B_\varepsilon(0)}, \qquad \int_{\partial B_\varepsilon(0)} \gamma_p\, t\, ds = -\varepsilon e_1$$

then one recovers the minimization problem (7.3) discussed above (in the special case of a single dislocation). In crystal plasticity one considers only those fields γ_p which correspond to a superposition of slips across a given set of slip planes and corresponding Burgers vectors, see, e.g., [AD15] or [CGO15]. Similarly one can consider multiple dislocations and their corresponding slip lines.

In some simple cases one can see easily that both approaches yield the same minimal energy. Consider, e.g., as above the geometrically linear two dimensional setting and let $\gamma : \Omega_\varepsilon \to \mathbb{R}^{2\times 2}$ be the elastic strain satisfying

$$\int_{\partial B_\varepsilon(x_i)} \gamma\, t\, ds = b_i, \qquad \operatorname{curl}\gamma = 0 \quad \text{in } \Omega_\varepsilon,$$

where x_i are the dislocation locations. Assume that the admissible slip lines are parallel to the two coordinate directions and that the Burgers vectors are of the form

$$b_i = \varepsilon z e_1 \quad \text{or} \quad b_i = \varepsilon z e_2 \quad \text{with } z \in \mathbb{Z}.$$

Then for a dislocation at x_i there exists a line segment Σ_i parallel to b_i (and hence to one of the coordinate axes) whose endpoints are x_i and a point on $\partial\Omega$ as well as a map $v^{(i)}$ which is smooth outside Σ_i and jumps by $-b_i$ across Σ_i. For $b_i = \varepsilon z e_1$ such a map was constructed above; for $b_i = \varepsilon z e_2$ one can rotate the previous construction by 90 degrees. Set $v = \sum_i v^{(i)}$. As above let ∇v denote the density of the absolutely continuous part of Dv and let γ_p denote the singular part of Dv. Then by construction γ_p is of the form required in crystal plasticity. Moreover $\operatorname{curl}(\gamma - \nabla u) = 0$ in Ω_ε and

$$\int_{\partial B_\varepsilon(x_i)} (\gamma - \nabla v)\, t\, ds = 0$$

for all i. Thus there exists a function $w \in SBV$ such that $Dw = \gamma - \nabla v$ in Ω_ε. Note that the singular part of Dw vanishes (and w is hence in the Sobolev space $W^{1,1}$). Set $u = v + w$. Then

$$Du - \gamma_p = Dv - \gamma_p + Dw = \nabla v + (\gamma - \nabla v) = \gamma.$$

Thus in this example every field $\gamma : \Omega_\varepsilon \to \mathbb{R}^{2\times 2}$ which satifies the circulation constraints can be written in the form $Du - \gamma_p$ where u is in SBV and γ_p corresponds to a superposition of single slips across the allowed slip lines.

Taking a slightly different point of view we can view the minimization problem in (7.3) (or the minimization of $\int_{\Omega^{sp}} \tilde{W}(G)\,dx$ subject to (7.10)) as a global minimization problem which involves only the elastic strain, but no global reference configuration. The maps G capture only the *local* deviation of the actual configuration from a perfect lattice and the strains γ are the linearized counterparts of G^{-1}. A 'plastic strain' is not required in this approach. By contrast, in crystal plasticity (and other versions of elastoplasticity which keep track of both the elastic and the plastic strain) one starts from a global defectless reference configuration and the plastic strain keeps track of global reordering of the lattice, while the elastic strain measures the local distorsion of the lattice. If one has an atomistic picture in mind the elastic strain only cares about the actual position of the atoms while the plastic strain keeps track of where they have come from. For a recent discussion in the context of the multiplicative decomposition in elastic and plastic strain see [RC14].

7.2 Key Mathematical Challenges

In the linearized setting one key difficulty is that the energy density $\mathbb{C}\gamma\cdot\gamma$ depends only on the symmetric part of γ. Hence control of the energy alone is not sufficient to get control on γ. If $\mu = \operatorname{curl}\gamma = 0$ (and if Ω is simply connected) then $\gamma = \nabla u$. Thus Korn's inequality allows one to estimate the L^2 norm of the skew-symmetric part of ∇u in terms of the L^2 norm of the symmetric part (up to an irrelevant skew-symmetric constant). If $\mu \neq 0$ one can use the Hodge decomposition $\gamma = \nabla u + \nabla^{\perp} v$ with $\Delta v = \mu$. Hence by standard elliptic regularity one obtains that if μ is in L^p (with $1 < p < 2$) then $\nabla^{\perp} v$ is in L^{p^*} with $\frac{1}{p^*} = \frac{1}{p} - \frac{1}{2}$ with a corresponding estimate.

Unfortunately we only have control of the total mass of the measure μ. This corresponds to the borderline case $p = 1$ and in this case standard elliptic regularity does not provide an estimate for $\nabla^{\perp} v$ in L^2 but only in the weaker Marcinkiewicz space $L^{2,\infty}$. This difficulty is overcome by a striking estimate of Bourgain & Brezis [BB07] which states that for $f \in L^1(\Omega;\mathbb{R}^2)$ with $\operatorname{div} f \in H^{-2}(\Omega)$

$$\|f\|_{H^{-1}} \leq C(\|f\|_{L^1} + \|\operatorname{div} f\|_{H^{-2}}). \tag{7.12}$$

This estimate extends to the case that f is a Radon measure with bounded mass and applying this estimate to $f = (\operatorname{curl}\gamma)^{\perp}$ and using that $\operatorname{div} f$ equals a linear combination of second derivatives of $\operatorname{sym}\gamma$ we get $f \in H^{-1}$ and

$$\|\operatorname{skw}\gamma\|_{L^2} \leq C(\|\operatorname{sym}\gamma\|_{L^2} + \|\operatorname{curl}\gamma\|_{\mathcal{M}}) \quad \text{if} \quad \int_{\Omega} \operatorname{skw}\gamma = 0, \tag{7.13}$$

see [GLP10], Theorem 11.

In the geometrically nonlinear setting the natural counterpart of Korn's inequality is the rigidity estimate of Friesecke et al. [FJM02]. If $U \subset \mathbb{R}^n$ is a bounded domain with Lipschitz boundary then there exists a constant $C(U)$ such that for every $u \in W^{1,2}(U;\mathbb{R}^n)$ there exists a rotation $R \in \mathrm{SO}(n)$ with

$$\|\nabla u - R\|_{L^2} \leq C(U)\, \|\operatorname{dist}(\nabla u, \mathrm{SO}(n))\|_{L^2}. \tag{7.14}$$

In our setting we do not have $\beta = \nabla u$, but we only have $\operatorname{curl}\beta = \mu$ where μ is a Radon measure with finite mass. A key idea in [MSZ15] is to show that one can combine (7.12) and (7.14) to establish the following result.

Theorem 1 (Rigidity estimate with dislocations, see [MSZ15], Theorem 3.3). *Let $\Omega \subset \mathbb{R}^2$ be a bounded and simply connected domain with Lipschitz boundary. Then there exists a constant $C(\Omega)$ with the following property. For every $\beta \in L^2(\Omega;\mathbb{R}^{2\times 2})$ with $\operatorname{curl}\beta \in \mathcal{M}(\Omega;\mathbb{R}^2)$ there exists an $R \in \mathrm{SO}(2)$ such that*

$$\boxed{\|\beta - R\|_{L^2} \le C(\Omega)\,(\|\operatorname{dist}(\beta,\mathrm{SO}(2))\|_{L^2} + \|\operatorname{curl}\beta\|_{\mathcal{M}})\,.} \tag{7.15}$$

This is the key ingredient to extend the analysis of Garroni et al. [GLP10] to the geometrically nonlinear setting. The proof of Theorem 1 proceeds in two steps. One first shows an estimate in the weak-L^2 space $L^{2,\infty}$ using [CDM14]. Then one improves this estimate to an L^2 estimate by a careful Taylor expansion.

A version of the estimate in the language for 1-forms and exterior derivatives and in dimensions $d \ge 2$ (with $\|\operatorname{curl}\beta\|_{\mathcal{M}}$ replaced by $\|\operatorname{curl}\beta\|_{L^p}$, $p = \frac{2d}{2+d}$ if $d \ge 3$) can be found in a recent preprint by Aumann [Aum15].

7.3 Heuristics for Scaling Regimes

7.3.1 The Core Energy of a Single Dislocation

A toy energy

Set $B_r = B_r(x_0)$ and assume that $\gamma \in L^2(B_\rho \setminus B_\varepsilon;\mathbb{R}^{2\times 2})$ and

$$\operatorname{curl}\gamma = 0 \quad \text{in } B_\rho \setminus \overline{B_\varepsilon}, \tag{7.16}$$

$$\int_{\partial B_\varepsilon} \gamma t\, ds = b. \tag{7.17}$$

Here the first identity is understood in the sense of distributions. It follows from $\operatorname{curl}\gamma = 0$ and $\gamma \in L^2$ that the tangential trace γt is well defined as a distribution in $H^{-1/2}(\partial B_\varepsilon)$ so that the second identity makes sense.

From (7.16) and (7.17) one easily obtains the lower bound

$$\int_{B_{2r}\setminus B_r} |\gamma|^2\, dx \ge \frac{1}{2\pi}\log 2\ |b|^2, \quad \forall r \in [\varepsilon, \rho/2] \tag{7.18}$$

and it follows that

$$\int_{B_\rho\setminus B_\varepsilon} |\gamma|^2\, dx \gtrsim \log\left(\frac{\rho}{\varepsilon}\right)\ |b|^2. \tag{7.19}$$

To prove (7.18) note that (7.16) and (7.17) imply that

$$\int_{\partial B_{r'}} \gamma t\, ds = b \tag{7.20}$$

for all $r' \in (\varepsilon, \rho)$. Since $|\gamma| \geq |\gamma t|$ it follows from Fubini's theorem and Jensen's inequality that

$$\begin{aligned}\int_{B_{2r}\setminus B_r} |\gamma|^2\,dx &\geq \int_r^{2r} \int_{\partial B_{r'}} |\gamma t|^2\,ds\,dr' \\ &\geq \int_r^{2r} \frac{1}{2\pi r'} \left| \int_{\partial B_{r'}} \gamma t\,ds \right|^2 dr' = \frac{1}{2\pi} \log 2\,|b|^2.\end{aligned}$$

This lower bound is in fact sharp as can be seen from the choice $\gamma = \frac{1}{2\pi} b \otimes \frac{x^\perp}{|x|^2}$.

A lower bound in linear elasticity

The estimate (7.19) suggests the natural scaling of the core energy, but is not immediately relevant since the linear elastic energy depends only on symmetric part $\operatorname{sym}\gamma$ and Jensen's inequality does not give a lower bound for $\int_{\partial B_{r'}} |\operatorname{sym}\gamma|^2\,ds$. We claim that nonetheless there exists a $c > 0$ such that

$$\int_{B_{2r}\setminus B_r} |\operatorname{sym}\gamma|^2\,dx \geq c\,|b|^2, \quad \forall r \in [\varepsilon, \rho/2]. \tag{7.21}$$

To see this, we follow [GLP10] and use the condition $\operatorname{curl}\gamma = 0$ to apply Korn's inequality. We first note that in view of the rescaling $\tilde{\gamma}(x) = \gamma(rx)$, $\tilde{b} = b/r$ it suffices to prove the estimate for $r = 1$. We claim that Korn's inequality holds in the following form

$$\begin{aligned}&\operatorname{curl}\gamma = 0 \quad \text{in } B_2 \setminus \overline{B_1} \quad \Longrightarrow \\ &\exists\, A \text{ skew-symmetric} \quad \|\operatorname{skw}\gamma - A\|_{L^2(B_2\setminus\overline{B_1})} \leq C \|\operatorname{sym}\gamma\|_{L^2(B_2\setminus\overline{B_1})}.\end{aligned} \tag{7.22}$$

To prove this set $\delta := \|\operatorname{sym}\gamma\|_{L^2(B_2\setminus\overline{B_1})}$ and write $B_2 \setminus \overline{B_1}$ as the union of two open, simply connected sets U^+ und U^- with Lipschitz boundaries. Then there exist $u^\pm \in H^1(U^\pm;\mathbb{R}^2)$ such that $\nabla u^\pm = \gamma$ in $U^\pm$. Thus by the usual Korn inequality there exist skew-symmetric matrices $A^\pm$ such that

$$\|\operatorname{skw}\gamma - A^\pm\|_{L^2(U^\pm)} \leq C \|\operatorname{sym}\gamma\|_{L^2(U^\pm)} \leq C\delta.$$

Since $U^+ \cap U^-$ has positive measure it follows that

$$|A^+ - A^-| \leq C\delta.$$

Thus we also have $\|\operatorname{skw}\gamma - A^-\|_{L^2(U^+)} \leq C\delta$ and this implies $\|\gamma - A^-\|_{L^2(B_2\setminus\overline{B_1})} \leq C\delta$ as claimed.

To prove (7.21) let A be as in (7.22) and note that $\int_{\partial B_{r'}} A t\,ds = 0$ and thus

$$b = \int_1^2 \int_{\partial B_{r'}} \gamma t\,ds\,dr' = \int_1^2 \int_{\partial B_{r'}} (\gamma - A) t\,ds\,dr' = \int_{B_2\setminus\overline{B_1}} (\gamma - A) \frac{x^\perp}{|x|}\,dx.$$

Hence

$$|b| \leq C\|\gamma - A\|_{L^2(B_2\setminus\overline{B_1})} \leq C\|\operatorname{sym}\gamma\|_{L^2(B_2\setminus\overline{B_1})}$$

which is the assertion.

Geometrically nonlinear setting

As observed in Scardia & Zeppieri [SZ12] one can extend the lower bound to the nonlinear setting if one replaces the Korn inequality by the rigidity estimate in [FJM02]. More precisely if $\beta \in L^2(B_\rho \setminus B_\varepsilon, \mathbb{R}^{2\times 2})$ and

$$\operatorname{curl}\beta = 0 \quad \text{in } B_\rho \setminus \overline{B_\varepsilon}, \qquad \int_{\partial B_\varepsilon} \beta\, t\, ds = b \tag{7.23}$$

then

$$\int_{B_{2r}\setminus B_r} \operatorname{dist}^2(\beta, SO(2))\, dx \geq c|b|^2 \quad \forall r \in [\varepsilon, \rho/2] \tag{7.24}$$

and thus

$$\int_{B_\rho\setminus B_\varepsilon} \operatorname{dist}^2(\beta, SO(2))\, dx \gtrsim \log\left(\frac{\rho}{\varepsilon}\right) |b|^2. \tag{7.25}$$

Note that in view of the rescaling $\tilde{\beta}(x) = \beta(rx)$ and $\tilde{b} = \frac{b}{r}$ it suffices to show the result for $r = 1$. To do so we argue exactly as in the case of linearized elasticity. First we note that

$$\begin{aligned} &\operatorname{curl}\beta = 0 \quad \text{in } B_2 \setminus \overline{B_1} \quad \Longrightarrow \\ &\exists R \in SO(2) \quad \|\beta - R\|_{L^2(B_2\setminus\overline{B_1})} \leq C\|\operatorname{dist}(\beta, SO(2))\|_{L^2(B_2\setminus\overline{B_1})}. \end{aligned} \tag{7.26}$$

This is deduced from (7.14) in the same way (7.22) was deduced from Korn's inequality. Now

$$b = \int_1^2 \int_{\partial B_{r'}} \beta\, t\, ds\, dr' = \int_1^2 \int_{\partial B_{r'}} (\beta - R)\, t\, ds\, dr' = \int_{B_2\setminus\overline{B_1}} (\beta - R)\frac{x^\perp}{|x|}\, dx.$$

Thus

$$|b| \leq C\|\beta - R\|_{L^2(B_2\setminus\overline{B_1})} \leq C\|\operatorname{dist}(\beta, SO(2))\|_{L^2(B_2\setminus\overline{B_1})}$$

which is the assertion.

7.3.2 The Core Energy of Many Dislocations

We now consider N_ε dislocations at points $x_i^\varepsilon \in \Omega$ and we assume that these points are well separated, i.e., there exists $s \in (0,1)$ such that

$$|x_i^\varepsilon - x_j^\varepsilon| \geq 2\varepsilon^s \quad \text{if } i \neq j \quad \text{and} \quad \operatorname{dist}(x_i^\varepsilon, \partial\Omega) \geq \varepsilon^s.$$

We assume that the Burgers vectors are of the order of the lattices spacing, i.e.,

$$b_i^\varepsilon = \varepsilon \xi_i^\varepsilon,$$

where ξ_i^ε belong to a given lattice of order 1. We assume that

$$\operatorname{curl} \gamma = 0 \quad \text{in } \Omega_\varepsilon := \Omega \setminus \bigcup_{i=1}^{N_\varepsilon} \overline{B_\varepsilon(x_i^\varepsilon)}$$

and

$$\int_{\partial B_\varepsilon(x_i^\varepsilon)} \gamma t \, ds = b_i^\varepsilon = \varepsilon \xi_i^\varepsilon \quad \forall i = 1, \dots, N_\varepsilon.$$

Then

$$\int_{\Omega_\varepsilon} |\operatorname{sym} \gamma|^2 \, dx \geq \sum_{i=1}^{N_\varepsilon} \int_{B_{\varepsilon^s}(x_i^\varepsilon) \setminus B_\varepsilon(x_i^\varepsilon)} |\operatorname{sym} \gamma|^2 \, dx \gtrsim N_\varepsilon \, s \, \varepsilon^2 \log\left(\frac{1}{\varepsilon}\right).$$

A similar lower bound holds in the geometrically nonlinear setting where $\operatorname{sym} \gamma$ is replaced by $\operatorname{dist}(\beta, SO(2))$. Thus the total core energy of the dislocations scales like

$$E_{\text{core}} \sim N_\varepsilon \, \varepsilon^2 \, |\log \varepsilon|.$$

7.3.3 The Interaction Energy

Here we want to understand the behaviour of γ (or β, in the geometrically nonlinear setting) away from the immediate neighbourhood of the dislocations. In this region γ is not dominated by the behaviour of the nearest dislocation but rather by the cumulative effect of all dislocations. At this point we are only interested in a rough estimate of orders of magnitude. We thus assume that Ω has area of order 1 and contains $N_\varepsilon \gg 1$ dislocations. We assume that the dislocations are roughly equally spaced, i.e., the typical distance between two neighbouring dislocations is

$$l_\varepsilon \sim \frac{1}{\sqrt{N_\varepsilon}}.$$

We also assume that each dislocation has a Burgers vector of order ε. The strain field of a single dislocation at x_i^ε is of order $\frac{\varepsilon}{|x - x_i^\varepsilon|}$. Now fix a point $x \in \Omega$ and let $r \geq l_\varepsilon$. Then the annulus $B_{2r}(x) \setminus B_r(x)$ contains approximately $r^2/l_\varepsilon^2 \sim N_\varepsilon r^2$ dislocations. The strain field of a single dislocation in this annulus generates a strain field of order $\frac{\varepsilon}{r}$ at x. If we ignore possible cancellations the total strain field of the dislocations in the annulus is given by $\varepsilon N_\varepsilon r$. Applying this with $r = 2^k l_\varepsilon$ for $k = 0, \dots, K_\varepsilon$ where $K_\varepsilon \sim \log_2 l_\varepsilon^{-1}$ and summing over k we see that the total field of all dislocations outside $B_{l_\varepsilon}(x)$ is of order $\varepsilon N_\varepsilon$. The disc $B_{l_\varepsilon}(x)$ contains only finitely many dislocations. Let x_i^ε be the position of the dislocation closest to x. Its strain field at x is of order $\frac{\varepsilon}{|x - x_i^\varepsilon|}$. Thus we have two cases

- If $|x-x_i^\varepsilon| \geq \frac{1}{N_\varepsilon} \sim l_\varepsilon^2$ then the strain field of the dislocations outside $B_{l_\varepsilon}(x)$ dominates and we have the bound

$$|\gamma(x)| \lesssim \varepsilon N_\varepsilon.$$

- If $|x-x_i^\varepsilon| < \frac{1}{N_\varepsilon} \sim l_\varepsilon^2$ then the strain field of the single dislocation at x_i^ε dominates $\gamma(x)$ and we have

$$|\gamma(x)| \gtrsim \varepsilon \frac{1}{|x-x_i^\varepsilon|} \geq \varepsilon N_\varepsilon.$$

Thus we get

$$\int_{\Omega_{l_\varepsilon^2}} |\gamma|^2 \, dx \lesssim \varepsilon^2 N_\varepsilon^2.$$

If we believe that the cancellations between the strain fields of different dislocations do not change the order of magnitude we expect that the interaction energy is of order

$$E_{\text{inter}} \sim \varepsilon^2 N_\varepsilon^2.$$

Thus we see that the core energy (concentrated near the dislocations, i.e., in $\bigcup_i B_{l_\varepsilon^2}(x_i^\varepsilon)$) and the interaction energy have the same scaling if

$$N_\varepsilon \sim |\log \varepsilon|.$$

This is the regime we will consider below. If $N_\varepsilon \ll |\log \varepsilon|$ then the core energy dominates, if $N_\varepsilon \gg |\log \varepsilon|$ then the interaction energy dominates.

For a slightly different view we consider the normalized dislocation measure

$$\lambda_\varepsilon = \frac{1}{\varepsilon N_\varepsilon} \mu_\varepsilon = \frac{1}{N_\varepsilon} \sum_{i=1}^{N_\varepsilon} \frac{b_i^\varepsilon}{\varepsilon} \delta_{x_i^\varepsilon} = \frac{1}{N_\varepsilon} \sum_{i=1}^{N_\varepsilon} \xi_i^\varepsilon \delta_{x_i^\varepsilon}$$

where the ξ_i^ε are of order 1. Then

$$\operatorname{curl} \frac{\gamma}{\varepsilon N_\varepsilon} = \lambda_\varepsilon.$$

If the x_i^ε are roughly equidistributed then λ_ε converges (in the weak* topology) to a uniform measure λ. Solving the equation $\operatorname{curl} \bar{\gamma} = \lambda$ we see that $\bar{\gamma}$ is of order 1. This again suggests that γ is of order $\varepsilon N_\varepsilon$. More generally one can show that the equation $\operatorname{curl} \gamma = \mu_\varepsilon$ has a solution such that $\frac{1}{\varepsilon N_\varepsilon}\gamma$ is globally bounded in the weak-L^2 space $L^{2,\infty}$. By removing small regions near the dislocations one can show that outside this regions one actually has a uniform bound in L^2 rather than $L^{2,\infty}$.

7.4 Main Result

7.4.1 Set-Up

Let $\Omega \subset \mathbb{R}^2$ be a simply connected, bounded, Lipschitz domain representing the horizontal cross section of an infinite cylindrical crystal. Let S be a set of admissible normalized Burgers vectors, e.g., $S = \{e_1, e_2\}$. Let

$$\mathbb{S} := \mathrm{Span}_{\mathbb{Z}} S = \{z_1 b_1 + z_2 b_2 : z_1, z_2 \in \mathbb{Z}\}.$$

For $0 < \varepsilon < 1$ let ρ_ε be such that

$$\lim_{\varepsilon \to 0} \frac{\rho_\varepsilon}{\varepsilon^s} = \infty \quad \forall s \in (0,1), \tag{7.27}$$

$$\lim_{\varepsilon \to 0} |\log \varepsilon| \rho_\varepsilon^2 = 0. \tag{7.28}$$

One possible choice is $\rho_\varepsilon = |\log \varepsilon|^{-1}$.

We denote by $\mathscr{M}(\Omega;\mathbb{R}^2)$ the set of $\mathbb{R}^2$-valued Radon measures on Ω with finite total mass. Given $\varepsilon > 0$ we define the set of admissible dislocation densities as

$$X_\varepsilon := \{\mu \in \mathscr{M}(\Omega;\mathbb{R}^2) : \mu = \sum_{i=1}^{M} \varepsilon \xi_i \delta_{x_i},\, M \in \mathbb{N},\, B_{\rho_\varepsilon}(x_i) \subset \Omega, \tag{7.29}$$

$$|x_j - x_k| \geq 2\rho_\varepsilon \text{ for every } j \neq k,\, \xi_i \in \mathbb{S} \setminus \{0\} \text{ for every } i\}. \tag{7.30}$$

For $\mu \in X_\varepsilon$ and $r > 0$ we define

$$\Omega_r(\mu) := \Omega \setminus \bigcup_{x_i \in \operatorname{supp}\mu} \overline{B_r(x_i)}. \tag{7.31}$$

Given $\mu \in X_\varepsilon$ the set of admissible strains is defined as

$$\mathscr{A}S_\varepsilon(\mu) := \{\beta \in L^2(\Omega;\mathbb{R}^{2\times 2}) : \beta = \mathrm{Id} \text{ in } B_\varepsilon(\operatorname{supp}\mu)$$
$$\operatorname{curl}\beta = 0 \text{ in } \Omega_\varepsilon(\mu), \quad \int_{\partial B_\varepsilon(y)} \beta t\, ds = \mu(y) \quad \forall y \in \operatorname{supp}\mu\}. \tag{7.32}$$

The choice $\beta = \mathrm{Id}$ in $B_\varepsilon(\operatorname{supp}\mu)$ is somewhat arbitrary (we could also set $\beta = 0$ in $B_\varepsilon(\operatorname{supp}\mu)$), but sufficient for our purposes. Note that for every measure $\mu \in X_\varepsilon$ there exists a unique finite set $\{x_1, \ldots, x_M\}$ and unique vectors $\xi_1, \ldots \xi_M$ such that $\mu = \sum_{i=1}^{M} \varepsilon \xi_i \delta_{x_i}$. Then the conditions in the definition of $\mathscr{A}S_\varepsilon(\mu)$ can be written more explicitly as

$$\beta = \mathrm{Id} \quad \text{in} \quad \bigcup_{i=1}^{M} B_\varepsilon(x_i), \qquad \int_{\partial B_\varepsilon(x_i)} \beta t\, ds = \varepsilon \xi_i \quad \forall i = 1, \ldots, M.$$

Let $W : \mathbb{R}^{2\times 2} \to [0,\infty]$ be an elastic energy density. Then we set

$$E_\varepsilon(\mu,\beta) := \int_\Omega W(\beta)\,dx.$$

Motivated by the heuristic considerations in Section 7.3 we define the rescaled energy functional as

$$\mathscr{E}_\varepsilon(\mu,\beta) := \begin{cases} \frac{1}{\varepsilon^2|\log\varepsilon|^2} E_\varepsilon(\mu,\beta) & \text{if } \mu \in X_\varepsilon,\ \beta \in \mathscr{AS}_\varepsilon(\mu), \\ +\infty & \text{otherwise in } \mathscr{M}(\Omega;\mathbb{R}^2)\times L^2(\Omega;\mathbb{R}^{2\times 2}). \end{cases} \tag{7.33}$$

The last ingredient in the set-up is a formula for the energy of a single dislocation and its relaxation. For $\xi \in \mathbb{R}^2$ and $\mathbb{C}$ which satisfies (7.1) we define

$$\psi(\xi) := \lim_{\delta\to 0} \frac{1}{|\log\delta|}\frac{1}{2}\int_{B_1\setminus\overline{B_\delta}} \mathbb{C}\eta_0\cdot\eta_0\,dx \tag{7.34}$$

where $\eta_0 : \mathbb{R}^2 \to \mathbb{R}^{2\times 2}$ is a distributional solution of

$$\begin{cases} \operatorname{curl}\eta = \xi\,\delta_0 & \text{in } \mathbb{R}^2, \\ \operatorname{div}\mathbb{C}\eta = 0 & \text{in } \mathbb{R}^2. \end{cases} \tag{7.35}$$

Note that two different solutions of (7.35) differ only by a smooth function and hence give the same limit in (7.34). Moreover there exists a unique -1 homogeneous distributional solution

$$\eta_{0,\xi}(x) = \frac{1}{r}\Gamma_\xi(\theta) \tag{7.36}$$

where (r,θ) are polar coordinates and the map $\xi \to \Gamma_\xi$ is linear (see, e.g., [BBS78]). For η_0 of the form (7.36) one sees immediately that the limit $\delta \to 0$ in (7.34) exists.

In [GLP10] it is shown that $\psi(\xi)$ corresponds to the minimal core energy of a dislocation, i.e.,

$$\psi(\xi) = \lim_{\varepsilon\to 0}\frac{1}{|\log\varepsilon|}\min\left\{\frac{1}{2}\int_{B_{\rho_\varepsilon}\setminus\overline{B_\varepsilon}} \mathbb{C}\eta\cdot\eta\,dx:\ \eta\in\mathscr{AS}_{\varepsilon,\rho_\varepsilon}(\xi)\right\}, \tag{7.37}$$

where

$$\mathscr{AS}_{\varepsilon,r}(\xi) := \left\{\eta \in L^2(B_r\setminus\overline{B_\varepsilon};\mathbb{R}^2) : \operatorname{curl}\eta = 0 \text{ in } B_r\setminus\overline{B_\varepsilon},\quad \int_{\partial B_\varepsilon}\eta\,t\,ds = \xi\right\}.$$

We want to approximate dislocation measures $\mu \in \mathscr{M}(\Omega;\mathbb{R}^2)$ which are absolutely continuous with respect to the Lebesgue measure by linear combinations of Dirac masses $\sum \varepsilon\xi_i\delta_{x_i}$ where $\xi_i \in \mathbb{S}$. Since there are many different ways to approximate a uniform measure $\xi\,dx$ it is natural to define the following relaxation of ψ. Here we also incorporate the effect of possible constant rotations which will emerge naturally in the proof. For $\xi \in \mathbb{R}^2$ and $R \in SO(2)$ define

$$\varphi(R,\xi) := \inf\left\{ \sum_{k=1}^{M} \lambda_k \psi(R^T \xi_k) : \sum_{k=1}^{M} \lambda_k \xi_k = \xi, M \in \mathbb{N}, \lambda_k \geq 0, \xi_k \in \mathbb{S} \right\}. \quad (7.38)$$

From this definition it follows easily that $\xi \mapsto \varphi(R,\xi)$ is convex and 1-homogeneous. Note also that ψ is 2-homogeneous and in particular $\psi(\xi) \geq c|\xi|^2$ with $c > 0$. From this one deduces easily that there exists a constant $\bar{M}$ (which only depends on the ratio c_2/c_1 of the constants in (7.1)) such that the minimum in the definition of $\varphi(R,\xi)$ does not change if we add the constraint $|\xi_k| \leq \bar{M}$. This argument also shows that the infimum in (7.38) is actually attained and we can write min instead of inf in (7.38).

7.4.2 Results

Theorem 2 (Compactness, see [MSZ15], Proposition 4.3). *Assume that*

$$\exists c > 0 \textit{ such that } \quad W(F) \geq c\,\mathrm{dist}^2(F, \mathrm{SO}(2)) \quad \forall F \in \mathbb{R}^{2\times 2}.$$

Assume that $\varepsilon_j \to 0$ *and let* $(\mu_j, \beta_j) \in \mathscr{M}(\Omega;\mathbb{R}^2) \times L^2(\Omega;\mathbb{R}^{2\times 2})$ *be such that*

$$\sup_j \mathscr{E}_j(\mu_j, \beta_j) < \infty.$$

Then

$$\sup_j (\|\beta_j\|_{L^2} + \|\mu_j\|_{\mathscr{M}}) < \infty. \quad (7.39)$$

Moreover there exist constant rotations $R_j \in \mathrm{SO}(2)$, *a measure* $\mu \in H^{-1}(\Omega;\mathbb{R}^2) \cap \mathscr{M}(\Omega;\mathbb{R}^2)$ *and a map* $\beta \in L^2(\Omega;\mathbb{R}^{2\times 2})$ *such that (after passage to subsequences)*

$$\frac{\mu_j}{\varepsilon_j |\log \varepsilon_j|} \overset{*}{\rightharpoonup} \mu \quad \textit{in } \mathscr{M}(\Omega;\mathbb{R}^2), \quad (7.40)$$

$$\frac{R_j^T \beta_j - \mathrm{Id}}{\varepsilon_j |\log \varepsilon_j|} \rightharpoonup \beta \quad \textit{in } L^2(\Omega;\mathbb{R}^{2\times 2}), \quad (7.41)$$

$$R_j \to R \quad \textit{in } \mathrm{SO}(2). \quad (7.42)$$

In addition we have

$$\mathrm{curl}\,\beta = R^T \mu.$$

Definition 1 ([MSZ15], Definition 4.5). Let $\varepsilon_j \to 0$. We say that a sequence of triplets $(\mu_j, \beta_j, R_j) \in \mathscr{M}(\Omega;\mathbb{R}^2) \times L^2(\Omega;\mathbb{R}^{2\times 2}) \times \mathrm{SO}(2)$ converges to a triplet (μ, β, R) if (7.40)–(7.42) hold.

We extend $\mathscr{E}_\varepsilon$ trivially to triplets, i.e., we set

$$\mathscr{E}_\varepsilon(\mu, \beta, R) := \mathscr{E}_\varepsilon(\mu, \beta) \quad \forall R \in \mathrm{SO}(2).$$

Theorem 3. *Assume that* $W : \mathbb{R}^{2\times 2} \to [0,\infty]$ *satisfies*

$$W(RF) = W(F) \quad \forall R \in \mathrm{SO}(2), \quad (7.43)$$

$$W \text{ is } C^2 \text{ in a neighbourhood of } \mathrm{SO}(2) \text{ and } W(\mathrm{Id}) = 0, \tag{7.44}$$

$$\exists c > 0 \text{ such that } \quad W(F) \geq c\,\mathrm{dist}^2(F, \mathrm{SO}(2)) \quad \forall F \in \mathbb{R}^{2\times 2}, \tag{7.45}$$

$$\text{for all } \delta > 0 \quad \sup\{W(F) : |F| \leq \delta^{-1}, \det F > \delta\} < \infty. \tag{7.46}$$

Then the functionals $\mathscr{E}_\varepsilon$ are Γ-convergent, with respect to the convergence in Definition 1, to a functional $\mathscr{E}$ on $\mathscr{M}(\Omega;\mathbb{R}^2) \times L^2(\Omega;\mathbb{R}^{2\times 2}) \times \mathrm{SO}(2)$ given by

$$\mathscr{E}(\mu,\beta,R) := \begin{cases} \frac{1}{2}\int_\Omega \mathbb{C}\beta\cdot\beta + \int_\Omega \varphi\left(R, \frac{d\mu}{d|\mu|}\right) d|\mu| & \text{if } \mu \in H^{-1}(\Omega;\mathbb{R}^2) \cap \mathscr{M}(\Omega;\mathbb{R}^2) \\ & \text{and } \mathrm{curl}\,\beta = R^T\mu, \\ +\infty & \text{otherwise,} \end{cases} \tag{7.47}$$

where $\mathbb{C} := D^2W(\mathrm{Id})$ and where φ is given by (7.38) *and* (7.37).
More specifically the following two inequalities hold true.

Γ-$\liminf$ *inequality:*
If $(\mu,\beta,R) \in \mathscr{M}(\Omega;\mathbb{R}^2) \times L^2(\Omega;\mathbb{R}^{2\times 2}) \times \mathrm{SO}(2)$, if $\varepsilon_j \to 0$ and if (μ_j,β_j,R_j) converge to the triplet (μ,β,R) then

$$\liminf_{j\to\infty} \mathscr{E}_{\varepsilon_j}(\mu_j,\beta_j,R_j) \geq \mathscr{E}(\mu,\beta,R).$$

Γ-$\limsup$ *inequality:*
If $(\mu,\beta,R) \in \mathscr{M}(\Omega;\mathbb{R}^2) \times L^2(\Omega;\mathbb{R}^{2\times 2}) \times \mathrm{SO}(2)$ with $\mathrm{curl}\,\beta = R^T\mu$ and if $\varepsilon_j \to 0$ then there exist (μ_j,β_j,R_j) which converge to the triplet (μ,β,R) and

$$\limsup_{j\to\infty} \mathscr{E}_{\varepsilon_j}(\mu_j,\beta_j,R_j) \leq \mathscr{E}(\mu,\beta,R).$$

In [MSZ15] the assumption (7.46) was replaced by the stronger assumption

$$W(F) \leq C\,\mathrm{dist}^2(F, \mathrm{SO}(2)),$$

for some $C > 0$. This makes it easier to handle the exceptional sets where the β_j are large, but for a usual elastic material we expect that the energy blows up for infinite compression, i.e., if $\det F \downarrow 0$. Therefore (7.46) is more realistic. Note that we also allow $W(F)$ to take the value ∞. This makes it possible to incorporate a constraint like $\det F > 0$ in the definition of the energy.

An interesting technical question is whether in the definition of the class of admissible measure X_ε the separation condition

$$|x_j - x_k| \geq 2\rho_\varepsilon$$

with ρ_ε satisfying $\rho_\varepsilon/\varepsilon^s \to \infty$ as $\varepsilon \to 0$ for *all* $s \in (0,1)$ can be relaxed. For the compactness result it certainly suffices to assume that $\rho_\varepsilon \geq C\varepsilon^s$ for *some* $s \in (0,1)$, but whether this suffices for a Gamma-convergence result is open in the nonlinear setting. In the case of linear elasticity, and in the *diluted*, core energy dominated regime $N_\varepsilon \ll |\log\varepsilon|$, De Luca et al. [DLGP12] relaxed the separation condition by extending the so called 'ball construction' for Ginzburg-Landau functionals [San98, San00, Jer99, SS11] to the setting of elasticity. Loosely speaking the Ginzburg-Landau functional corresponds to the toy energy $\int_\Omega |\beta|^2\,dx$ where β is now a vector field (not a matrix field) and thus the Burgers vector is replaced by scalar and the lattice $\mathbb{S} \subset \mathbb{R}^2$ is replaced by $2\pi\mathbb{Z} \subset \mathbb{R}$. Thus for the Ginzburg-Landau functionals Korn's inequality is not needed.

7.5 Ideas of Proof

Proof (of Theorem 2). The compactness result follows from the lower bound (7.25) and the new rigidity estimate (7.15). Indeed if we apply (7.25) around each of the N_{ε_j} points $x_i^{\varepsilon_j} \in \operatorname{supp}\mu_j$ we get

$$\mathscr{E}_j(\mu_j,\beta_j) \geq \frac{c}{\varepsilon_j^2|\log\varepsilon_j|^2}\log\left(\frac{\rho_{\varepsilon_j}}{\varepsilon_j}\right)\varepsilon_j^2\sum_{i=1}^{N_{\varepsilon_j}}|\xi_i^{\varepsilon_j}|^2 \geq \frac{c}{|\log\varepsilon_j|}\sum_{i=1}^{N_{\varepsilon_j}}|\xi_i^{\varepsilon_j}|^2.$$

The $\xi_i^{\varepsilon_j}$ belong to $\mathbb{S}\setminus\{0\}$ and are thus bounded from below. Hence

$$\|\mu_j\|_{\mathscr{M}} = \varepsilon_j\sum_{i=1}^{N_{\varepsilon_j}}|\xi_i^{\varepsilon_j}| \leq C\varepsilon_j\sum_{i=1}^{N_{\varepsilon_j}}|\xi_i^{\varepsilon_j}|^2 \leq C\varepsilon_j|\log\varepsilon_j|\,\mathscr{E}_j(\mu_j,\beta_j).$$

This proves the bound for $\frac{1}{\varepsilon_j|\log\varepsilon_j|}\mu_j$ and hence the weak* compactness. Since $|\xi_i^{\varepsilon_j}|$ is bounded from below we also obtain a bound on the number N_{ε_j} of dislocations

$$N_{\varepsilon_j} \leq C\frac{1}{\varepsilon_j}\|\mu_j\|_{\mathscr{M}} \leq C|\log\varepsilon_j|. \tag{7.48}$$

For β_j we have by (7.45)

$$\int_\Omega \operatorname{dist}^2(\beta_j,\mathrm{SO}(2))\,dx \leq C\varepsilon_j^2|\log\varepsilon_j|^2\,\mathscr{E}_j(\mu_j,\beta_j)$$

and the L^2 bound for β_j follows from (7.15). Actually one has to be a little bit careful here since $\operatorname{curl}\beta_j$ and μ_j are closely related but not identical. It can be shown, however, that there exists $\tilde\beta_j$ such that $\|\operatorname{curl}\tilde\beta_j\|_{\mathscr{M}} = \|\mu_j\|_{\mathscr{M}}$ and $\|\beta_j - \tilde\beta_j\|_{L^2}$ can be controlled, see [MSZ15], pp. 1380–1382. In this argument one uses (7.24). □

Proof (of Theorem 3). Γ-lim inf inequality: Using the Taylor expansion

$$W(\beta_j) = W(R_j^T \beta_j) \approx \frac{1}{2}\mathbb{C}(R_j^T \beta_j - \mathrm{Id}) \cdot (R_j^T \beta_j - \mathrm{Id})$$

one can essentially reduce the proof to the argument in the linear theory [GLP10]. One subtlety is that the convergence (7.41) does not guarantee that $R_j^T \beta_j - \mathrm{Id}$ converges uniformly to zero and therefore Taylor expansion cannot be used at all points. To get the lower bound by the linear interaction energy $\int_\Omega \mathbb{C}\beta \cdot \beta \, dx$ we can use that the measure of the set where $|R_j^T \beta_j - \mathrm{Id}| \geq \varepsilon_j^{1/2}$ goes to zero and argue as in [FJM02] to apply Taylor expansion as if we had uniform convergence. For this argument we can actually also remove the set $\bigcup_{i=1}^{N_{\varepsilon_j}} B_{\rho_{\varepsilon_j}}(x_i^{\varepsilon_j})$ since by (7.48) and (7.28) the measure of this set goes to zero. For further details, see [MSZ15], pp. 1385–1387.

To show that the limes inferior of $\int_{\Omega_{\varepsilon_j} \setminus \Omega_{\rho_{\varepsilon_j}}} W(\beta_j) \, dx$ is bounded from below by $\int_\Omega \varphi\left(R, \frac{d\mu}{d|\mu|}\right) d|\mu|$ one establishes a nonlinear counterpart of (7.37), using again Taylor expansion with the exception of a small set, see [MSZ15], pp. 1386–1387 and [SZ12], Proposition 3.11 for the details.

Γ-lim sup inequality: Here we provide more details since compared to [MSZ15] we no longer assume an upper bound $W(F) \leq C\mathrm{dist}^2(F, \mathrm{SO}(2))$. The proof still follows closely the argument in [MSZ15]. The main difference is that we change the definition of the recovery sequence in the regions $B_{L\varepsilon_j}(x_i^{\varepsilon_j}) \setminus B_{\varepsilon_j}(x_i^{\varepsilon_j})$ to ensure that the nonlinear energy remains controlled. Specifically we use the following results.

Lemma 1 (Existence of a core field which is compatible with the nonlinear energy). *For every $\xi \in \mathbb{R}^2$ there exists a -1 homogeneous smooth map $\gamma : \mathbb{R}^2 \setminus \{0\} \to \mathbb{R}^{2\times 2}$ such that*

$$\mathrm{curl}\, \gamma = 0 \quad \text{in} \quad \mathbb{R}^2 \setminus \{0\}, \qquad \int_{\partial B_r(0)} \gamma t \, ds = \xi \quad \forall r > 0, \tag{7.49}$$

$$|\gamma(x)| \leq \frac{1}{|x|} |\xi| \quad \forall x \in \mathbb{R}^2 \setminus \{0\} \tag{7.50}$$

and

$$\det(\mathrm{Id} + \gamma) \geq 1. \tag{7.51}$$

Proof (of Lemma 1). We first show that it suffices to prove the result for vectors ξ of the form ae_1 with $a > 0$. To see this, note that if $R \in \mathrm{SO}(2)$ and $\hat{\gamma}(x) = \gamma(Rx)R$ then

$$\mathrm{curl}\, \hat{\gamma}(x) = \mathrm{curl}\, \gamma(Rx) \quad \text{and} \quad \int_{\partial B_r} \hat{\gamma} t \, ds = \int_{\partial B_r} \gamma t \, ds.$$

Thus if γ has the desired properties for $\xi = ae_1$ then $\tilde{\gamma}(x) = R^T \hat{\gamma}(x)$ has the desired properties for $\xi = R^T(ae_1)$ because

$$\det(\mathrm{Id} + \tilde{\gamma}(x)) = \det(R^T(\mathrm{Id} + \gamma(Rx))R) = \det(\mathrm{Id} + \gamma(Rx)) \geq 1.$$

Thus assume that $\xi = ae_1$ with $a > 0$. We will define γ using Polar coordinates. Let $\delta \in (0, \frac{1}{2})$ and let $g : \mathbb{R} \to \mathbb{R}$ be a smooth 2π-periodic function such that

$$
\begin{aligned}
g &= 0 \quad \text{in } [0,\pi+\delta]\cup[2\pi-\delta,2\pi],\\
0 &\le g \le 1,\\
\int_0^{2\pi} g(t)\,dt &= 1.
\end{aligned}
$$

Set

$$G(t) := \int_0^t g(s)\,ds \quad \text{for } t \in [0,2\pi].$$

With $x = (r\cos\theta, r\sin\theta)$ define

$$\gamma(x) := ae_1 \otimes \frac{1}{r^2} g(\theta)x^{\perp} = ag(\theta)\frac{1}{r^2}\begin{pmatrix} -x_2 & x_1 \\ 0 & 0 \end{pmatrix}.$$

Note that γ is well-defined and smooth in $\mathbb{R}^2 \setminus \{0\}$ since g is 2π-periodic. We also have $|\gamma| \le a/r = |\xi|/r$. Moreover for $\theta \neq 0$

$$\gamma(x) = ae_1 \otimes g(\theta)\nabla\theta = ae_1 \otimes \nabla(G\circ\theta)$$

and hence $\operatorname{curl}\gamma = 0$ in the region in $\mathbb{R}^2 \setminus [0,\infty)\times\{0\}$. Since γ is smooth it follows that $\operatorname{curl}\gamma = 0$ in $\mathbb{R}^2\setminus\{0\}$. Moreover

$$\int_{\partial B_r} \gamma t\,ds = ae_1[G(2\pi) - G(0)] = ae_1.$$

The explicit form of γ gives

$$\det(\mathrm{Id}+\gamma) = 1 - \frac{a}{r^2} g(\theta)x_2.$$

Now $g(\theta) = 0$ for $\theta \in [0,\pi]$ and hence $g(\theta) = 0$ if $x_2 > 0$. Thus $\det(\mathrm{Id}+\gamma) \ge 1$. □

Lemma 2 (Interpolation between core field and far field). *Assume that $\mathbb{C}$ is a linear symmetric operator on $\mathbb{R}^{2\times 2}$ which satisfies*

$$c_1|\operatorname{sym}\xi|^2 \le \mathbb{C}\xi\cdot\xi \le c_2|\operatorname{sym}\xi|^2$$

with $c_1 > 0$. Then there exists a constant $\bar{C} \ge 1$, which only depends on c_1 and c_2, with the following property. Let $\xi \in \mathbb{R}^2$ and as in (7.36) let $\eta_{0,\xi}(x) = \frac{1}{r}\Gamma_\xi(\theta)$ be the -1 homogeneous distributional solution of

$$\operatorname{curl}\eta_{0,\xi} = \xi\delta_0, \qquad \operatorname{div}\mathbb{C}\eta_{0,\xi} = 0 \qquad \text{in } \mathbb{R}^2.$$

Let $L \ge 1$. Then there exists a smooth map $\eta_\xi : \mathbb{R}^2\setminus\{0\} \to \mathbb{R}^{2\times 2}$ such that

$$\operatorname{curl} \eta_\xi = 0 \quad \text{in } \mathbb{R}^2 \setminus \{0\}, \tag{7.52}$$

$$\int_{\partial B_r} \eta_\xi \, t \, ds = \xi \quad \forall r > 0, \tag{7.53}$$

$$\eta_\xi = \eta_{0,\xi} \quad \text{in } \mathbb{R}^2 \setminus B_{2L}(0), \tag{7.54}$$

$$|\eta_\xi(x)| \leq \frac{\bar{C}}{|x|}|\xi| \quad \text{in } \mathbb{R}^2 \setminus B_L(0), \tag{7.55}$$

$$|\eta_\xi(x)| \leq \frac{1}{|x|}|\xi| \quad \text{in } B_L(0) \setminus \{0\}, \tag{7.56}$$

$$\det(\mathrm{Id} + \eta_\xi) \geq 1 \quad \text{in } B_L(0) \setminus \{0\}. \tag{7.57}$$

Proof (of Lemma 2). Since Γ_ξ is smooth by elliptic regularity and $\xi \to \Gamma_\xi$ is linear there exists a constant C' (which depends only on c_1 and c_2) such that

$$|\eta_{0,\xi}(x)| \leq \frac{C'}{|x|}|\xi|.$$

Let γ be the function in Lemma 1. Then

$$\operatorname{curl}(\gamma - \eta_{0,\xi}) = 0 \quad \text{in } \mathbb{R}^2 \setminus \{0\}, \qquad \int_{\partial B_r(0)} (\gamma - \eta_{0,\xi}) \, t \, ds = 0 \quad \forall r > 0.$$

Thus there exists a smooth function $u : \mathbb{R}^2 \setminus \{0\} \to \mathbb{R}^2$ such that

$$\gamma - \eta_{0,\xi} = \nabla u.$$

We have $|\nabla u| \leq (1 + C')|\xi|/|x|$ and we may assume that $u(Le_1) = 0$. Thus

$$|u| \leq (1 + C')(\pi + 1)|\xi| \quad \text{in } B_{2L}(0) \setminus B_L(0).$$

Let $\alpha \in C_0^\infty(\mathbb{R}^2)$ be a cut-off function with $\operatorname{supp} \alpha \subset B_{2L}(0)$, $0 \leq \alpha \leq 1$, $\alpha = 1$ on $B_L(0)$, $|\nabla \alpha| \leq 2/L$ and define

$$\eta_\xi := \eta_{0,\xi} + \nabla(\alpha u) \quad \text{on } \mathbb{R}^2 \setminus \{0\}.$$

Thus (7.52) and (7.53) follow. Moreover $\eta_\xi = \gamma$ in $B_L(0)$ and $\eta_\xi = \eta_{0,\xi}$ in $\mathbb{R}^2 \setminus B_{2L}(0)$. This yields (7.54), (7.56) and (7.57). Moreover the bound (7.55) holds for $x \notin B_{2L}(0) \setminus B_L(0)$. To verify that (7.55) holds true also in $B_{2L}(0) \setminus B_L(0)$ note that

$$\eta_\xi = \alpha\gamma + (1 - \alpha)\eta_{0,\xi} + u \otimes \nabla \alpha.$$

Now the bounds on γ, $\eta_{0,\xi}$ and u imply that for $x \in B_{2L}(0) \setminus B_L(0)$

$$|\eta_\xi(x)| \leq \frac{(1 + C')}{L}(2\pi + 3)|\xi|.$$

Thus (7.55) holds with $\bar{C} = (1 + C')(2\pi + 3)$. □

Proof of Γ-lim sup inequality (continued):
Step 1. We show the result first for $\beta \in L^\infty(\Omega;\mathbb{R}^{2\times 2})$ and piecewise constant measures of the form

$$\mu = \xi\, \chi_U\, dx$$

where dx denotes the Lebesgue measure, U is a cube with $\overline{U} \subset \Omega$ and χ_U is the characteristic function of U. To simplify the notation slightly we only consider the case $R = \mathrm{Id}$.

As in [MSZ15] we first undo the effect of the minimization in the definition of φ and reduce the problem to vectors ξ_k in the lattice $\mathbb{S}$. By definition (7.38) of φ and the comments following (7.38) there exist $M \in \mathbb{N}$, $\lambda_k \geq 0$ and $\xi_k \in \mathbb{S}\setminus\{0\}$ such that

$$\varphi(\mathrm{Id},\xi) = \sum_{k=1}^{M} \lambda_k \psi(\xi_k), \qquad \xi = \sum_{i=k}^{M} \lambda_k \xi_k \quad \text{and} \quad |\xi_k| \leq \bar{M},$$

where $\bar{M}$ depends only on $\mathbb{C} = D^2W(\mathrm{Id})$. We set

$$\Lambda := \sum_{k=1}^{M} \lambda_k, \quad r_j := \frac{1}{2\sqrt{\Lambda |\log \varepsilon_j|}}.$$

Since $\varphi(\mathrm{Id},\xi) \leq C|\xi|$ and $\psi(\xi_k) \geq c|\xi_k|^2 \geq c' > 0$ we see that $\Lambda \leq C$ where C depends only on $\mathbb{C}$ (or, more generally, only on the constants c_1 and c_2 in (7.1)). Note also that by (7.28)

$$r_j \gg \rho_{\varepsilon_j}.$$

By [GLP10], Lemma 14, there exists a sequence of admissible measures $\mu_j \in X_{\varepsilon_j}$ of the form

$$\mu_j = \sum_{k=1}^{M} \varepsilon_j \xi_k \mu_j^k, \quad \text{where} \quad \mu_j^k = \sum_{i=1}^{M_k^j} \delta_{x_{i,k}^j}$$

with the following properties:

$$B_{r_j}(x_{i,k}^j) \subset U, \qquad |x_{i,k}^j - x_{i',k'}^j| \geq 2r_j \quad \text{if } (i,k) \neq (i',k')$$

and

$$\frac{\mu_j^k}{|\log \varepsilon_j|} \overset{*}{\rightharpoonup} \lambda_k \chi_U\, dx \quad \text{in } \mathcal{M}(\Omega,\mathbb{R}^2) \quad \text{for } k = 1,\dots,M \text{ as } j \to \infty, \tag{7.58}$$

$$\frac{\mu_j}{\varepsilon_j |\log \varepsilon_j|} \overset{*}{\rightharpoonup} \mu \quad \text{in } \mathcal{M}(\Omega,\mathbb{R}^2) \text{ as } j \to \infty. \tag{7.59}$$

In the following it is useful to combine the two summations in the definition of μ_j into a single sum and to rewrite μ_j as $\mu_j = \sum_{i=1}^{M_j} \varepsilon_j \xi_i^j \delta_{x_i^j}$.

By assumption there exists a $\delta > 0$ such that W is C^2 in the set $\{F : \mathrm{dist}(F,\mathrm{SO}(2)) < 3\delta\}$. It follows that

$$W(F) \le C\mathrm{dist}^2(F, \mathrm{SO}(2)) \qquad \text{if } \mathrm{dist}(F, \mathrm{SO}(2)) \le 2\delta \tag{7.60}$$

with

$$C = \max\left\{\frac{1}{2}\|D^2W(G)\| : \mathrm{dist}(G, \mathrm{SO}(2)) \le 2\delta\right\}. \tag{7.61}$$

Let $\bar{C}$ denote the constant in Lemma 2 and choose

$$L = \max\left(\frac{\bar{C}\bar{M}}{\delta}, 1\right)$$

in Lemma 2. Then the function η_ξ in Lemma 2 satifies

$$|\eta_\xi| \le \delta \quad \text{in } \mathbb{R}^2 \setminus B_L(0) \qquad \text{whenever } |\xi| \le \bar{M}.$$

Define

$$\eta_i^j(x) := \eta_{\xi_i^j}\left(\frac{x - x_i^j}{\varepsilon_j}\right)\alpha\left(\frac{x - x_i^j}{r_j}\right), \qquad \eta^j = \sum_{i=1}^{M_j} \eta_i^j,$$

where $\alpha \in C_0^\infty(\mathbb{R}^2)$ with $\mathrm{supp}\,\alpha \subset B_1(0)$ is a standard cut-off function, i.e., $0 \le \alpha \le 1$, $\alpha = 1$ in $B_{1/2}(0)$ and $|\nabla\alpha| \le 4$. By the properties of $\eta_{\xi_i^j}$ and the change of variables $x \mapsto x_i^j + \varepsilon_j x$ we have

$$\int_{\partial B_{\varepsilon_j}(x_i^j)} \eta_i^j t\, ds = \varepsilon_j \int_{\partial B_1(0)} \eta_{\xi_i^j} t\, ds = \varepsilon_j \xi_i^j.$$

Moreover $\mathrm{curl}\,\eta_i^j$ vanishes outside $\{x_i^j\} \cup B_{r_j}(x_i^j) \setminus B_{r_j/2}(x_i^j)$ and

$$\int_{B_{r_j}(x_i^j) \setminus B_{\varepsilon_j}(x_i^j)} \mathrm{curl}\,\eta_i^j\, dx = -\int_{\partial B_{r_j/2}(x_i^j)} \eta_i^j t\, ds = -\varepsilon_j \xi_i^j, \tag{7.62}$$

$$|\mathrm{curl}\,\eta_i^j(x)| \le \frac{C}{r_j^2}\varepsilon_j|\xi_i^j| \le C\varepsilon_j|\log\varepsilon_j| \quad \text{for } x \ne x_i^j. \tag{7.63}$$

Set

$$v^j := \chi_{\Omega_{\varepsilon_j}}\, \mathrm{curl}\,\eta^j. \tag{7.64}$$

We would like to define $\beta_j \approx \mathrm{Id} + \eta^j + \varepsilon_j|\log\varepsilon_j|\beta$, so that η^j captures the behaviour near the dislocations and β the behaviour away from the dislocations. The right hand side is, however, not curl-free in Ω_{ε_j}, indeed $\mathrm{curl}(\eta^j + \varepsilon_j|\log\varepsilon_j|\beta) = v^j + \varepsilon_j|\log\varepsilon_j|\mu$. We will show below that

$$\frac{1}{\varepsilon_j|\log\varepsilon_j|}v^j + \mu \overset{*}{\rightharpoonup} 0 \quad \text{in } L^\infty(\Omega; \mathbb{R}^2). \tag{7.65}$$

We define

$$\beta_j := \begin{cases} \mathrm{Id} + \eta^j + \varepsilon_j |\log \varepsilon_j| \beta + \tilde{\beta}_j & \text{in } \overline{\Omega_{\varepsilon_j}} \\ \mathrm{Id} & \text{in } \bigcup_i B_{\varepsilon_j}(x_i^j), \end{cases} \tag{7.66}$$

with $\tilde{\beta}_j = \nabla w_j J$, where J is the 90 degree anticlockwise rotation and w_j is the solution of

$$\begin{cases} \Delta w_j = \varepsilon_j |\log \varepsilon_j| \mu + \nu^j & \text{in } \Omega, \\ w_j \in H_0^1(B_R(0)) \end{cases} \tag{7.67}$$

for some ball $B_R(0) \supset \Omega$. Now $\operatorname{curl} \tilde{\beta}_j = -\Delta w_j$ and thus

$$\operatorname{curl} \beta_j = 0 \quad \text{in } \Omega_{\varepsilon_j}$$

and

$$\operatorname{curl}(\varepsilon_j |\log \varepsilon_j| \beta + \tilde{\beta}_j) = \varepsilon_j |\log \varepsilon_j| \mu_j - \varepsilon_j |\log \varepsilon_j| \mu_j - \nu^j = 0 \quad \text{in } \bigcup_i B_{\varepsilon_j}(x_i^j).$$

It follows that

$$\int_{\partial B_{\varepsilon_j}(x_i^j)} \beta_j t \, ds = \int_{\partial B_{\varepsilon_j}(x_i^j)} \eta_i^j t \, ds + \int_{B_{\varepsilon_j}(x_i^j)} \operatorname{curl}(\varepsilon_j |\log \varepsilon_j| \beta + \tilde{\beta}_j) \, dx = \varepsilon_j \xi_i^j,$$

thus $\beta_j \in \mathscr{A}S_{\varepsilon_j}(\mu_j)$.

We next show that

$$\frac{\beta_j - \mathrm{Id}}{\varepsilon_j |\log \varepsilon_j|} \rightharpoonup \beta \quad \text{in } L^2(\Omega; \mathbb{R}^{2\times 2}). \tag{7.68}$$

Since $|\eta_\xi|(x) \leq C|\xi|/|x|$ and since the ξ_i^j are uniformly bounded it follows that

$$\|\eta^j\|_{L^1} \leq C M_j \varepsilon_j r_j \leq C |\log \varepsilon_j| \, \varepsilon_j r_j, \quad \|\eta^j\|_{L^2(\Omega_{\varepsilon_j})}^2 \leq C M_j \varepsilon_j^2 |\log \varepsilon_j| \leq C |\log \varepsilon_j|^2 \, \varepsilon_j^2.$$

Thus $\eta^j \chi_{\Omega_{\varepsilon_j}} / (\varepsilon_j |\log \varepsilon_j|)$ is bounded in L^2 and converges to zero in L^1. Hence

$$\frac{\eta^j}{\varepsilon_j |\log \varepsilon_j|} \rightharpoonup 0 \quad \text{in } L^2(\Omega; \mathbb{R}^{2\times 2}).$$

By elliptic regularity, from (7.65) and (7.67) we get $w_j / (\varepsilon_j |\log \varepsilon_j|) \rightharpoonup 0$ in $W^{2,p}(B_R(0))$ for all $p < \infty$. Hence by the compact Sobolev embedding $W^{1,p} \hookrightarrow C^0$ (for $p > 2$)

$$\frac{\tilde{\beta}_j}{\varepsilon_j |\log \varepsilon_j|} \to 0 \quad \text{uniformly in } \Omega. \tag{7.69}$$

Together with the weak convergence of $\frac{\eta^j \chi_{\Omega_{\varepsilon_j}}}{\varepsilon_j |\log \varepsilon_j|}$ this implies (7.68).

Before we turn to the upper bound for $\mathscr{E}_{\varepsilon_j}(\mu_j, \beta_j, \mathrm{Id})$ we first prove (7.65). By the assumption $\mu \in L^\infty$ and by (7.63) we know that $\frac{1}{\varepsilon_j |\log \varepsilon_j|} \nu^j + \mu$ is bounded in L^∞. It thus suffices to show that

$$\left\langle \frac{1}{\varepsilon_j|\log\varepsilon_j|}\nu^j+\mu,\Phi\right\rangle = \frac{1}{\varepsilon_j|\log\varepsilon_j|}\langle \nu^j+\mu_j,\Phi\rangle + \left\langle \mu - \frac{1}{\varepsilon_j|\log\varepsilon_j|}\mu_j,\Phi\right\rangle \to 0$$

for every Lipschitz continuous function Φ. The second term converges to zero by (7.59). To estimate the first term we use (7.62). This yields

$$\begin{aligned}
|\langle \nu^j+\mu_j,\Phi\rangle| &= \left|\sum_i \int_{B_{r_j}(x_i^j)\setminus B_{\varepsilon_j}(x_i^j)} \operatorname{curl}\eta_i^j(x)\Phi(x)\,dx + \varepsilon_j\xi_i^j\Phi(x_i^j)\right| \\
&\underset{(7.62)}{=} \left|\sum_i \int_{B_{r_j}(x_i^j)\setminus B_{\varepsilon_j}(x_i^j)} \operatorname{curl}\eta_i^j(x)(\Phi(x)-\Phi(x_i^j))\,dx\right| \\
&\underset{(7.63)}{\leq} C\sum_i \varepsilon_j|\xi_i^j|\, r_j \operatorname{Lip}\Phi \leq C r_j \operatorname{Lip}\Phi\,\|\mu_j\|_{\mathscr{M}}
\end{aligned}$$

and the assertion follows since $r_j \to 0$.

The upper bound

$$\limsup_{j\to\infty} \mathscr{E}_{\varepsilon_j}(\mu_j,\beta_j,\mathrm{Id}) \leq \mathscr{E}(\mu,\beta,\mathrm{Id})$$

can now be proved as in [MSZ15] if one uses the following observations

- In the core regions $B_{L\varepsilon_j}(x_i^j)\setminus B_{\varepsilon_j}(x_i^j)$ we have $\det(\mathrm{Id}+\eta^j)\geq 1$ and $|\eta^j|\leq C$. Since $\beta\in L^\infty$ and since $\tilde\beta_j$ converges uniformly to zero we also have $\det\beta_j\geq\frac{1}{2}$ and $|\beta_j|\leq C$ in this region if j is sufficiently large. Thus by assumption (7.46) we have $W(\beta_j)\leq C$ and hence

$$\int_{B_{L\varepsilon_j}(x_i^j)\setminus B_{\varepsilon_j}(x_i^j)} W(\beta_j)\,dx \leq CL^2\varepsilon_j^2$$

  which implies that

$$\frac{1}{\varepsilon_j^2|\log\varepsilon_j|^2}\sum_i \int_{B_{L\varepsilon_j}(x_i^j)\setminus B_{\varepsilon_j}(x_i^j)} W(\beta_j)\,dx = 0.$$

- In the region $\Omega_{L\varepsilon_j}(\mu_j)$ we have $|\eta^j|\leq\delta$ and hence $|\beta_j-\mathrm{Id}|\leq 2\delta$ if j is large enough. Thus in this region we can use the estimate (7.60) which was assumed in [MSZ15].
- Finally the construction above differs slightly from the one in [MSZ15] in the region $B_{r_j}(x_i^j)\setminus B_{r_j/2}(x_i^j)$ (cut-off with a smooth function α vs. cut-off with a characteristic function). Since $|\eta_\xi(x)|\leq C|\xi|/|x|$ the total energy contribution from these regions is bounded by $|\log\varepsilon_j|\varepsilon_j^2 \ll |\log\varepsilon_j|^2\varepsilon_j^2$ and hence irrelevant.

If we define $\mathscr{E}_{\varepsilon_j}(\mu,\beta,\mathrm{Id};U)$ and $\mathscr{E}(\mu,\beta,\mathrm{Id};U)$ by replacing Ω by U we also get

$$\limsup_{j\to\infty} \mathscr{E}_{\varepsilon_j}(\mu_j,\beta_j,\mathrm{Id};U) \le \mathscr{E}(\mu,\beta,\mathrm{Id};U).$$

In fact the same arguments as above show that if in addition $\frac{1}{\varepsilon_j|\log\varepsilon_j|}\alpha_j \to 0$ uniformly in U and $\operatorname{curl}\alpha_j = 0$ in U then

$$\limsup_{j\to\infty} \mathscr{E}_{\varepsilon_j}(\mu_j,\beta_j+\alpha_j,\mathrm{Id};U) \le \mathscr{E}(\mu,\beta,\mathrm{Id};U). \tag{7.70}$$

Step 2. Assume that $\beta \in L^\infty$ and that there exist finitely many cubes U_k with disjoint interior such that

$$\mu = \sum_{k=1}^{K} \mu_k, \qquad \mu_k = \xi_k\, \chi_{U_k}\, dx.$$

Let $\mu_{k,j}$, η_k^j and $\tilde\beta_{k,j}$ be the sequences corresponding to μ_k constructed in Step 1. Note that all sequences depend only on μ_k and not on β. Then

$$\eta_k^j = 0 \quad \text{in } \Omega\setminus U_k, \tag{7.71}$$

$$\lim_{j\to\infty} \frac{1}{\varepsilon_j|\log\varepsilon_j|}\sup|\tilde\beta_{k,j}| = 0 \quad \forall k\in\{1,\dots,K\}, \tag{7.72}$$

$$\operatorname{curl}\tilde\beta_{k,j} = 0 \quad \text{in } \Omega\setminus U_k. \tag{7.73}$$

Let

$$\beta_j := \mathrm{Id} + \varepsilon_j|\log\varepsilon_j|\beta + \sum_{k=1}^{K}\left(\eta_k^j + \tilde\beta_{k,j}\right).$$

Then $\beta_j \in \mathscr{A}S_{\varepsilon_j}(\mu_j)$ (here we use Step 1 as well as (7.71) and (7.73)). Since $\eta_k^j \rightharpoonup 0$ in L^2 as $j\to\infty$ we get from (7.72)

$$\frac{\beta_j - \mathrm{Id}}{\varepsilon_j|\log\varepsilon_j|} \rightharpoonup \beta \quad \text{in } L^2(\Omega;\mathbb{R}^{2\times 2}).$$

Using (7.72) and (7.70) we get

$$\limsup_{j\to\infty} \mathscr{E}_{\varepsilon_j}(\mu_j,\beta_j,\mathrm{Id};U_k) \le \mathscr{E}(\mu,\beta,\mathrm{Id};U_k)$$

for all $k = 1,\dots K$. Moreover by (7.71) and (7.72) and Taylor expansion

$$\lim_{j\to\infty} \mathscr{E}_{\varepsilon_j}(\mu_j,\beta_j,\mathrm{Id};\Omega\setminus\bigcup_{k=1}^{K}U_k) = \int_{\Omega\setminus\bigcup_{k=1}^{K}U_k}\frac{1}{2}\mathbb{C}\beta\cdot\beta\,dx = \mathscr{E}(\mu,\beta,\mathrm{Id};\Omega\setminus\bigcup_{k=1}^{K}U_k)$$

since $|\mu|(\Omega\setminus\bigcup_{k=1}^{K}U_k) = 0$. Thus

$$\limsup_{j\to\infty} \mathscr{E}_{\varepsilon_j}(\mu_j, \beta_j, \mathrm{Id}) \leq \mathscr{E}(\mu, \beta, \mathrm{Id}).$$

Step 3. The general case follows by standard approximation and diagonalization arguments, see [MSZ15] for the details. In general one has to be careful in using diagonalization arguments in connection with the weak or weak* topology since these topologies are not metrizable on the whole space but only on bounded sets. We can, however, use the bound (7.39) to ensure that we work in bounded sets of L^2 and $\mathscr{M}$, so that weak and weak* convergence are metrizable. □

Acknowledgements. This work was supported by the DFG Research Unit FOR 797 *Analysis and computation of microstructures in finite plasticity*, subproject P6. The authors are grateful to Sergio Conti, Georg Dolzmann, Patrick Dondl and Alexander Mielke for many inspiring discussions.

References

[AD14a] Anguige, K., Dondl, P.W.: Optimal energy scaling for a shear experiment in single-crystal plasticity with cross-hardening. Zeitschrift für Angewandte Mathematik und Physik. ZAMP. Journal of Applied Mathematics and Physics. Journal de Mathématiques et de Physique Appliquées 65(5), 1011–1030 (2014)

[AD14b] Anguige, K., Dondl, P.W.: Relaxation of the single-slip condition in strain-gradient plasticity. arXiv.org (February 2014)

[AD15] Anguige, K., Dondl, P.W.: Energy estimates, relaxation, and existence for strain-gradient plasticity with cross-hardening. In: Hackl, K., Conti, S. (eds.) Analysis and Computation of Microstructure in Finite Plasticity. LNACM, vol. 78, pp. 157–174. Springer, Heidelberg (2015)

[AO05] Ariza, M.P., Ortiz, M.: Discrete crystal elasticity and discrete dislocations in crystals. Archive for Rational Mechanics and Analysis 178(2), 149–226 (2005)

[Aum15] Aumann, S.: Spontaneous breaking of rotational symmetry with arbitrary defects and a rigidity estimate. Preprint, pp. 1–40 (January 2015)

[BB07] Bourgain, J., Brezis, H.: New estimates for elliptic equations and Hodge type systems. Journal of the European Mathematical Society (JEMS) 9(2), 277–315 (2007)

[BBS78] Bacon, D., Barnett, D., Scattergood, R.: Anisotropic continuum theory of lattice defects. Progress in Materials Science 23, 51–262 (1978)

[CDM14] Conti, S., Dolzmann, G., Müller, S.: Korn's second inequality and geometric rigidity with mixed growth conditions. Calculus of Variations and Partial Differential Equations 50(1-2), 437–454 (2014)

[Cer99] Cermelli, P.: Material Symmetry and Singularities in Solids. Proceedings of The Royal Society of London. Series A. Mathematical, Physical and Engineering Sciences 455, 299–322 (1999)

[CGM14] Conti, S., Garroni, A., Massaccesi, A.: Modeling of dislocations and relaxation of functionals on 1-currents with discrete multiplicity. Calc. Var. PDE (2015), doi:10.1007/s00526-015-0846-x

[CGO15] Conti, S., Garroni, A., Ortiz, M.: The line-tension approximation as the dilute limit of linear-elastic dislocations. Preprint, pp. 1–57 (February 2015)

[DLGP12] De Luca, L., Garroni, A., Ponsiglione, M.: Gamma-convergence analysis of systems of edge dislocations: the self energy regime. Archive for Rational Mechanics and Analysis 206(3), 885–910 (2012)

[DMNP02] Dal Maso, G., Negri, M., Percivale, D.: Linearized elasticity as Γ-limit of finite elasticity. Set-Valued Analysis. An International Journal Devoted to the Theory of Multifunctions and its Applications 10(2-3), 165–183 (2002)

[DP91a] Davini, C., Parry, G.P.: A Complete List of Invariants for Defective Crystals. Proceedings of the Royal Society. London. Series A. Mathematical, Physical and Engineering Sciences 432, 341–365 (1991)

[DP91b] Davini, C., Parry, G.P.: Errata: A Complete List of Invariants for Defective Crystals. Proceedings of the Royal Society. London. Series A. Mathematical, Physical and Engineering Sciences 434, 735 (1991)

[DRMD15] Dmitrieva, O., Raabe, D., Müller, S., Dondl, P.W.: Microstructure in plasticity, A comparison between theory and experiment. In: Hackl, K., Conti, S. (eds.) Analysis and Computation of Microstructure in Finite Plasticity. LNACM, vol. 78, pp. 205–218. Springer, Heidelberg (2015)

[FH93] Fleck, N.A., Hutchinson, J.W.: A phenomenological theory for strain gradient effects in plasticity. Journal of the Mechanics and Physics of Solids 41(12), 1825–1857 (1993)

[FJM02] Friesecke, G., James, R.D., Müller, S.: A theorem on geometric rigidity and the derivation of nonlinear plate theory from three-dimensional elasticity. Communications on Pure and Applied Mathematics 55(11), 1461–1506 (2002)

[GLP10] Garroni, A., Leoni, G., Ponsiglione, M.: Gradient theory for plasticity via homogenization of discrete dislocations. Journal of the European Mathematical Society (JEMS) 12(5), 1231–1266 (2010)

[GPPS13] Geers, M.G.D., Peerlings, R.H.J., Peletier, M.A., Scardia, L.: Asymptotic Behaviour of a Pile-Up of Infinite Walls of Edge Dislocations. Archive for Rational Mechanics and Analysis 209(2), 495–539 (2013)

[Hut00] Hutchinson, J.W.: Plasticity at the micron scale. International Journal of Solids and Structures 37(1-2), 225–238 (2000)

[Jer99] Jerrard, R.: Lower bounds for generalized Ginzburg-Landau functionals. SIAM Journal of Mathematical Analysis 30(4), 721–746 (1999)

[Kon64] Kondo, K.: On the analytical and physical foundations of the theory of dislocations and yielding by the differential geometry of continua. Int. J. Eng. Sci. 2, 219–251 (1964)

[Krö58] Kröner, E.: Kontinuumstheorie der Versetzungen und Eigenspannungen. Springer, Heidelberg (1958)

[KT77] Kléman, M., Toulouse, G.: Classification of topologically stable defects in ordered media. J. Physique Lett. 38, L195–L197 (1977)

[LM10] Luckhaus, S., Mugnai, L.: On a mesoscopic many-body Hamiltonian describing elastic shears and dislocations. Continuum Mechanics And Thermodynamics 22, 251–290 (2010)

[LW14] Luckhaus, S., Wohlgemuth, J.: Study of a model for reference-free plasticity. arXiv.org (August. 2014)

[Mer77] Mermin, N.: Classification of topologically stable defects in ordered media. J. Physique Lett. 38, L195–L197 (1977)

[Mic80] Michel, L.: Symmetry defects and broken symmetry. Rev. Mod. Phys. 52, 617–651 (1980)

[Mie15] Mielke, A.: Variational approaches and methods for dissipative material models with multiple scales. In: Hackl, K., Conti, S. (eds.) Analysis and Computation of Microstructure in Finite Plasticity. LNACM, vol. 78, pp. 125–156. Springer, Heidelberg (2015)

[MM06] Mielke, A., Müller, S.: Lower semicontinuity and existence of minimizers in incremental finite-strain elastoplasticity. ZAMM. Zeitschrift für Angewandte Mathematik und Mechanik. Journal of Applied Mathematics and Mechanics 86(3), 233–250 (2006)

[MP08] Müller, S., Palombaro, M.: Existence of minimizers for a polyconvex energy in a crystal with dislocations. Calculus of Variations and Partial Differential Equations 31(4), 473–482 (2008)

[MS13] Mielke, A., Stefanelli, U.: Linearized plasticity is the evolutionary Γ-limit of finite plasticity. Journal of the European Mathematical Society (JEMS) 15(3), 923–948 (2013)

[MSZ15] Müller, S., Scardia, L., Zeppieri, C.I.: Geometric rigidity for incompatible fields, and an application to strain gradient plasticity. Indiana University Mathematics Journal 63, 1365–1396 (2015)

[Pon07] Ponsiglione, M.: Elastic Energy Stored in a Crystal Induced by Screw Dislocations: From Discrete to Continuous. SIAM Journal on Mathematical Analysis 39, 449–469 (2007)

[RC14] Reina, C., Conti, S.: Kinematic description of crystal plasticity in the finite kinematic framework: a micromechanical understanding of $F = F^e F^p$. Journal of the Mechanics and Physics of Solids 67, 40–61 (2014)

[San98] Sandier, E.: Lower bounds for the energy of unit vector fields and applications. Journal of Functional Analysis 152(2), 379–403 (1998)

[San00] Sandier, E.: Erratum: Lower bounds for the energy of unit vector fields and applications. Journal of Functional Analysis 171(1), 233 (2000)

[SPPG14] Scardia, L., Peerlings, R.H.J., Peletier, M.A., Geers, M.G.D.: Mechanics of dislocation pile-ups: A unification of scaling regimes. Journal of the Mechanics and Physics of Solids 70, 42–61 (2014)

[SS11] Sandier, E., Serfaty, S.: Improved lower bounds for Ginzburg-Landau energies via mass displacement. Analysis & PDE 4(5), 757–795 (2011)

[SvG14] Scala, R., van Goethem, N.: Dislocations at the continuum scale: functional setting and variational properties. Preprint, pp. 1–28 (November 2014)

[SZ12] Scardia, L., Zeppieri, C.I.: Line-tension model for plasticity as the Γ-limit of nonlinear dislocation energy. SIAM Journal on Mathematical Analysis 44(4), 2372–2400 (2012)

[TK76] Toulouse, G., Kléman, M.: Principles of a classification of defects in ordered media. J. Physique 37, 149–151 (1976)

Chapter 8
Microstructure in Plasticity, a Comparison between Theory and Experiment

Olga Dmitrieva, Dierk Raabe, Stefan Müller, and Patrick W. Dondl

Abstract. We review aspects of pattern formation in plastically deformed single crystals, in particular as described in the investigation of a copper single crystal shear experiment in [DDMR09]. In this experiment, the specimen showed a band-like microstructure consisting of alternating crystal orientations. Such a formation of microstructure is often linked to a lack of convexity in the free energy describing the system. The specific parameters of the observed bands, namely the relative crystal orientation as well as the normal direction of the band layering, are thus compared to the predictions of the theory of kinematically compatible microstructure oscillating between low-energy states of the non-convex energy. We conclude that this theory is suitable to describe the experimentally observed band-like structure. Furthermore, we link these findings to the models used in studies of relaxation and evolution of microstructure.

8.1 Introduction

Plastically deformed single crystals often exhibit the formation of complex microstructure [BHHK92, Han90, HBO10], where dislocations in the crystal

Olga Dmitrieva · Dierk Raabe
Max-Planck-Institut fuer Eisenforschung, Duesseldorf, Germany
e-mail: {dmitrieva,raabe}@mpie.de

Stefan Müller
Hausdorff Center for Mathematics and Institute for Applied Mathematics, University Bonn, Bonn, Germany
e-mail: stefan.mueller@hcm.uni-bonn.de

Patrick W. Dondl
Department of Mathematical Sciences, Durham University, Durham, UK
e-mail: patrick.dondl@durham.ac.uk

S. Conti and K. Hackl (eds.), *Analysis and Computation of Microstructure in Finite Plasticity*, Lecture Notes in Applied and Computational Mechanics 78, DOI: 10.1007/978-3-319-18242-1_8

arrange in intricate patterns. Since the creation and propagation of dislocations ultimately mediates the plastic behavior of crystals, and these dislocations interact through the elastic field and the local lattice distortion they generate over many length scales, understanding the macroscopic plastic behavior of crystalline specimen is strongly dependent on understanding the microstructure formation.

The seminal work by Ortiz and Repetto [OR99] introduced an incremental implicit time-stepping approach in order to study the evolution of plastic deformation. In this approach, in each time-step, the sum of the stored elastic energy and an incremental dissipation is minimized. The elastic energy is usually assumed to be polyconvex, the dissipation in single crystals, however, is naturally non-convex: plastic deformation is easier in single-slip since otherwise sessile atomistic products of dislocations on different slip planes—so-called Lomer-Cottrell locks—form [RP99]. This non-convexity of the dissipation potential due to latent- (or cross-)hardening is a candidate for the description of plastic microstructure. Basically, the homogeneously deformed state becomes energetically unfavourable, so the plastic strain oscillates between favourable energy wells in the dissipation potential, i.e., states of single slip. This, however, can create long-range elastic effects which can only be avoided if the generated microstructure is kinematically compatible, in the sense that the plastic strain is the gradient of a continuous deformation. This links the study of plastic microstructure to the study of Martensitic phase transformations in shape-memory materials, where a non-convexity in the elastic energy (due to the underlying phase transformation) is the source of microstructure [BJ87]. In this context of phase transformations, sharply delineated, laminated, zones of alternating deformation states are often observed. At first glance, these laminates bear a striking resemblance to some of the microstructure seen in experiments in single-crystal plasticity.

In [OR99], there is already a number of experimental studies referenced in order to link such kinematically compatible laminates to observed plastic microstructure in a more rigorous manner [RP80, JW84]. Many of these experiments, however, were performed in fatigue, i.e., using repeated oscillating small amplitude plastic deformation. The goal in [DDMR09] was to examine the pattern formation in a well-controlled single pass shear experiment, thus bringing it closer to applications in deep-drawing and related industrial deformation methods.

The remainder of this chapter is organized as follows. In section 8.2 we will briefly review some approaches to the continuum modeling of microstructure in crystal plasticity. Section 8.3 describes the experiment from [DDMR09]. We then offer some conclusions in the final section 8.4.

8.2 Modeling Continuum Plasticity

The goal of this chapter is to test the underlying modeling assumptions that are used, for example, in the studies of energy relaxation in [CDK13b, CDK13a, CDK11, ACD09, CDK09] and in the studies of plastic evolution [HHK12, HK11, KH11] in continuum plasticity. We thus consider, in the framework of multiplicative continuum plasticity, a deformation $y\colon \Omega \to \mathbb{R}^3$, for $\Omega \subset \mathbb{R}^3$, with appropriate boundary conditions. The deformation gradient $F = \nabla y$ is decomposed multiplicatively into an elastic contribution and a purely plastic part (neither of which necessarily have to be kinematically compatible) as

$$F = F_{\mathrm{el}} F_{\mathrm{pl}}.$$

For further discussion on the subject of multiplicative strain decomposition see [RC14]. As mentioned in the introduction, we follow the approach by Ortiz and Repetto [OR99] in studying a time-discrete problem instead of a continuous time evolution problem. We will furthermore restrict ourselves here to the analysis of microstructure formation in a single time step. A suitable functional for such a time step now reads, with a suitable elastic energy density W_{el} and a plastic dissipation W_{pl},

$$E(y, F_{\mathrm{pl}}) = \int_\Omega W_{\mathrm{el}}(F_{\mathrm{el}}) + \int_\Omega W_{\mathrm{pl}}(F_{\mathrm{pl}}). \tag{8.1}$$

The assumption of strong latent hardening now leads to the assumption that the plastic deformation necessarily has to occur in single slip only. Given a set $\{m_j\}_{j=1}^M$ slip plane normals in the crystal and a set $\{b_{ij}\}_{i=1}^{N_j}$ Burgers' vectors (orthogonal to m_j, respectively) in each slip plane, we thus assume that

$$W_{\mathrm{pl}} = \begin{cases} 0 & \text{if } F_{\mathrm{pl}} = \mathrm{Id} + \sum_{j=1}^M \sum_{i=1}^{N_j} \gamma_{ij} b_{ij} \otimes m_j, \\ & \text{with } \gamma_{ij}\gamma_{kl} = 0 \text{ for } i \neq k \text{ or } j \neq l, \\ +\infty & \text{otherwise,} \end{cases}$$

where we have (for simplicity) disregarded dissipation, which is small for plastic deformation in single slip and introduced the single-slip side condition as an infinite penalty on the energy. We remark that we expect that our results in section 8.3 below would not change substantially if this side condition was made somewhat less strict by introducing a hardening matrix with large off-diagonal entries instead.

Commonly, a higher-gradient curl-type term is introduced to such an energy in order to account for geometrically necessary dislocations [Nye53, Kon52, CG01]. Even with such an additional regularization (certainly also without), one relaxation of the model that can immediately be performed is that of considering a single-plane side condition instead of the single-slip condition above. See [AD14] for more information. After this relaxation, the plastic dissipation can be written as

$$W_{\mathrm{pl}} = \begin{cases} 0 & \text{if } F_{\mathrm{pl}} = \mathrm{Id} + \sum_{j=1}^{M} s_j \otimes m_j, \\ & \text{with } s_j \in m_j^{\perp} \text{ and } |s_j|\,|s_l| = 0 \text{ for } j \neq l, \\ +\infty & \text{otherwise}, \end{cases} \tag{8.2}$$

The investigations in the aforementioned articles usually start with an energy of this kind, often assuming a limited number of slip systems (i.e., one or two) and then studying further relaxation, evolution, or computational problems [MRF10].

Our approach, as mentioned above, was somewhat different but complementary: we wanted to check whether a configuration admitting low energy can be found that reproduces observed microstructure. The experiment performed and its outcome are described in the following section.

8.3 A Single-Pass Shear Deformation Experiment and the Resulting Microstructure

This section is a review of material published in [DDMR09] and [DSDR10]. We briefly recapitulate the experimental methods and observations and then answer the question posed at the end of section 8.2 by explicitly constructing a low-energy microstructure that reproduces the experimentally observed parameters.

8.3.1 Sample Preparation and Shear Deformation Experiments

The specimens were cut by spark erosion from 99.98% pure copper single crystals produced by the melt-grow method to a dimension of 3 mm × 2 mm with a height of 10 – 15 mm. They were then polished, first mechanically and finally electrolytically. The specimen is illustrated in Fig. 8.1a.

The shear experiments were performed on a special miniaturized testing device made by Kammrath Weiss GmbH (44141 Dortmund, Germany). The specimen is fixed in a stable, centered position between two movable cross-heads in the device. These cross-heads were sheared with respect to each other at a rate of 5 μm/s as measured by the machine extensometer, the load was controlled to a maximal load value of 1 kN by the device's load cell. For a schematic of the position of the specimen in the device see Fig. 8.1a. As seen in the figure, the freestanding part of the specimen has a length of 2.4 mm.

To an accuracy of 0.5°, the orientation of the undeformed single crystal specimen in the device is $(101)[12\bar{1}]$ with the shear load applied along the $[12\bar{1}]$ direction. Fig. 8.1b shows the shear load with respect to the crystal orientation in the context of the relevant f.c.c. slip systems in the $\{111\}$-plane given by the normal of the applied shear load. Under the given loading conditions, there are two primarily active coplanar slip systems with Schmid factors of maximal magnitude.

8.3.2 Digital Image Correlation for Strain Mapping and EBSD for Texture Mapping

In the course of the shear deformation, the strain on the surface of the sample was measured using the digital image correlation (DIC) method. The basic idea of this method is that an optical pattern (graphite spray for optical decoration on white acrylic spray) is applied to the surface of the sample and geometrical changes of this pattern are recognized by means of digital image analysis. For DIC we used a GOM Inc. Aramis System (version 6.0.0-3) with two digital cameras (CCD-1300, maximal resolution 1280 × 1024 pixels) placed behind the testing device. The recording time for each frame was 1 s.

After the deformation of the specimen, the surface of the samples was characterized structurally and crystallographically. In order to perform this characterization, a scanning electron microscope (SEM)[1] with a field emission gun operated at 15 kV was used. The microscope was equipped with a detector for the imaging of backscattered electrons (BSE imaging). The EBSD patterns were then recorded and evaluated by an EDAX/TSL EBSD

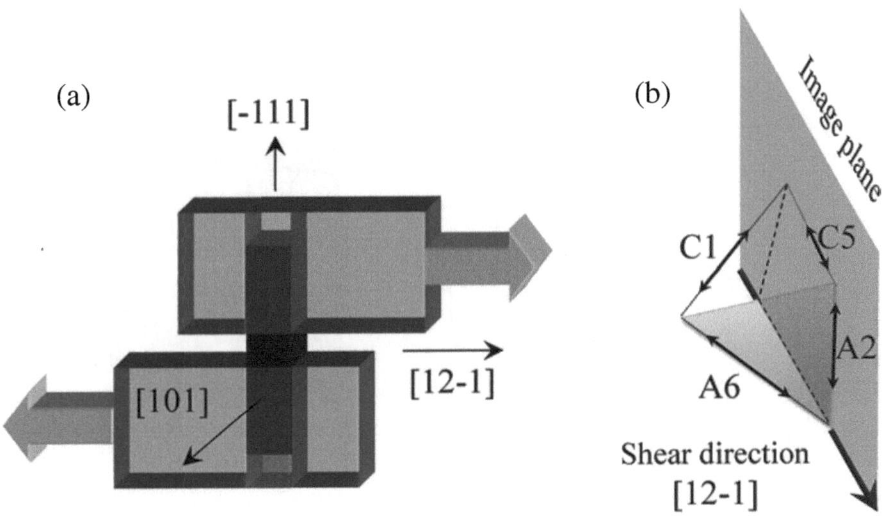

Fig. 8.1 **a:** Specimen in the specimen holder. **b:** Illustration of the crystallographic orientation of the specimen, displaying the slip systems (according to the Schmid-Boas nomenclature (also used in [OR99]). From [DSDR10], © IOP Publishing. Reproduced by permission of IOP Publishing. All rights reserved.

[1] The device used was a JEOL JSM 6500F microscope.

System equipped with a Digiview camera. In the high-resolution EBSD measurements the exposure time for each frame was set to about 0.5 s at the smallest binning size, and for the calculation of the Hough transformation a binned pattern size 240 × 240 and an angular spacing control of 0.5° were chosen.

8.3.3 Outcome of the Single Crystal Shear Deformation Experiments

Shear Deformation and DIC Analysis: The specimens were deformed in simple shear up to a deformation of $\gamma = 0.23$ as measured in the machine extensometer. The load/displacement dependence was recorded during the shear deformation, the data can be seen in Fig. 8.2a. The digital images of the initial and final deformation state as captured for DIC are shown in Fig. 8.2b. A closer inspection of the DIC data of the deformed sample revealed some strain concentration in the sample near the clamps and a large central region of homogeneously deformed material.

This homogeneously deformed part of the sample was then examined more closely. In particular, we found that it can be very well approximated by a completely homogeneous deformation in simple shear with a shear magnitude of $\gamma = 0.20$, but with a normal of shear rotated clockwise by $\varphi = 4.5°$ from the vertical direction.

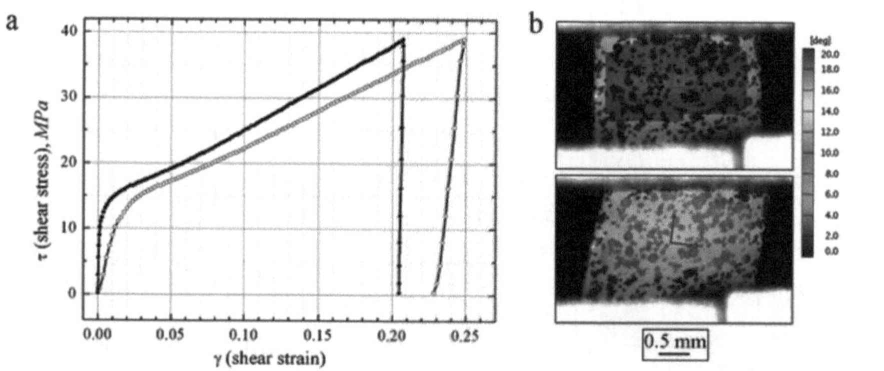

Fig. 8.2 a: Stress/shear angle dependence obtained from the load/displacement measurement (grey curve) and using the DIC method (black curve). **b:** Digital images of the sample surface decorated for the DIC method before (upper image) and after the deformation (lower image). Image from [DDMR09], © Elsevier. Reproduced by permission of Elsevier. All rights reserved.

Microstructural Characterization of the Sample: As mentioned above, the specimen was analyzed in an SEM after deformation. The BSE micrograph of the single crystal sample is shown in Fig. 8.3a, with the direction of the applied shear load indicated by the white arrows. Some glide bands can be observed; their orientation in relation to the nearest plane trace of a {111}-plane is indicated by a grey line. The central area containing the glide bands was then analyzed using EBSD.

Fig. 8.3b shows the EBSD image of the deformed part of the crystal recorded with a step size 2 μm. The mismatch of orientation to the orientation of an arbitrarily chosen point in the image is plotted. One can observe a variation of the orientation within 3°. The formation of a microscopic band structure with a different orientation compared to the material in between the bands can be observed from this map. The averaged orientation of the sample in the image was used to calculate the the plane traces of the {111}-slip planes displayed in the upper right hand corner of the image. The plane trace of the {111}-slip plane containing the two slip systems with maximal modulus of the Schmid factor (as illustrated in Fig. 8.1 above) is shown as a black line in the center of the image. Comparing this direction with the orientation of the microbands (illustrated by the grey line), a deviation of approximately 7° can be determined. Note that the orientation of the microbands is *not* crystallographic. A further feature of the generated bands is

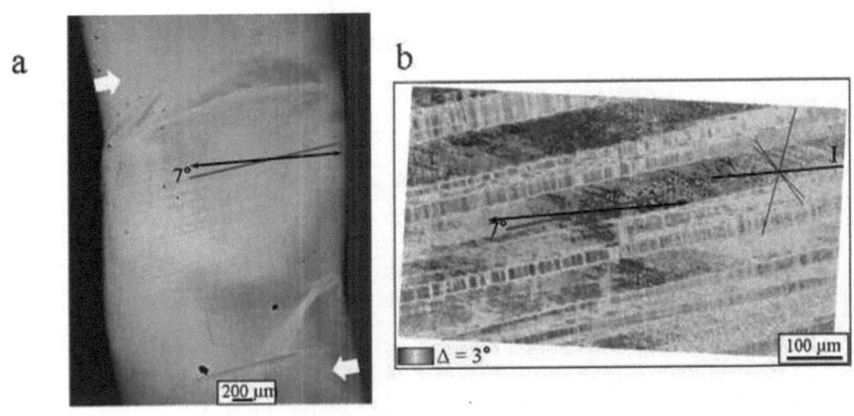

Fig. 8.3 a: BSE overview micrograph with white arrows indicating the applied shear load. The black line is a plane trace of a slip-plane (i.e., a {111}-plane). **b:** EBSD characterization of the local lattice orientation. The band-like structure can be clearly observed here. In the upper right corner of the image, the set of {111}-plane traces is displayed, and in the center the orientation of the laminates with respect to the nearest {111}-plane trace is shown. The 7° mismatch between the two is characteristic for the microstructure. From [DDMR09], used with permission.

the additional substrcuture inside of the microbands (which are subdivided by orthogonal lines). For an analysis of this substructure see section 8.3.5.

A high-resolution EBSD image is shown in Fig. 8.4a, taken with a step-size of 0.1 μm. Again, this image presents the local deviation of the orientation with respect to a reference orientation. The grey scale displays that the crystal lattice inside of the microbands is rotated by 3° in comparison to the material outside. Fig. 8.4b, on the other hand, demonstrates the variation of the the crystal orientation out of the defined crystallographic direction, which is nearly parallel to the normal of the frontal face. The grey scale range is the same as in Fig. 8.4a and displays no significant tilt on that order of magnitude. We therefore conclude that the microbands' orientation is rotated 3° clockwise with respect to the outside material and with a rotation axis that is perpendicular to the front face of the specimen.

8.3.4 Energy Minimizing Microstructure

Our hypothesis is that a non-convexity in the energy landscape is the basic mechanism underlying the formation of the patterns described in the previous section. The idea, as pioneered by Ball and James [BJ87], is the following: Consider the deformation of a body from a reference configuration $\Omega \subset \mathbb{R}^3$

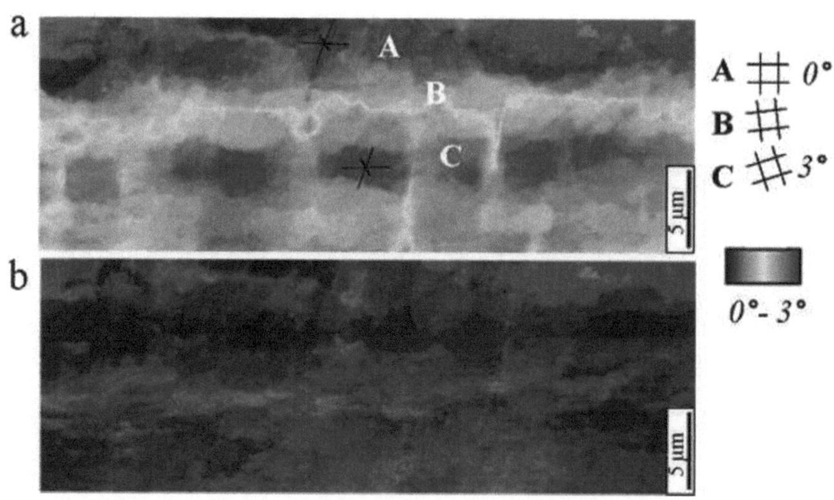

Fig. 8.4 **a:** High resolution EBSD map showing the edge of a microband. The grey scale indicates the the crystal orientation relative to a reference point. A graphical illustration of the observed local lattice rotation is shown on the right hand side. **b:** EBSD map of the same area showing the variation of the orientation away from the normal direction (same scale as above). From [DDMR09], © Elsevier. Reproduced by permission of Elsevier. All rights reserved.

by a sufficiently smooth function $y\colon \Omega \to \mathbb{R}^3$ mapping each point in the reference configuration to its current location. The free energy density of this continuum body is given by a frame indifferent function W depending on the deformation gradient $F = \nabla y$. Assume now that the deformation admits two preferred states (energy minima) F_1 and F_2. Due to material frame indifference, the energy must not change under rigid body motions, i.e., $W(F) = W(QF)$ with $Q \in SO(3)$. Assuming the body is elastically rigid, a good free energy functional can be written as

$$W(F) = \begin{cases} 0 & \text{if } F = QF_1 \text{ or } F = QF_2 \text{ for } Q \in SO(3), \\ \infty & \text{otherwise.} \end{cases} \tag{8.3}$$

A natural question to ask now is which boundary conditions such a material can accommodate in an averaged sense. After relaxing the elastic rigidity to a strong growth away from the minima, one can also introduce small boundary layers in the deformation. In general, this is an open question. However, it is possible to give an interesting upper bound for the relaxation of such a non-convex W. If one assumes that there exists an invariant plane between the two minimizers of the energy, i.e., a plane that is deformed in the same way by both deformation gradients, then one can alternate these two deformation gradients to form a fine scale mixture known as a laminate. The condition for this can be written in the following way: There must exist $Q \in SO(3)$ and $a, n \in \mathbb{R}^3$ such that

$$QF_1 - F_2 = a \otimes n. \tag{8.4}$$

In other words, modulo a rigid body motion, the difference between the deformation gradients must be a rank one matrix. Under these conditions, it is possible to find a *continuous, piecewise affine* deformation y whose gradient is at any point given either by QF_1 or F_2. Alternating these deformations with volume fraction λ and $1 - \lambda$ results in affine boundary conditions of the form $\lambda QF_1 + (1 - \lambda)F_2$ that can be accommodated by the material. This situation is illustrated in Fig. 8.5.

To this end, we first determine the macroscopic strain of the sample, which will act a side condition to minimizing an elasto-plastic energy functional. An analysis of the DIC-measurement illustrated in Fig. 8.2 revealed a central, homogeneously deformed section of the specimen with measured macroscopic strain

$$U = \begin{pmatrix} 0.9555 & -0.0198 & 0.0445 \\ & 1.1036 & 0.0198 \\ & & 0.9555 \end{pmatrix} \tag{8.5}$$

in the final deformation state. This calculation of the average strain assumed that the deformation in the direction normal to the crystal surface was the identity. Apart from this we only performed a basis change from the basis used in the DIC measurement to the basis of the f.c.c. lattice.

Subsequently, a MATLAB program was employed to find a deformation that satisfies the twinning equation (8.4) with the two deformation gradients

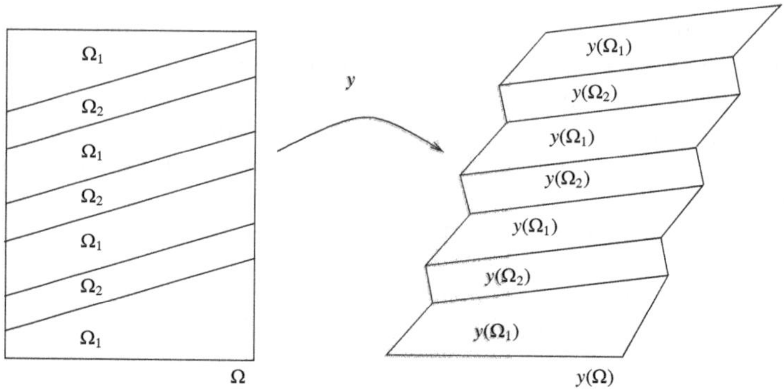

Fig. 8.5 Lamination of a piecewise affine deformation: the two subdomains, Ω_1 and Ω_2, are deformed in an affine manner such that the resulting deformation y is continuous. From [DDMR09], © Elsevier. Reproduced by permission of Elsevier. All rights reserved.

being simple shear in f.c.c. slip systems, while averaging to the strain (8.5) from the measurement. The program simply employs a gradient flow method for the magnitudes of slip in two specifically chosen slip systems with respect to the elasto-plastic energy

$$E = \left\| \sqrt{(UP^{-1})^{\mathrm{T}} \cdot (UP^{-1})} - \mathrm{Id} \right\|^2 , \tag{8.6}$$

where $P = (1-\lambda)Q(\gamma_1 P_1 + \mathrm{Id}) + \lambda(\gamma_2 P_2 + \mathrm{Id})$. Here, λ is the volume fraction of one part of the laminate, P_1 and P_2 are the displacement gradients of the two chosen slip systems, and Q is the lattice rotation as calculated in the twinning equation (8.4). Equation (8.6) can be seen as a relaxation of the energy in (8.3) allowing for elastic deformation.

As the main component P_1 of plastic deformation we choose an equal activity in the A2 and A6 slip systems (in the following, we refer to slip systems by the Schmid-Boas nomenclature, also used in [OR99]). We note that these slip systems are naturally compatible without a lattice rotation—they are coplanar slip systems. Following [AD14], we thus consider coplanar slip systems to be lumped into one.

For the secondary component of plastic deformation, there are a number of different choices for a kinematically compatible complementary slip activity. Noting that we are solely looking for a low energy state, we restrict our investigation to two particular possibilities, which guarantee kinematic compatibility independently of the choice of γ_1 and γ_2: first, an equal activity in C1 and C5, and second, an equal activity in B4 and D4.

Which volume fraction is attributed to which slip is a further degree of freedom in our construction. Using the histogram of the distribution of lattice rotations on the whole homogeneously deformed part of the face of the crystal, we determined that one component of the microstructure occupies approximately two thirds of the total area in the picture. This larger area can now be associated with either the primary component or with the secondary component of the plastic deformation. Note that from the experiments alone, we can not determine this directly, since only the lattice orientation, and not the shear strain was measured microscopically.

We note that if we find such a laminate as described above that also results in a small energy E in (8.6), we immediately have found a deformation state that admits a small energy in the plastic deformation model (8.1) with non-convex dissipation of type (8.2) which averages to the given macroscopic deformation. The reason is the following: the laminate of consisting of two different plastic deformation states as constructed above is automatically admissible with vanishing energy for (8.2). Furthermore, since the laminate is itself a gradient (modulo rigid body motions), it does not require any elastic strain to be made compatible. The only elastic strain appearing in (8.1) is thus the strain from the inexact recovery of the given average strain, i.e., the strain in the energy E in (8.6).

The parameters of the energy minimizers determined by the gradient flow are displayed in Table 8.1. It can be observed that some of the slip-system combinations yield lamination parameters (i.e., direction of the lamination normal, and relative lattice orientation) matching very well the experimental data. The resulting deformations are illustrated in Fig. 8.6, to be compared with the EBSD-results.

Table 8.1 Results from the energy minimization algorithm used to find a low energy laminate of slip systems recovering the average strain in (8.5). The value of λ is the volume fraction of slip in P_2. The values γ_i are the amount of slip in the respective slip system, α is the (clockwise) angle of orientation of the lamination normal with respect to the nearest $\langle 111\rangle$-direction. The value β is the (clockwise) angle of lattice misorientation. The value E is the energy at the minimizing state. All rotations are exactly in the plane with normal [101].

Choice of slip systems	λ	γ_{P1}	γ_{P2}	α	β	E
$P_2 =$ B4+D4	1/3	0.18	0.082	$-7.5°$	$3.8°$	$5.0 \cdot 10^{-6}$
$P_2 =$ C1−C5	2/3	0.28	-0.040	$-6.3°$	$3.3°$	$5.0 \cdot 10^{-6}$

8.3.5 An Analysis of the Substructure Within the Lamination Bands

In Fig. 8.3 one can clearly see the formation of a substructure within the microbands of an orientation variation of 1°. The reasons for the formation of this substructure were examined in [DSDR10]. There, a discrete dislocation dynamics model was used to determine the equilibrium distribution of

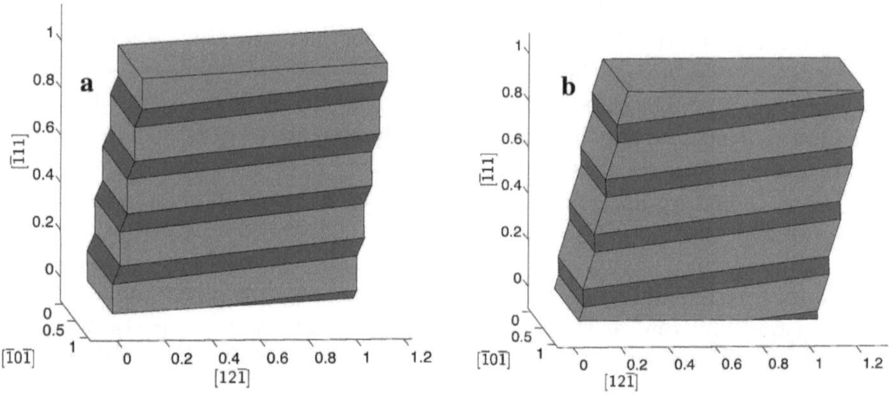

Fig. 8.6 Illustration of the two laminates from Table 8.1 that agree well with the EBSD result from Fig. 8.3b. From [DDMR09],

dislocations within the microbands. Given the boundary conditions of bending near the clamped edges of the specimen, the formation of domain walls within the microbands in quantitative agreement with the observed orientation variation was found.

8.4 Conclusions

Here we presented the findings of a copper single crystal shear experiment as published in [DDMR09, DSDR10], in relation to modeling plasticity microstructure by variational approaches. These experiments show that the theory of kinematically compatible microstructures, in particular laminate microstructures, can indeed be used to predict the formation of microstructure in plasticity. We demonstrated that there exist low-energy states consisting of plastic laminates whose macroscopic deformation averages to the measured macroscopic strain, while at the same time their microstructural properties do match the measured properties of the local lattice orientation patterning. In this sense, our work justifies the energy minimization approach to plasticity used for example in [CDK13b, CDK13a, CDK11, ACD09, CDK09, HHK12, HK11, KH11].

A main question that remains open regards the evolution of such laminate microstructures, in particular, whether those structures can arise in finite-strain deformation, where the lamination normal depends on the amount of shear. Some ideas for modeling the evolution of laminates have been explored in [KH11]. In this particular case, we see that during the course of the deformation, only a minor change in the lamination normal was necessary [DDMR09].

Acknowledgements. The authors gratefully acknowledge support from the DFG through the research unit FOR 797 *Analysis and computation of microstructure in finite plasticity*, subprojects P6 and P7.

References

[ACD09] Albin, N., Conti, S., Dolzmann, G.: Infinite-order laminates in a model in crystal plasticity. Proc. Roy. Soc. Edinburgh Sect. A 139(4), 685–708 (2009)

[AD14] Anguige, K., Dondl, P.W.: Relaxation of the single-slip condition in strain-gradient plasticity. Proceedings of the Royal Society of London A: Mathematical, Physical and Engineering Sciences, 470(2169) (2014)

[BHHK92] Bay, B., Hansen, N., Hughes, D., Kuhlmann-Wilsdorf, D.: Overview no-96 – evolution of FCC deformation structures in polyslip. Acta Metallurgica et Materialia 40(2), 205–219 (1992)

[BJ87] Ball, J.M., James, R.D.: Fine phase mixtures as minimizers of energy. Arch. Rational Mech. Anal. 100(1), 13–52 (1987)

[CDK09] Conti, S., Dolzmann, G., Klust, C.: Relaxation of a class of variational models in crystal plasticity. Proc. R. Soc. Lond. Ser. A Math. Phys. Eng. Sci. 465(2106), 1735–1742 (2009)

[CDK11] Conti, S., Dolzmann, G., Kreisbeck, C.: Asymptotic behavior of crystal plasticity with one slip system in the limit of rigid elasticity. SIAM J. Math. Anal. 43(5), 2337–2353 (2011)

[CDK13a] Conti, S., Dolzmann, G., Kreisbeck, C.: Relaxation and microstructure in a model for finite crystal plasticity with one slip system in three dimensions. Discrete Contin. Dyn. Syst. Ser. S 6(1), 1–16 (2013)

[CDK13b] Conti, S., Dolzmann, G., Kreisbeck, C.: Relaxation of a model in finite plasticity with two slip systems. Math. Models Methods Appl. Sci. 23(11), 2111–2128 (2013)

[CG01] Cermelli, P., Gurtin, M.E.: On the characterization of geometrically necessary dislocations in finite plasticity. Journal of the Mechanics and Physics of Solids 49(7), 1539–1568 (2001)

[DDMR09] Dmitrieva, O., Dondl, P.W., Müller, S., Raabe, D.: Lamination microstructure in shear deformed copper single crystals. Acta Materialia 57(12), 3439–3449 (2009)

[DSDR10] Dmitrieva, O., Svirina, J.V., Demir, E., Raabe, D.: Investigation of the internal substructure of microbands in a deformed copper single crystal: experiments and dislocation dynamics simulation. Modelling and Simulation in Materials Science and Engineering 18(8) (December 2010)

[Han90] Hansen, N.: Cold deformation microstructures. Materials Science and Technology 6(11), 1039–1047 (1990)

[HBO10] Hansen, B.L., Bronkhorst, C.A., Ortiz, M.: Dislocation subgrain structures and modeling the plastic hardening of metallic single crystals. Modelling and Simulation in Materials Science and Engineering 18(5), 055001 (2010)

[HHK12] Hackl, K., Hoppe, U., Kochmann, D.M.: Generation and evolution of inelastic microstructures—an overview. GAMM-Mitt. 35(1), 91–106 (2012)

[HK11] Hackl, K., Kochmann, D.M.: An incremental strategy for modeling laminate microstructures in finite plasticity—energy reduction, laminate orientation and cyclic behavior. In: de Borst, R., Ramm, E. (eds.) Multiscale Methods in Computational Mechanics. LNACM, vol. 55, pp. 117–134. Springer, Heidelberg (2011)

[JW84] Jin, N., Winter, A.: Dislocation structures in cyclically deformed [001] copper crystals. Acta Metallurgica 32(8), 1173–1176 (1984)

[KH11] Kochmann, D.M., Hackl, K.: The evolution of laminates in finite crystal plasticity: a variational approach. Contin. Mech. Thermodyn. 23(1), 63–85 (2011)

[Kon52] Kondo, K.: On the geometrical and physical foundations of the theory of yielding. In: Proc. 2nd Japan Nat. Congr. Applied Mechanics, pp. 41–47 (1952)

[MRF10] Miehe, C., Rosato, D., Frankenreiter, I.: Fast estimates of evolving orientation microstructures in textured bcc polycrystals at finite plastic strains. Acta Materialia 58(15), 4911–4922 (2010)

[Nye53] Nye, J.F.: Some geometrical relations in dislocated crystals. Acta Metallurgica 1(2), 153–162 (1953)

[OR99] Ortiz, M., Repetto, E.A.: Nonconvex energy minimization and dislocation structures in ductile single crystals. J. Mech. Phys. Solids 47(2), 397–462 (1999)

[RC14] Reina, C., Conti, S.: Kinematic description of crystal plasticity in the finite kinematic framework: a micromechanical understanding of $F = F^{\mathrm{e}} F^{\mathrm{p}}$. J. Mech. Phys. Solids 67, 40–61 (2014)

[RP80] Rasmussen, K., Pedersen, O.: Fatigue of copper polycrystals at low plastic strain amplitudes. Acta Metallurgica 28(11), 1467–1478 (1980)

[RP99] Rodney, D., Phillips, R.: Structure and strength of dislocation junctions: An atomic level analysis. Physical Review Letters 82(8), 1704–1707 (1999)

Chapter 9
Construction of Statistically Similar RVEs

Lisa Scheunemann, Daniel Balzani, Dominik Brands, and Jörg Schröder

Abstract. In modern engineering, micro-heterogeneous materials are designed to satisfy the needs and challenges in a wide field of technical applications. The effective mechanical behavior of these materials is influenced by the inherent microstructure and therein the interaction and individual behavior of the underlying phases. Computational homogenization approaches, such as the FE^2 method have been found to be a suitable tool for the consideration of the influences of the microstructure. However, when real microstructures are considered, high computational costs arise from the complex morphology of the microstructure. Statistically similar RVEs (SSRVEs) can be used as an alternative, which are constructed to possess similar statistical properties as the real microstructure but are defined by a lower level of complexity. These SSRVEs are obtained from a minimization of differences of statistical measures and mechanical behavior compared with a real microstructure in a staggered optimization scheme, where the inner optimization ensures statistical similarity and the outer optimization problem controls the mechanical comparativity of the SSRVE and the real microstructure. The performance of SSRVEs may vary with the utilized statistical measures and the parameterization of the microstructure of the SSRVE. With regard to an efficient construction of SSRVEs, it is necessary to consider statistical measures which can be computed in reasonable time and which provide sufficient information of the real microstructure. Minkowski functionals are analyzed as possible basis for statistical descriptors of microstructures and compared with other well-known statistical measures to investigate the performance. In order to emphasize the general importance of considering microstructural features by more sophisticated measures than basic ones, i.e. volume fraction, an analysis of upper bounds on the error of statistical measures and mechanical response is presented.

Lisa Scheunemann · Dominik Brands · Jörg Schröder
Institut für Mechanik, Abteilung Bauwissenschaften,
Univeristät Duisburg-Essen, 45117 Essen, Germany
e-mail: {lisa.scheunemann,dominik.brands,j.schroeder}@uni-due.de

Daniel Balzani
Institut für Mechanik und Flächentragwerke, TU Dresden, 01062 Dresden, Germany
e-mail: daniel.balzani@tu-dresden.de

S. Conti and K. Hackl (eds.), *Analysis and Computation of Microstructure in Finite Plasticity*,
Lecture Notes in Applied and Computational Mechanics 78, DOI: 10.1007/978-3-319-18242-1_9

9.1 Introduction

The description of the effective mechanical behavior of micro-heterogeneous materials is a large field of interest in research. This behavior is determined by the underlying microstructure and therein the behavior of the individual phases and their interaction. From the viewpoint of computational simulation, homogenization approaches such as the direct micro-macro transition approach, also known as multilevel Finite Element (FE) method or FE^2 method, provide a framework for the direct incorporation of the microstructure, see e.g. [GTK97], [SBM98], [MS98], [MSS99], [FS99], [Fey99], [FC00], [KBB01], [GKB03] and [Sch14]. In this approach, a microscopic FE boundary value problem is attached at every gauss-point of the macroscopic FE boundary value problem with the macroscopic quantities being obtained from suitable volumetric averages over their microscopic counterparts. The microscopic problem can be more efficiently solved using spectral methods, see e.g. [LLR11] and [EDLR13]. The solution of the microscopic boundary value problem replaces the evaluation of an analytical material law. At the microscale a representative volume element (RVE), is considered. There exist various definitions of RVEs in the literature, for an overview see e.g. [Zem03] and the references therein. Especially the estimation of the size of an RVE is important to ensure representativity and minimize computational costs. An approach for the computation of an appropriate size of an RVE for composites is presented in [KFG$^+$03] and was later on extended by [PBMP09], lowering computational costs for the determination of RVEs. For dual phase steel, represented by inclusion-matrix microstructures with 35% martensite, suitable sizes for RVEs are determined in [RMPB12] based on equiaxed and banded inclusions by comparing different RVE sizes under varying boundary conditions. A suitable size of 24 μm with a necessary number of 12 and 19 martensite inclusions was found for equiaxed and banded microstructures, respectively. An overview on different methods to analyze RVEs can be found in [ZW05].

For RVEs defined as subsections of real microstructures, typically large portions are necessary to obtain representativity, leading to high computational costs. Even if smaller subsections are admissible as RVEs, the complex nature of a real microstructure's morphology leads to disadvantages with respect to computational expenses. An alternative use of artificial microstructures governed by a lower complexity while maintaining similarities with respect to the original microstructure is desirable. Thereby aside from advantages in computational homogenization, other method defined on microscale level, e.g analysis of dislocation patterns, could benefit from the use of artificial microstructures. The concept of statistically similar RVEs (SSRVEs) is a possible method for the construction of such artificial microstructures. Therein a least-square functional is minimized, which compares statistical measures evaluated for a real microstructure and the SSRVE. The method was proposed for two-dimensional microstructures in [BSB09], see also [SBB11], [BBSC10] and [BBS14a], the extension to three-dimensional microstructures is proposed in [BSBS14]. Similar methods are utilized in [Pov95] and [KBC06] to reconstruct real microstructures. The latter one shows a good representation of stress distribution, peak stresses as well as nucleation strains and hot spot regions in the reconstructed microstructures compared with the original microstructures. The concept of SSRVEs was adapted in [RPBP11] and is applied in [RKP14] for a simulation of stamping. In [ABRP12] the authors have used SSRVEs in crash box stamping using scalar statistical descriptors for the construction.

The reconstruction of real microstructures to obtain samples for an analysis of microstructure properties is a topic of current research in order to circumvent costly microscopy analyses. In [KBC06], a least-square functional considering lineal-path function and spectral density is minimized in a simulated annealing process to reconstruct 2D complex microstructures, which are found to possess similar mechanical properties as the original microstructure. A Monte-Carlo approach is used for the reconstruction of three-dimensional microstructures based on a minimization of a least-square functional taking into account two-point correlation function and two-point cluster function achieving a comparability of statistical and mechanical properties of the reconstructed microstructures compared with the real one in [BMH$^+$12]. When aiming for a description of a real microstructure for reconstruction purposes, the statistical measures used play a crucial role. The quality of reconstructed microstructures and SSRVEs in terms of comparability with the original one is strongly influenced by the choice of statistical measures used in the construction process. There exist numerous statistical descriptors in the literature, an overview can be found in e.g. [OM00], [Tor02], [Zem03] and [EH86]. It is obvious that hybrid approaches utilizing combinations of statistical measures perform better in the reconstruction of microstructures, as shown in [YT98], providing that the different measures capture different aspects of the microstructure. A sensitivity analysis presented in [BBSC10] showed that a combination of phase fraction and spectral density as statistical measures performs well in the construction of 2D SSRVEs compared to using solely classical scalar descriptors. The class of Minkowski functionals, which describe geometric objects using integral geometry, offers a set of descriptors for the characterization of microstructure morphologies. The Minkowski functionals, forming scalar, vectorial or tensorial measures, are defined as integrals over curvatures, position vectors and normal vectors on the volume or surface of a geometric object, for detailed information see [MS00]. While the scalar measures are correlated to basic measures describing area, boundary length and Euler characteristics in 2D and volume, surface area, mean curvature and Euler characteristics in 3D, the vectorial measures are e.g. related to center of mass and the tensorial measures are related to the well know mechanical measure tensor of inertia. In [AKM10], scalar Minkowski functionals are applied for the characterization of three-dimensional structures while in [MJM08] they are utilized in the field of 2D image analysis. Minkowski functionals of tensorial form have been used in [STKB$^+$10] and [STMK$^+$11] for a characterization of multiphase structures, i.e. planar microscopy data, granular materials and foam structures, focusing on morphological anisotropy properties. Local anisotropy in fluids has been characterized using Minkowski tensors in [KMS$^+$10]. Since Minkowski tensors are defined for individual geometrical objects, a consideration of probability density functions based on specific measures derived for single inclusions seems reasonable to obtain a descriptor for a two-phase microstructure consisting of a matrix phase with multiple embedded inclusions. The analysis of such measures for the construction of SSRVEs is a main objective of this contribution.

The outline of the paper is as follows: Section 9.2, the concept of SSRVEs is described, first focusing on the general method and lower and upper bounds on RVEs regarding statistical similarity and similar mechanical behavior compared with a real microstructure. Furthermore, a summary of the statistical measures used later on in the construction of SSRVEs is presented. Here, a detailed elaboration of the statistical measures given by probability density functions based on Minkowski tensors is presented. Section 9.3 discusses different aspects of the construction of SSRVEs, including parameterization of SSRVEs and size determination as well as giving specific objective functions and briefly resuming the FE^2 method. The second part of the section presents SSRVEs based on a minimization of the specific objective functions and discusses the results. An analysis of the microscopic mechanical response is performed for selected SSRVEs. Exemplarily, a construction of upper and lower bounds for RVEs is carried out. Section 9.4 concludes and summarizes the main outcome.

9.2 Statistically Similar RVEs

In computational homogenization, the choice of a representative volume element (RVE) is an important aspect for an appropriate consideration of the microstructure of the material. There exist several definitions of an RVE in the literature, an overview can be found in e. g. [Hil63], [DW96] and [Zem03]. However, the definition of an RVE obtained from experimental measurements of a real material's microstructure is constrained by the limitations of microscopy techniques, as the measurable volume is limited. Typically, the largest measurable portion of a microstructure is assumed as a representative portion, thus an RVE. Due to the highly complex composition of micro-constituents on this scale, the resultant discretization necessary for computational purposes leads to very high costs making efficient computations unachievable. If a smaller subsection of this cutout still exhibits similar overall mechanical properties, it could be instead considered as an RVE. The definition of statistically similar RVEs (SSRVEs) in this context is an alternative to circumvent the issue of high computational costs. These artificial structures possess a microstructural morphology which is statistically similar to a real microstructure while being smaller in size and governed by a lower complexity. Still, they exhibit a similar mechanical behavior. The method resumes the idea proposed in [Pov95] and [KBC06], where a special class of microstructures is taken into account in the former. The construction of SSRVEs is proposed in [BSB09] and [SBB11] for two-dimensional microstructures and extended to 3D in [SBBS13] and [BSBS14]. The method is adapted in [RPBP11] and [RKP14]. In [ABRP12], SSRVEs are used for the simulation of crash box stamping. The general premises for the application of this method are the representativity of the real microstructure by a periodic microstructure consisting of SSRVEs as periodic subvolumes, as illustrated in Figure 9.1. Furthermore, the macroscopic material behavior must be solely depending on the local microscopic mechanical behavior and the microstructure morphology. The behavior of the individual constituents of the microstructure is

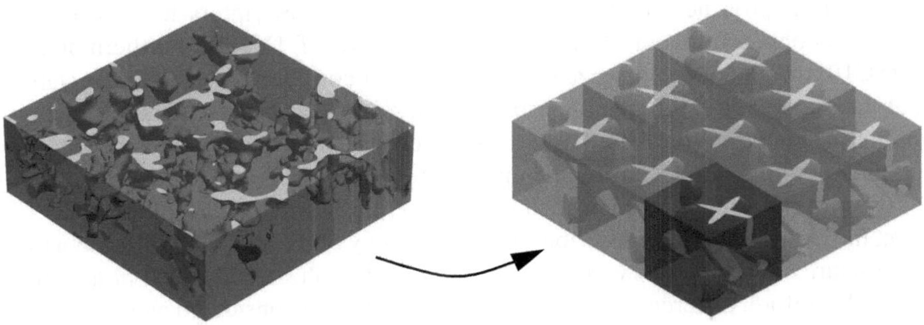

Fig. 9.1 Random target microstructure (left) of an inclusion-matrix microstructure (phase reconstruction of a 3D EBSD/FIB measurement of a DP steel performed at Max-Planck Institute for Iron Research, Düsseldorf, Prof. D. Raabe, details see [BBSR11], [BBS+14b]) and associated periodic microstructure (right) with periodic SSRVEs. Taken from [BSBS14].

assumed to be known. Note that real microstructures and RVEs obtained therefrom are mostly non-periodic by nature. This section presents the method of construction of SSRVEs and the statistical descriptors used therein. Furthermore some aspects on bounds of RVEs will be discussed.

9.2.1 Method

The construction of SSRVEs is based on the minimization of a least-square functional which compares the differences of statistical measures computed for the real microstructure and the SSRVE. The SSRVE's artificial inclusion morphology is described by γ_i which considers different types i of descriptions for the inclusion morphology. The least-square functional comparing a set of statistical measures $\mathscr{G}$ reads

$$\mathscr{E}_{\mathscr{G},\omega}(\gamma_i) := \sum_{L\in\mathscr{G}} \omega_L \mathscr{L}_L(\gamma_i) \qquad \text{with} \qquad \mathscr{L}_L(\gamma_i) := \left(\mathscr{P}_L^{\text{real}} - \mathscr{P}_L^{\text{SSRVE}}(\gamma_i)\right)^2 \tag{9.1}$$

with the least-square functional $\mathscr{L}_L(\gamma_i)$ for an individual statistical measure L. These functionals are weighted with weighting factors ω_L of the set ω. $\mathscr{P}_L^{\text{real}}$ denotes the evaluation of a statistical measure L for the real microstructure and $\mathscr{P}_L^{\text{SSRVE}}(\gamma_i)$ gives the counterpart for the SSRVE. For a fixed set of statistical measures, a fixed set of weighting factors and a specific parameterization, the ideal inclusion morphology $\tilde{\gamma}_i$, in a statistical sense, can be found by solving the optimization problem

$$\tilde{\gamma}_i = \arg\left[\min_{\gamma_i}\left[\mathscr{E}_{\mathscr{G},\omega}(\gamma_i)\right]\right]. \tag{9.2}$$

The construction of an optimal SSRVE is influenced by many factors. Depending on the statistical measures taken into account, different properties of the microstructure are captured and similarities thereon are enforced in the SSRVE. It was found that hybrid approaches combining multiple statistical descriptors lead to beneficial results opposed to considering only one measure, cf. [YT98]. Furthermore, in [BBSC10] the authors emphasize the importance of considering higher order statistical measures. For an efficient construction of SSRVEs, the individual evaluation time of the statistical measure has to be considered, since the measures have to be computed multiple times during the optimization process. Additionally, the different types of parameterization of inclusion morphology affect the resulting SSRVE. Here, the parameterization describes the morphology of the inclusion phase in a microstructure consisting of two materials, an inclusion phase embedded in a matrix phase. Therefore, different shapes of inclusions could be considered as well as different number of inclusions. Permitting a parameterization leading to an arbitrary inclusion morphology would be beneficial for the quality of the SSRVE, since features of the real microstructure can be adjusted more easily. However, it would also imply a high effort regarding discretization and computation. In order to evaluate the performance of different sets of statistical measures and weighting factors as well as types of parameterization, an optimization problem related to (9.2) can be used. Therein, the obtained parameterizations $\tilde{\gamma}_i$ are compared with the real microstructure regarding their mechanical response. In view of this staggered optimization process, (9.2) can be considered as an inner optimization problem and the evaluation considering the mechanical performance represents an outer optimization problem. The full optimization problem to be solved reads

$$\tilde{\gamma} = \underbrace{\arg\left\{\min_{\tilde{\gamma}_i}\left[\tilde{r}_{\varnothing}\left(\arg\underbrace{\left[\min_{\gamma_i}\left[\mathscr{E}_{\mathscr{G},\omega}(\gamma_i)\right]\right]}_{\text{inner problem, compare (9.2)}}\right)\right]\right\}}_{\text{outer problem}}, \tag{9.3}$$

with the overall optimal inclusion morphology $\tilde{\gamma}$ and the objective function comparing the mechanical responses given by $\tilde{r}_{\varnothing}$. In this sense, the staggered optimization process ensures the best possible statistical similarity of the inclusion morphology compared with the real microstructure morphology and then evaluates the obtained morphologies regarding their mechanical performance. This scheme is favorable if the inner optimization problem can be solved at less computational costs compared to the outer problem, because it limits the number of evaluations of $\tilde{r}_{\varnothing}$. For an efficient solution of the inner optimization problem, a high efficiency regarding the statistical measures is important. In Sec. 9.2.3, different sets of statistical measures are presented.

9.2.2 Lower and Upper Bounds of RVEs

Using the inner optimization problem in (9.3), bounds related to the statistical similarity of the microstructure morphology can be computed. These bounds represent the "best" and "worst" scenario of a microstructure morphology in a volume element for a fixed parameterization i, a fixed set of weighting factors ω and a fixed set of statistical descriptors $\mathscr{G}$. Note that a considered volume portion is here denoted as volume element (VE), since a representativity for the "worst" case is not given. They are obtained by

$$\mathscr{E}^{\text{best}} = \min_{\gamma_i} \left[\mathscr{E}_{\mathscr{G},\omega}(\gamma_i) \right] \quad \text{and} \quad \mathscr{E}^{\text{worst}} = \max_{\gamma_i} \left[\mathscr{E}_{\mathscr{G},\omega}(\gamma_i) \right], \tag{9.4}$$

with the "best" and "worst" possible scenario of a microstructure morphology defined by

$$\tilde{\gamma}_i^{\text{best}} = \arg \left[\mathscr{E}^{\text{best}} \right] \quad \text{and} \quad \tilde{\gamma}_i^{\text{worst}} = \arg \left[\mathscr{E}^{\text{worst}} \right]. \tag{9.5}$$

Note that here $\tilde{\gamma}_i^{\text{best}}$ coincides with the definition of an SSRVE. These cases of a microstructure describe extrema with respect to similarity of the VE and the real microstructure. It can be expected that for an infinite flexibility of microstructure morphology and a set of statistical measures describing the microstructure perfectly, the mechanical response of the resulting "best" and "worst" VEs becomes extremal, thus in the "best" case, the mechanical response should coincide with the one of the real microstructure under the assumptions made in the end of Section 9.2 on the representativity of the real microstructure. The bounds contain valuable information since they describe the range of a VE which could be obtained for a given parameterization and given set of statistical descriptors. Under the consideration of a sufficient set of statistical descriptors for the description of the microstructure morphology, the bounds are expected to expand with respect to greater flexibility of the parameterization of the microstructure morphology. A greater flexibility of parameterization in this sense allows for more complex microstructure morphologies. For the lower bound, representing the SSRVE, a decreasing value of the objective function $\mathscr{E}^{\text{best}}$ should be observed with increasing flexibility. Only together with the general premises of Section 9.2 for the application of SSRVEs, a suitable set of statistical descriptors and a reasonable parameterization for the microstructure morphology, the increase of flexibility is expected to coincide with a qualitative decrease of the mechanical error compared to the real microstructure, again due to a wider range of possible microstructure morphologies. Additionally, a converging behavior to optimal bounds is expected with continuing increase of flexibility of microstructure morphology.

9.2.3 Statistical Measures

There exist various statistical descriptors for the characterization of microstructures, an overview can be found in [EH86], [OM00], [Tor02] and [Zem03]. The detailed description of microstructures is indispensable for high-performance materials, for which a rapid development has taken place in the last decades. For characterizing position, size, orientation and shape of a microstructural constituent, distribution densities of specific properties go beyond the traditional approach of average values. Here, only the statistical descriptors used for the construction of SSRVEs in Section 9.3 are recapitulated.

9.2.3.1 Spectral Density and Lineal-Path Function

The notion of n-point probability functions is an important concept for the characterization of microstructures, see [Ber68] and [Bro55]. Let $D^{(P)}$ be a domain occupied by the microscopic phase P and $\boldsymbol{x}$ a position vector to a material point in the microstructure, then the indicator function

$$\chi^{(P)}(\boldsymbol{x}) := \begin{cases} 1 \text{ if } \boldsymbol{x} \in D^{(P)}, \\ 0 \text{ otherwise,} \end{cases} \tag{9.6}$$

describes the respective phase P, thus in every point $\boldsymbol{x}$ the indicator function fulfills the condition $\sum_P \chi^{(P)}(\boldsymbol{x}) = 1$. The n-point probability function for a specific phase P is defined as the average of the products of indicator functions at different positions $\boldsymbol{x}_1,...\boldsymbol{x}_n$ over an ensemble of individual samples α as

$$\mathscr{S}_n^{(P)}(\boldsymbol{x}_1,...,\boldsymbol{x}_n) := \overline{\chi^{(P)}(\boldsymbol{x}_1,\alpha)\,\chi^{(P)}(\boldsymbol{x}_2,\alpha)\ldots\chi^{(P)}(\boldsymbol{x}_n,\alpha)}, \tag{9.7}$$

where $\overline{(\bullet)}$ denotes the ensemble average of $(\bullet)$. This function is also referred to as n-point correlation function and represents the probability of n points $\boldsymbol{x}_1...\boldsymbol{x}_n$ being located in the same phase P. The ensemble average over a set of samples can be replaced by the average over one sample $\mathscr{B}$ with infinite sample size providing that the microstructure fulfills ergodicity assumptions. The definition then transforms to

$$\mathscr{S}_n^{(P)}(\boldsymbol{x}_1,...,\boldsymbol{x}_n) := \lim_{V\to\infty} \frac{1}{V(\mathscr{B})} \int_{\mathscr{B}} \chi^{(P)}(\boldsymbol{y}+\boldsymbol{x}_1)\chi^{(P)}(\boldsymbol{y}+\boldsymbol{x}_2)\ldots\chi^{(P)}(\boldsymbol{y}+\boldsymbol{x}_n)\,\mathrm{d}\boldsymbol{y}, \tag{9.8}$$

where $V(\bullet)$ describes the volume of a domain $\bullet$ and $\boldsymbol{y}$ denotes a position vector to material points located in the phase P. Note that for practicable application the integral over an infinite volume is replaced by the integral over a sufficiently large amount of a microstructure. For most application, it is feasible to consider the 1-point and 2-point probability function. The 1-point probability function

$$\mathscr{S}_1^{(P)}(\boldsymbol{x}) := \lim_{V\to\infty} \frac{1}{V(\mathscr{B})} \int_{\mathscr{B}} \chi^{(P)}(\boldsymbol{y}+\boldsymbol{x})\,\mathrm{d}\boldsymbol{y}, \tag{9.9}$$

describes the probability of finding a material point in the inclusion phase, thus it is equal to the phase fraction

$$\mathscr{P}_V^{(P)} := V(\mathscr{B}^{(P)})/V(\mathscr{B}), \tag{9.10}$$

with $\mathscr{B}^{(P)}$ characterizing the domain of phase P. The 2-point probability function, also denoted as second-order correlation function or autocorrelation function, is described by

$$\mathscr{S}_2^{(P)}(\boldsymbol{x}_1,\boldsymbol{x}_2) := \lim_{V\to\infty} \frac{1}{V(\mathscr{B})} \int_{\mathscr{B}} \chi^{(P)}(\boldsymbol{y}+\boldsymbol{x}_1)\chi^{(P)}(\boldsymbol{y}+\boldsymbol{x}_2)\mathrm{d}\boldsymbol{y}, \tag{9.11}$$

and gives the probability of two points $\boldsymbol{x}_1$ and $\boldsymbol{x}_2$ both being located in the same phase P for an infinite volume $V(\mathscr{B})$. The two-point probability is strongly correlated to the spectral density, which can be used as an alternative descriptor. The spectral density for a discrete three-dimensional data set is calculated based on the discrete Fourier transform

$$\mathscr{F}^{(P)}(k_x,k_y,k_z) := \sum_{n_x=1}^{N_x}\sum_{n_y=1}^{N_y}\sum_{n_z=1}^{N_z} \exp\left(-\pi\mathrm{i}\left(\frac{n_x k_x}{N_x}+\frac{n_y k_y}{N_y}+\frac{n_z k_z}{N_z}\right)\right)\chi^{(P)}(n_x,n_y,n_z), \tag{9.12}$$

where a discrete material point in the data set of size N_x, N_y and N_z in the three respective directions is given by $\boldsymbol{x} = [n_x\ n_y\ n_z]$ with $n_x = 1...N_x$, $n_y = 1...N_y$ and $n_z = 1...N_z$. The coordinates in the frequency domain are defined by k_x, k_y and k_z, respectively. The spectral density is defined by

$$\mathscr{P}_{\mathrm{SD}}^{(P)} := \frac{(\mathscr{F}^{(P)})^*\mathscr{F}^{(P)}}{2\pi\, N_x\, N_y\, N_z} \tag{9.13}$$

with the conjugate complex $(\mathscr{F}^{(P)})^*$ of the Fourier transform $\mathscr{F}^{(P)}$. The spectral density and two-point probability function capture information in the sense of periodicity of a microstructure. The spectral density gains its popularity in many fields, e.g. image analysis, from the fact that fast numerical algorithms, e.g. the "FFTW" ("Fastest Fourier Transform in the West"), developed at the Massachusetts Institute of Technology by M. Frigo and S.G. Johnson (www.fftw.org), can be used to perform the computation of the discrete Fourier transform for discrete data sets.

Another statistical measure for the description of microstructures is the lineal-path function, cf. [LT92]. The function describes the probability that an entire line segment $\overrightarrow{z} := \overrightarrow{\boldsymbol{x}_1\boldsymbol{x}_2}$ connecting the points $\boldsymbol{x}_1$ and $\boldsymbol{x}_2$ is located in the same phase. The modified indicator function

$$\chi_{\mathrm{LP}}^{(P)}(\overrightarrow{z}) := \begin{cases} 1 \text{ if } \overrightarrow{z} \in D^{(P)}, \\ 0 \text{ otherwise,} \end{cases} \tag{9.14}$$

examines whether the line segment $\overrightarrow{z}$ is located in the domain $D^{(P)}$ of phase P. The lineal-path function is then characterized by the ensemble average over a series of samples α by

$$\mathscr{P}_{\mathrm{LP}}^{(P)}(\overrightarrow{z}) := \overline{\chi_{\mathrm{LP}}^{(P)}(\overrightarrow{z},\alpha)}, \tag{9.15}$$

whereas for ergodic microstructures the volumetric average can be considered instead of the ensemble average, thus leading to

$$\mathscr{P}_{\mathrm{LP}}^{(P)}(\overrightarrow{z}) := \lim_{V\to\infty}\int_{\mathscr{B}} \chi_{\mathrm{LP}}^{(P)}(\boldsymbol{y}+\overrightarrow{z})\,\mathrm{d}\boldsymbol{y}. \tag{9.16}$$

Therein, $\boldsymbol{y}+\overrightarrow{z}$ denotes a shift of the line segment $\overrightarrow{z}$ by the position vector $\boldsymbol{y}$ relative to an origin of the considered domain $\mathscr{B}$. Again, note that the infinite volume $\mathscr{B}$

can be replaced by a sufficiently large portion of microstructure for practical reasons. The lineal-path function captures information about the connectedness of a phase in a microstructure and therefore accounts for some kind of long-range properties of a phase.

9.2.3.2 Tensorial Minkowski Functionals

For a description of spatial structures, which can e. g. be of cellular, periodic or porous type, the class of Minkowski tensors (MTs) offers a robust alternative to the concept of correlation functions. Due to their tensorial nature, anisotropy and orientational aspects are particularly captured. MTs are defined as integrals of normal and position vectors as well as surface curvatures. A detailed description on the topic can be found in [MS00] and [BDMW02]. Let $\mathbb{R}^3$ be a three-dimensional space, then MTs are defined as

$$\mathscr{W}_\nu^{a,b} = \frac{1}{3}\int_{\partial\Omega} \mathscr{G}_\nu\, \boldsymbol{r}^a \otimes \boldsymbol{n}^b \,\mathrm{d}S \quad \text{with} \quad \nu = 1,2,3. \tag{9.17}$$

The special case $\nu = 0$ is given by

$$\mathscr{W}_0^{a,0} = \int_{\Omega} \mathbf{r}^a \,\mathrm{d}\Omega \tag{9.18}$$

for a convex body Ω with the boundary surface $\partial\Omega$ and the scalar functions $\mathscr{G}_\nu$ are given by $\mathscr{G}_1 = 1$, the mean curvature $\mathscr{G}_2 = (\kappa_1 + \kappa_2)/2$ and the Gaussian curvature $\mathscr{G}_3 = \kappa_1\kappa_2$, with the principle curvatures κ_1 and κ_2 in 3D. The tensor products of position vectors $\boldsymbol{r}$ and normal vectors $\boldsymbol{n}$ on the boundary $\partial\Omega$ are defined as

$$\boldsymbol{r}^a \otimes \boldsymbol{n}^b := \underbrace{\boldsymbol{r}\otimes\ldots\otimes\boldsymbol{r}}_{a \text{ times}} \otimes \underbrace{\boldsymbol{n}\otimes\ldots\otimes\boldsymbol{n}}_{b \text{ times}} \tag{9.19}$$

with the symmetric tensor product $(a\otimes b)_{ij} = (a_i b_j + a_j b_i)/2$. In the case of MTs of rank $a+b=2$, ten second order tensor measures are obtained with the respective tensor products $\boldsymbol{r}\otimes\boldsymbol{r}$, $\boldsymbol{r}\otimes\boldsymbol{n}$ and $\boldsymbol{n}\otimes\boldsymbol{n}$. Linear dependencies exist between these tensors, cf. [STKB$^+$10]. A number of six linearly independent MTs result, cf. [STMK$^+$11], i.e.

$$\mathscr{W}_0^{2,0} := \int_{\Omega} \boldsymbol{r}\otimes\boldsymbol{r}\,\mathrm{d}V, \qquad \mathscr{W}_1^{2,0} := \frac{1}{3}\int_{\partial\Omega} \boldsymbol{r}\otimes\boldsymbol{r}\,\mathrm{d}A, \tag{9.20}$$

$$\mathscr{W}_2^{2,0} := \frac{1}{3}\int_{\partial\Omega} \mathscr{G}_2(\boldsymbol{r})\boldsymbol{r}\otimes\boldsymbol{r}\,\mathrm{d}A, \qquad \mathscr{W}_3^{2,0} := \frac{1}{3}\int_{\partial\Omega} \mathscr{G}_3(\boldsymbol{r})\boldsymbol{r}\otimes\boldsymbol{r}\,\mathrm{d}A, \tag{9.21}$$

$$\mathscr{W}_1^{0,2} := \frac{1}{3}\int_{\partial\Omega} \boldsymbol{n}\otimes\boldsymbol{n}\,\mathrm{d}A, \qquad \mathscr{W}_2^{0,2} := \frac{1}{3}\int_{\partial\Omega} \mathscr{G}_2(\boldsymbol{r})\boldsymbol{n}\otimes\boldsymbol{n}\,\mathrm{d}A. \tag{9.22}$$

Higher order MTs are well defined but will not be considered here, for further information the reader is referred to [SGM08]. MTs can be related to the well-known engineering measure tensor of inertia $\mathbb{I}$ by

$$\mathbb{I}(\Omega) = \int_{\Omega} (-\boldsymbol{r} \otimes \boldsymbol{r} + |\boldsymbol{r}|^2 \mathbf{1}) \, \mathrm{d}V = -\mathscr{W}_0^{2,0} + \mathrm{tr}(\mathscr{W}_0^{2,0})\mathbf{1}, \tag{9.23}$$

with the three dimensional unit tensor $\mathbf{1}$, cf. [STMK+11]. From the above MTs (9.20) to (9.22), motion invariant and motion covariant measures can be distinguished considering a pure translational shift of the considered body. Obviously due to the consideration of position vectors, $(9.20)_1$, $(9.20)_2$, $(9.21)_1$ and $(9.21)_2$, change under a pure translational shift of body $\boldsymbol{\Omega}$. However, $(9.22)_1$ and $(9.22)_2$ remain unchanged under a shift, since only normal vectors on the surface contribute to the tensor product. A change in orientation, i. e. rotation of the considered body would result in changes in all measures (9.20) through (9.22).

MTs are generally defined for convex bodies. Due to the additivity theorem, their definition can also be applied to non-convex bodies, cf. [STMK+11]. In the analysis of spatial structures, one typically uses discrete data in the form of images or 3D data sets. MTs can be evaluated for this type of data, e. g. by considering a triangulation as an approximation of the convex body. The specific formulas for the computation of MTs for convex and non-convex polytopes are given in [STMK+13].

9.2.3.3 Anisotropy Measure Based on MT

For the application as a descriptor of microstructures, the aim is to deduce relevant measures from a MT. A measure describing the anisotropy of a body can be defined following [KMS+10] by considering the eigenvalue ratio of the respective MT. Using the minimal and maximal eigenvalue given by $\mu_{\min}$ and $\mu_{\max}$, respectively, with $|\mu_{\max}| \geq |\mu_{\min}|$, one obtains

$$\beta_{\nu}^{a,b} := \frac{|\mu_{\min}|}{|\mu_{\max}|} \in [0,1] \tag{9.24}$$

for a considered MT $\mathscr{W}_{\nu}^{a,b}$. The different MTs in (9.20) to (9.22) describe different properties of the underlying body. While $\mathscr{W}_0^{2,0}$ describes the distribution of mass inside the body, $\mathscr{W}_1^{0,2}$ characterizes the surface orientation. In [STMK+11] two-dimensional boolean models of overlapping ellipses with varying values of preferred orientation are analyzed regarding $\beta_1^{0,2}$. A constant value of the average value $\left\langle \beta_1^{0,2} \right\rangle = 1$ is found for a sufficiently large number of particles in the isotropic case, whereas the average value $\left\langle \beta_1^{0,2} \right\rangle \approx 0.44$ results from the analysis of aligned ellipses, which equals the result from the analysis of a single pixelized ellipse in the analysis. An analysis of local properties of a turing pattern system is carried out in [STKB+10], where regions of interest are the ones with non-lamellar concentration of one chemical. Here, the Minkowski tensor $\mathscr{W}_1^{0,2}$ can identify the local non-lamellar regions in the pattern, corresponding to regions with an isotropic value of $\beta_1^{0,2}$. Based on these observations, the MT $\mathscr{W}_1^{0,2}$ is used for the description of microstructures in this sequel. Certainly, the procedure can be adapted for other MTs.

9.2.3.4 Orientation Measure Based on $\mathscr{W}_1^{0,2}$

From the eigenvalue analysis of the MT $\mathscr{W}_1^{0,2}$, the eigenvectors carry information about the orientation of the analyzed body. In the case of ellipsoidal bodies in three-dimensional space, the eigenvectors correspond to the direction of orientation axes of the ellipsoid. For a more detailed analysis, the bodies of ellipsoidal shape can be separated into three cases regarding their symmetry properties: **(a)** ellipsoids without any axial rotational symmetry, i. e. three differing radii, **(b)** ellipsoids exhibiting rotational symmetry about one axis, i. e. two radii are equal and **(c)** spheres, where all three radii are equal. The eigenvalues and eigenvectors show specific properties related to the cases **(a)**, **(b)** and **(c)**.

(a) An ellipsoidal body with no rotational symmetry properties yields a MT $\mathscr{W}_1^{0,2}$ with three distinct eigenvalues $\mu_1 \neq \mu_2 \neq \mu_3$, with μ_i for $i = 1,2,3$ being the eigenvalues of $\mathscr{W}_1^{0,2}$ in ascending order, thus $\mu_1 \geq \mu_2 \geq \mu_3$, one eigenvector associated to each distinct eigenvalue. The directions of these eigenvectors $\boldsymbol{v}_1$, $\boldsymbol{v}_2$ and $\boldsymbol{v}_3$ correspond to the three axis of the ellipsoid, which represent the main directions of orientation of the body. This case is depicted in Figure 9.2 (a).

(b) Bodies of ellipsoidal shape which possess one axis of rotational symmetry result in a MT $\mathscr{W}_1^{0,2}$ with one double and one single eigenvalue, i. e. $\mu_1 = \mu_2 \neq \mu_3$ or $\mu_1 \neq \mu_2 = \mu_3$ and associated eigenvectors. It is observed that the direction of the eigenvector associated to the distinct eigenvalue coincides with the direction of rotational symmetry axis. Thus, this eigenvector gives information about a preferred direction of orientation of the ellipsoidal body. This is illustrated in Figure 9.2(b). The eigenvectors associated to the double eigenvalue span a plane perpendicular to the axis of symmetry which represents the space of all possible eigenvectors associated to the double eigenvalue.

(c) Ellipsoids of spherical shape show an isotropic distribution of surface orientation, thus a value of $\beta_1^{0,2} = 1$ results due to the fact that a triple eigenvalue $\mu_1 = \mu_2 = \mu_3$ is found. Consequently, three ambiguous eigenvectors are obtained. This observation is in accordance with the arbitrariness of rotational symmetry axis in a sphere, where an infinite number of symmetry axes exist due to spherical symmetry. In this special case no distinct axis of symmetry or direction of orientation of the body can be detected. An illustration of this case is given in Figure 9.2(c). The three eigenvectors span a space of all permissible eigenvectors, i.e. any vector originating in the center of the sphere.

The above described observations for the analysis of eigenvalues and eigenvectors suggest that the direction of eigenvectors associated to distinct eigenvalues correspond to preferred orientation directions of an ellipsoidal body. Thus, this class of eigenvectors is considered in the analysis of the microstructure together with their separation due to the different types of ellipsoids **(a)**, **(b)** and **(c)**. The direction of the distinct eigenvectors can be described in spherical coordinates by two angles, θ

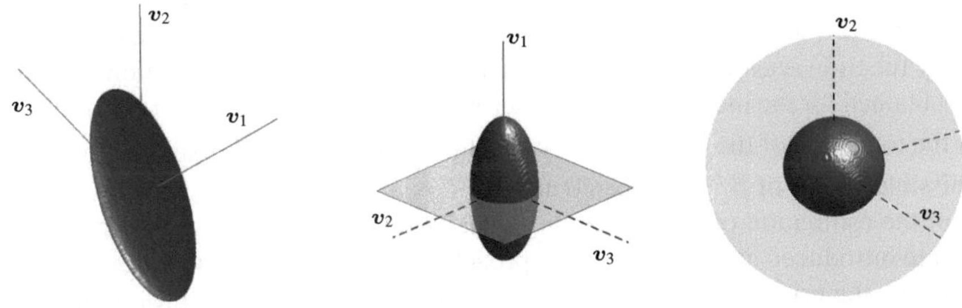

(a) Ellipsoid without rotational symmetry: three distinct eigenvalues $\mu_1 \neq \mu_2 \neq \mu_3$ with associated eigenvectors $\boldsymbol{v}_1$, $\boldsymbol{v}_2$ and $\boldsymbol{v}_3$.

(b) Ellipsoid with rotational symmetry: one distinct eigenvalue and a double eigenvalue $\mu_1 \neq \mu_2 = \mu_3$ with a distinct eigenvector $\boldsymbol{v}_1$ and a plane of possible eigenvectors $\boldsymbol{v}_2$ and $\boldsymbol{v}_3$

(c) Sphere: triple eigenvalue $\mu_1 = \mu_2 = \mu_3$ and three ambiguous vectors $\boldsymbol{v}_1, \boldsymbol{v}_2$ and $\boldsymbol{v}_3$ spanning the associated eigenspace.

Fig. 9.2 Illustration of possible cases for an ellipsoid and associated eigenvectors of $\mathscr{W}_1^{0,2}$. Dashed lines denote directions of eigenvectors associated to multiple eigenvalues, which span the possible space of all eigenvectors associated to these eigenvalues (gray plane or space). Red solid lines denote directions of eigenvectors associated to distinct eigenvalues, describing distinct orientation directions of the ellipsoidal bodies. Taken from [SBBS14].

and φ. The admissible range of θ and φ can be reduced due to joint consideration of a direction and its inversion with $\theta, \varphi \in [-\pi/2, \pi/2]$.

9.2.3.5 Microstructural Descriptors Based on Probability Density Functions

In order to analyze the microstructure of a two-phase material consisting of an inclusion phase embedded in a matrix phase, the afore-mentioned measures based on MT $\mathscr{W}_1^{0,2}$ will be used. Since this MT is defined as the integral over a single body and based on the above described interpretation of the measures of anisotropy and orientation drawn from $\mathscr{W}_1^{0,2}$, the following procedure is proposed to compute statistical descriptors for the complete microstructure. First, the measures $\beta_1^{0,2}$ and the orientation given by θ and φ are computed for each inclusion in the microstructure. Next, a probability density function is considered in order to take into account the distribution of the respective measure in the complete microstructure. The definition of the probability density function $\mathscr{P}_{\text{MA}}$ for the scalar anisotropy measure $\beta_m = \beta_1^{0,2}$ for an inclusion Ω_m is straight forward, hence

$$\mathscr{P}_{\text{MA}}(i_\beta) := \frac{1}{d_\beta \; n_{\text{incl}}} \sum_{m=1}^{n_{\text{incl}}} \xi(i_\beta) \qquad \text{with} \qquad \xi := \begin{cases} 1 \text{ if } \beta_m = \beta_1^{0,2} \in c_{i_\beta} \\ 0 \text{ else,} \end{cases} \tag{9.25}$$

with a number n_β of categories c_{i_β}, $i_\beta = 1...n_\beta$ with an equal category size $d_\beta = 1/n_\beta$, since $\beta \in [0,1]$ and a total number of inclusions given by n_{incl}. The probability density function is estimated by the use of a histogram over the distribution of values $\beta_1^{0,2}$ obtained for the inclusions in the microstructure, which is then normalized such that the total area of the histogram is equal to one to satisfy $\int_0^1 (\mathscr{P}_{\text{MA}}) = 1$ with the admissible values of $\beta_1^{0,2}$ between 0 and 1.

For the estimation of the probability density function describing the orientation measure introduced above, a similar approach is applied. Since the random variable describing the orientation is two-dimensional (described by two angles θ and φ), a number of $n_\theta \times n_\varphi$ categories c_{i_θ,i_φ} with $i_\theta = 1...n_\theta$ and $i_\varphi = 1...n_\varphi$ and an equal category size $d_\theta = \pi/n_\theta$ and $d_\varphi = \pi/n_\varphi$, due to $\theta, \varphi \in [-\pi/2, \pi/2]$, is obtained. The probability density function then reads

$$
\begin{aligned}
\mathscr{P}_{\text{MO},k}(i_\theta, i_\varphi) &:= \frac{1}{d_\theta\, d_\varphi\, n_{\text{incl}}} \sum_{m=1}^{n_{\text{incl}}} \xi(i_\theta, i_\varphi) \\
\text{with} \quad \xi &:= \begin{cases} 1 \text{ if } \Omega_m \in \mathscr{B}_k \cup \{\theta_m \in c_{i_\theta}, \varphi_m \in c_{i_\varphi}\} \\ 0 \text{ else,} \end{cases}
\end{aligned} \tag{9.26}
$$

with individual sets of inclusions $\mathscr{B}_k$ and an individual inclusion Ω_m . These sets are determined by the composition of eigenvalues, according to the characterization of ellipsoids described in Section 9.2.3.4 as

$$
\begin{aligned}
\mathscr{B}_{k=1,2,3} &:= \{\Omega_m | \mu_1 > \mu_2 > \mu_3\} &&\rightarrow \quad \text{case } \mathbf{(a)}, \\
\mathscr{B}_{k=4} &:= \{\Omega_m | \mu_1 = \mu_2 > \mu_3\} &&\rightarrow \quad \text{case } \mathbf{(b)}, \\
\mathscr{B}_{k=5} &:= \{\Omega_m | \mu_1 > \mu_2 = \mu_3\} &&\rightarrow \quad \text{case } \mathbf{(b)}, \\
\mathscr{B}_{k=6} &:= \{\Omega_m | \mu_1 = \mu_2 = \mu_3\} &&\rightarrow \quad \text{case } \mathbf{(c)}.
\end{aligned} \tag{9.27}
$$

In detail, the different types of ellipsoids and their orientations represented by direction of eigenvectors are treated separately. In the case of an ellipsoid without rotational symmetry, case **(a)**, three distinct eigenvectors are obtained. The according eigenvectors to each eigenvalue are represented by $k = 1$ for μ_1, $k = 2$ for μ_2 and $k = 3$ for μ_3. For case **(b)**, where a rotationally symmetric ellipsoid about one axis is treated, only one distinct eigenvalue exists, with the according eigenvector treated by $k = 4$ for $\mu_1 = \mu_2 > \mu_3$ and $k = 5$ for $\mu_1 > \mu_2 = \mu_3$. For case **(c)**, i.e. a spherical inclusion, no distinct eigenvalue and thus no distinct direction of eigenvectors can be observed. In this case simply the existence and not the direction of an eigenvector is captured in $\mathscr{P}_{\text{MO},6}$. Hence, in this special case there exists only one category c_{i_θ,i_φ} with $n_\theta = n_\varphi = 1$ and $d_\theta = d_\varphi = \pi$.

Using the presented microstructural descriptors based on the Minkowski tensor $\mathscr{W}_1^{0,2}$, a real DP-steel microstructure is analyzed in the following. It was reconstructed from 3D Electron Backscatter Diffraction measurements at the Max-Planck Institute for Iron Research in Düsseldorf, Germany (Prof. D. Raabe), cf. Figure 9.3(a), for details see [ZRS+06], [BBS+14b], [BBSR11]. The respective probability density distribution of the anisotropy measure is depicted in

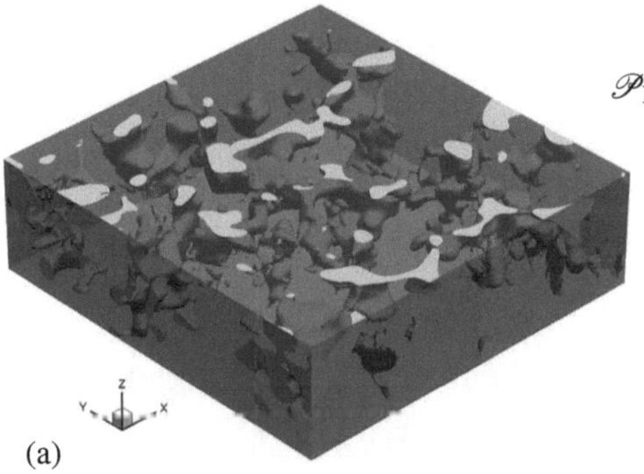

(a)

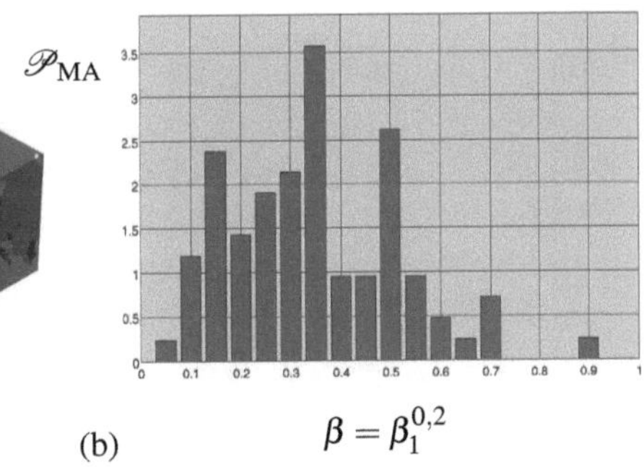

(b)

Fig. 9.3 (a) Realistic DP-steel microstructure (green indicating the martensitic inclusion phase and red indicating the ferritic matrix phase) and (b) probability density function $\mathscr{P}_{MA}$ versus $\beta = \beta_1^{0,2}$, see (9.25). Taken from [SBBS14].

Figure 9.3(b). Figure 9.4 shows the probability distribution of orientation for the relevant cases ($k = 1,2,3$). The other cases were rarely found in the analysis.

9.3 Construction and Analysis of SSRVEs

In this section, SSRVEs based on combinations of different statistical measures will be constructed and analyzed for a real DP steel microstructure. Therefore, specific objective functions are proposed based on different sets of statistical descriptors as well as an objective function for the evaluation of the mechanical response. The DP steel microstructure is obtained from 3D EBSD measurements (electron backscatter diffraction combined with focused ion beam sectioning), cf. [ZRS$^+$06], [BBS$^+$14b], [BBSR11]. The physical size of the microstructure portion is 15.9μm×16.45μm×5.0μm. For the solution of the optimization problem, a differential evolution algorithm incorporated in the optimization framework Mystic, see [MHA09, MSS$^+$11], is used. Based on each objective function, SSRVEs for different parameterization types are constructed. Here, the inclusion morphology of the SSRVEs is described by different numbers of ellipsoidal inclusions which are defined in terms of generalized ellipsoids, whose outer surface is defined by

$$\sum_{l=1}^{3} \left(\frac{|\boldsymbol{v}_l \cdot (\boldsymbol{X} - \boldsymbol{X}_c)|}{r_l} \right)^{p_l} = 1. \tag{9.28}$$

The shape of the generalized ellipsoid is defined by p_l, here a value of $p_l = 2$ is chosen for (classical) ellipsoids. The orientation of semi-axis of the ellipsoid is defined by the vectors $\boldsymbol{v}_l$, the radii on the semi-axis are defined by r_l and the ellipsoid's center position is given by $\boldsymbol{X}_c = [X_c \ Y_c \ Z_c]$. With a restriction to an orthogonal semi-axis coordinate system, the direction of this system can be represented in spherical

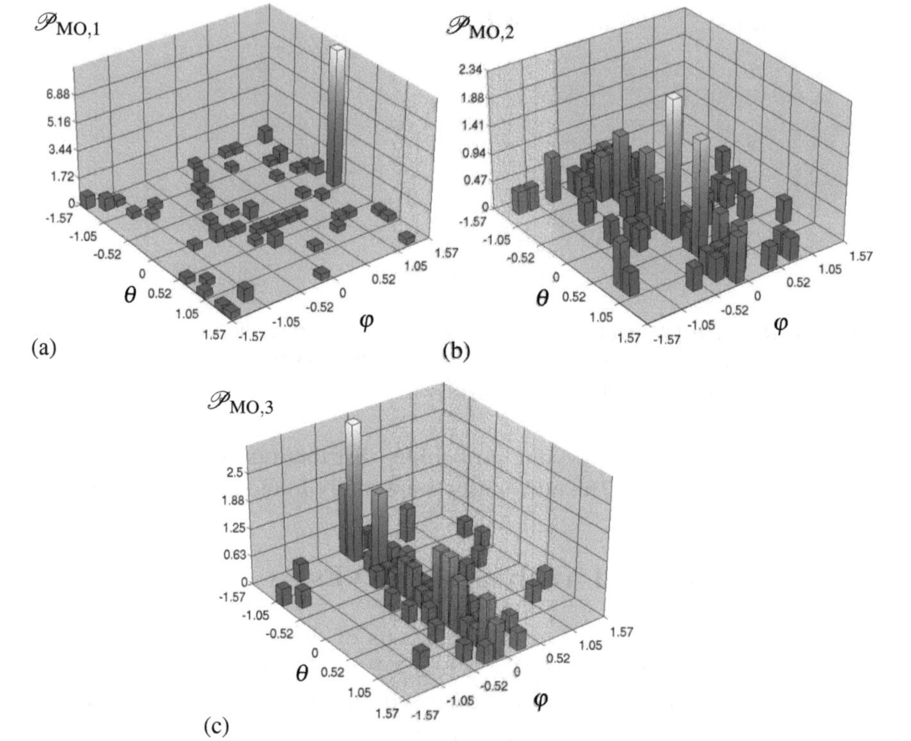

Fig. 9.4 Relevant cases of probability distribution of orientation: (a) $\mathscr{P}_{\mathrm{MO},1}$, (b) $\mathscr{P}_{\mathrm{MO},2}$ and (c) $\mathscr{P}_{\mathrm{MO},3}$ for realistic DP-steel microstructure. Taken from [SBBS14].

coordinates by three angles, i. e. ϑ, φ and ζ. All position $\boldsymbol{X}$ satisfying (9.28) describe points on the boundary of the ellipsoid. Thereby, every ellipsoidal inclusion can be described by

$$\gamma^{(j)} = \left[X_c^{(j)}\ Y_c^{(j)}\ Z_c^{(j)}\ \vartheta^{(j)}\ \varphi^{(j)}\ \zeta^{(j)}\ r_1^{(j)}\ r_2^{(j)}\ r_3^{(j)}\right]^{\mathrm{T}}. \tag{9.29}$$

The different types of parameterization considered here describe SSRVEs with different numbers of ellipsoidal inclusions, thus the vector for parameterization type i is given by

$$\gamma_i = \left[(\gamma^{(1)})^T\ (\gamma^{(2)})^T\ \dots\ (\gamma^{(i)})^T\right]^{\mathrm{T}}. \tag{9.30}$$

The SSRVEs are constructed such that periodic extensibility is provided. Note that the center coordinates of the first ellipsoid are fixed in order to avoid redundancy due to pure translational movement of the inclusion phase. Here, SSRVEs with one to four ellipsoidal inclusions, i.e. $i = 1, 2, 3, 4$, are constructed.

9.3.1 Objective Functions

The different statistical measures presented in Sec. 9.2.3.1 and Sec. 9.2.3.5 have the ability to capture different features of a microstructure morphology. For the construction of SSRVEs for a real (target) microstructure it is beneficial to incorporate multiple measures to account for more statistical information. In order to compare the performance of the statistical measures describing the real microstructure, three objective functions are defined based on weighted sums of least-square functionals of different statistical descriptors. A first objective function taking into account the volume fraction and spectral density is given by

$$\mathscr{E}_I = \omega_\text{V} \mathscr{L}_\text{V} + \omega_\text{SD} \mathscr{L}_\text{SD} \tag{9.31}$$

with the weighting factors for the volume fraction and spectral density given by ω_V and ω_SD, respectively. The least-square functional

$$\mathscr{L}_\text{V}(\gamma_i) = \left(1 - \frac{\mathscr{P}_\text{V}^\text{SSRVE}(\gamma_i)}{\mathscr{P}_\text{V}^\text{target}}\right)^2 \tag{9.32}$$

describes the deviation of the volume fraction of the SSRVE, $\mathscr{P}_\text{V}^\text{SSRVE}$, and the one of the real (target) microstructure, $\mathscr{P}_\text{V}^\text{target}$. The least-square functional of the spectral density reads

$$\mathscr{L}_\text{SD}(\gamma_i) = \frac{1}{\tilde{N}^\text{SD}} \sum_{m=1}^{\tilde{N}^\text{SD}} \left(\mathscr{P}_\text{SD}^\text{target}(\boldsymbol{y}_m) - \mathscr{P}_\text{SD}^\text{SSRVE}(\boldsymbol{y}_m, \gamma_i)\right)^2 \tag{9.33}$$

accounting for the difference of the spectral density of the SSRVE, $\mathscr{P}_\text{SD}^\text{SSRVE}$, and the real (target) microstructure, $\mathscr{P}_\text{SD}^\text{target}$, respectively. In contrast to the comparison of the volume fraction, where scalar values are compared, the spectral density requires a more sophisticated handling. Reasons for this are firstly the higher dimensionality (here a three-dimensional data field) and secondly the difference in data resolution for the SSRVE and the real (target) microstructure. In a preprocessing step, the complete spectral density of the real (target) microstructure is computed in the original resolution of the input data. Using a threshold value to identify the relevant entries of the spectral density, a relevant section of the spectral density is identified and stored. In the subsequent optimization process, only this relevant section of the SSRVE is compared to the one stored for the real (target) microstructure. This requires a rebinning process[1], since the real microstructures input data and the SSRVE do not necessarily have the same resolution. It was noted by [Pov95] that the degree of rebinning strongly influences the accuracy of the overall results and should be performed such that the main characteristics of the spectral density are still captured. The number of relevant entries in (9.33) is denoted by $\tilde{N}^\text{SD}$ with the vector $\boldsymbol{y}_m$ defining the position vector of the individual entry.

---

[1] Here, rebinning stands for the standard procedure to change the resolution of data, which is well-known from image processing.

Another objective function can be considered as an extension of (9.31) by additionally considering the lineal-path function. The resulting objective function reads

$$\mathscr{E}_{II} = \omega_{V}\mathscr{L}_{V} + \omega_{SD}\mathscr{L}_{SD} + \omega_{LP}\mathscr{L}_{LP} \tag{9.34}$$

with the least square functional computed in a similar manner as for (9.33) by

$$\mathscr{L}_{LP}(\gamma_i) = \frac{1}{\tilde{N}^{LP}} \sum_{m=1}^{\tilde{N}^{LP}} \left(\mathscr{P}_{LP}^{target}(y_m) - \mathscr{P}_{LP}^{SSRVE}(y_m, \gamma_i) \right)^2 \tag{9.35}$$

with the lineal-path function computed for the SSRVE and the real microstructure given by $\mathscr{P}_{LP}^{SSRVE}$ and $\mathscr{P}_{LP}^{target}$, respectively. Since the computation of complete lineal-path functions is rather costly, only a subset of line segments is evaluated here, provided that this set of line segments still characterizes the important properties of the microstructure adequately. As for the spectral density, a threshold value is used to determine a relevant section of the lineal-path function, which is later on compared to the respective part of the lineal-path function of the SSRVE in the optimization process. This is again done in a preprocessing step, since the relevant part has to be determined only once for the real (target) microstructure. The relevant section of the lineal-path function is then defined by a total number of $\tilde{N}^{LP}$ entries and the position vector to the individual entries is given by y_m.

In order to analyze the performance of statistical measures computed based on the Minkowski functional $\mathscr{W}_1^{0,2}$ proposed in Section 9.2.3.5, a third objective function is defined by

$$\mathscr{E}_{III} = \omega_{V}\mathscr{L}_{V} + \omega_{SD}\mathscr{L}_{SD} + \omega_{MA}\mathscr{L}_{MA} + \omega_{MO}\mathscr{L}_{MO} \tag{9.36}$$

with the respective least-square functionals

$$\mathscr{L}_{MA} = \frac{1}{N_{MA}} \sum_{m=1}^{N_{MA}} \left(\frac{V_{MA}^{target}(m)}{V^{target}} \mathscr{P}_{MA}^{target}(m) - \frac{V_{MA}^{SSRVE}(m)}{V^{SSRVE}} \mathscr{P}_{MA}^{SSRVE}(m, \gamma_i) \right)^2, \tag{9.37}$$

$$\mathscr{L}_{MO} = \frac{1}{\sum_{k=1}^{6} N_{MO,k}} \sum_{k=1}^{6} \sum_{m=1}^{N_{MO,k}} \left(\frac{V_{MO,k}^{target}(m)}{V_k^{target}} \mathscr{P}_{MO,k}^{target}(m) - \frac{V_{MO,k}^{SSRVE}(m)}{V_k^{SSRVE}} \mathscr{P}_{MO,k}^{SSRVE}(m, \gamma_i) \right)^2. \tag{9.38}$$

Here, the number of categories in the probability density function describing aniso-tropy is given by N_{MA}, whereas the number of categories in the probability density function characterizing the orientation measure $\mathscr{P}_{MO,k}$ is denoted by $N_{MO,k}$ for the different cases $k = 1...6$ described in (9.27). The individual categories in the functions are denoted by m. In the prefactors of the probability density function, the quantities $V_{MA}^{target|SSRVE}(m)$ and $V^{target|SSRVE}$ denote the volume of all inclusions associated to the category m for the anisotropy measure and the volume of the inclusion phase in the considered microstructure, respectively. In the case of the orientation measure, the quantity $V_{\mathrm{MO},k}^{\mathrm{target}|\mathrm{SSRVE}}(m)$ describes the volume of inclusions associated to category m of case k and $V_k^{\mathrm{target}|\mathrm{SSRVE}}$ gives the total volume of inclusions of case k in the real (target) microstructure and the SSRVE. Subsequently, the prefactors $V_{\mathrm{MA}}^{\mathrm{target}|\mathrm{SSRVE}}(m)/V^{\mathrm{target}|\mathrm{SSRVE}}$ and $V_{\mathrm{MO},k}^{\mathrm{target}|\mathrm{SSRVE}}(m)/V_k^{\mathrm{target}|\mathrm{SSRVE}}$ define phase fractions of inclusion phase for individual categories. These factors enable a higher weighting for categories associated to large fractions of inclusion phase and thus enforce the influence of larger inclusions compared to smaller ones.

The objective function $\tilde{r}_\varnothing$ for the comparison of the mechanical response is based on the evaluation of a set of virtual experiments performed for the real (target) microstructure and the SSRVE. Here, the macroscopic stress-strain response is compared and the deviation $r_{e,j}$ is defined by

$$r_{e,j} = \frac{\overline{\sigma}_{e,j}^{\mathrm{target}}(\overline{\varepsilon}_{e,j}) - \overline{\sigma}_{e,j}^{\mathrm{SSRVE}}(\overline{\varepsilon}_{e,j})}{\overline{\sigma}_{e,j}^{\mathrm{target}}(\overline{\varepsilon}_{e,j})}, \tag{9.39}$$

with the individual evaluation point $j = 1...n_{\mathrm{ep}}$ and a total number of evaluation points n_{ep} for each virtual experiment e, $\overline{\sigma}$ denotes the macroscopic Cauchy stress evaluated at the macroscopic engineering strain $\overline{\varepsilon}$. Only values with non-zero denominator are considered in (9.39). An average error for each virtual experiment is computed by

$$\tilde{r}_e = \sqrt{\frac{1}{n_{\mathrm{ep}}} \sum_{j=1}^{n_{\mathrm{ep}}} \left[r_{e,j}(\overline{\varepsilon}_{e,j})\right]^2} \qquad \text{with} \qquad \overline{\varepsilon}_{e,j} = \frac{j}{n_{\mathrm{ep}}} \overline{\varepsilon}_e^{\max}, j = 1...n_{\mathrm{ep}} \tag{9.40}$$

where the maximum strain in each experiment is denoted by $\overline{\varepsilon}_e^{\max}$. As an overall error measure to compare the individual SSRVEs to each other, the average

$$\tilde{r}_\varnothing = \sqrt{\frac{1}{n_{\exp}} \sum_{e=1}^{n_{\exp}} \tilde{r}_e^2} \tag{9.41}$$

for a total number of virtual experiments $n_{\exp}$ is defined. (9.41) also represents the objective function comparing the mechanical response of the SSRVE and the real (target) microstructure. In this contribution, the following virtual experiments are considered: two uniaxial tension tests in x- and z- direction, denoted by subscript x and z in the following, and two simple shear tests with a displacement of the xy-plane in x-direction and in y- direction, respectively, denoted by subscript xy and yx.

In view of the staggered optimization scheme, cf. (9.3), it has to be noted that a fully automated procedure is expensive. To evaluate the outer optimization problem in a reasonable amount of time, the following procedure is used for the construction of the SSRVEs:

1.) Initialize the parameterization type $i = 1$, i.e. consider one inclusion.
2.) Solve the inner optimization problem with the respective objective function given in Sec.9.3.1 in order to identify $\tilde{\gamma}_i$, which reproduces the real microstructure best in terms of statistical measures.
3.) Evaluate $\tilde{r}_{\varnothing}(\tilde{\gamma}_i)$. If the value does not deceed a certain tolerance, increase the complexity of the SSRVE's microstructure morphology by increasing the number of inclusions i, go back to step 2.

The maximum number of inclusions in the SSRVEs was limited to $i = 4$. For the evaluation of the mechanical response using coupled two-scale simulations, the real microstructure and the SSRVEs are discretized using quadratic tetrahedral elements. For the real microstructure, approximately 8.5 million degrees of freedom were obtained. For the SSRVEs, the resulting number of elements is given in the respective sections. Note that mesh convergence studies have been performed to verify the use of a suitable mesh density. Furthermore, in the case of the SSRVEs, which are governed by a periodic geometry, periodic boundary conditions are applied. As the real microstructure does not show a periodic geometry, linear displacement boundary conditions are applied here.

In the description of the staggered optimization problem in (9.3), an optimization with respect to a suitable set of weighting factors is proposed. The weighting factors can be used to emphasize one statistical measure against another and trigger the optimization process. Subsequently, different weighting factors determine different objective functions which pronounce the importance of a certain feature of the function. Since a thorough analysis of the weighting factors is costly, suitable sets of weighting factors are used for the construction of the SSRVEs instead. An analysis of their relevance can be found in [SBBS14] from which suitable sets of weighting factors are taken. These sets focus on a good fitting of the statistical measures of the SSRVE to the ones of the real microstructure. The specific weighting factors are $\omega_{\rm V} = 1$, $\omega_{\rm SD} = 1$, $\omega_{\rm LP} = 1000$, $\omega_{\rm MA} = \omega_{\rm MO} = 1$.

9.3.1.1 Definition of SSRVE Size

The size of the SSRVE plays an important role, because the statistical measures are not necessarily dimensionless. For example, the lineal-path function contains information about the characteristic size of an inclusion in the microstructure and is observed to interfere with the inclusion phase fraction if the size of the SSRVE is chosen too small. Consequently, the SSRVE size is not variable and has to be defined apriori. Therefore, for an estimation of the size of the SSRVE, the following approach is used to circumvent a contradiction between phase fraction and lineal-path function:

1.) Based on the lineal-path function, an estimation for an average inclusion volume in the real (target) microstructure can be made. For this purpose, a threshold $p^{\rm thres}$ is defined and all entries/voxel in the lineal-path function smaller than $p^{\rm thres}$ are deleted, leaving a number of $n_{\rm LP}^{\rm thres}$ entries/voxel. The remaining entries are associated to line segments with a relevant probability and

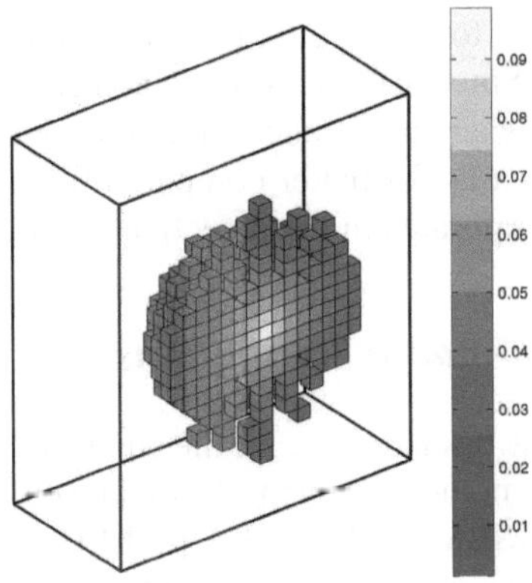

Fig. 9.5 Section with lineal-path function values higher than the threshold of 0.02 representing an estimation for the average inclusion size computed for the microstructure shown in Figure 9.3 (a). Taken from [BSBS14]

therefore define the average inclusion. The volume of this inclusion is computed by $\overline{V}_{\text{inc}}^{\text{target}} = \sum_{i=1}^{n_{\text{LP}}^{\text{thres}}} V_i$, where V_i denotes the unit volume of one voxel in the lineal-path function, cf. Figure 9.5.

2.) With a prescribed number of inclusions i in the SSRVE and the same average size of inclusions as in the real (target) microstructure,i.e. $V_{\text{inc}}^{\text{SSRVE}} = \overline{V}_{\text{inc}}^{\text{target}}$, the SSRVE's inclusion phase fraction can be estimated by

$$\mathscr{P}_{\text{V}}^{\text{SSRVE}} = \frac{i\, V_{\text{inc}}^{\text{SSRVE}}}{V^{\text{SSRVE}}} = \frac{i\, \overline{V}_{\text{inc}}^{\text{target}}}{V^{\text{SSRVE}}} \tag{9.42}$$

with the total volume of the SSRVE denoted by V_{SSRVE}. By claiming the same inclusion phase fraction in the SSRVE and in the real (target) microstructure, i.e. $\mathscr{P}_{\text{V}}^{\text{SSRVE}} \equiv \mathscr{P}_{\text{V}}^{\text{target}}$, and rearranging (9.42) with respect to V_{SSRVE} an estimation of the required SSRVE volume is obtained by

$$V^{\text{SSRVE}} = \frac{i\, \overline{V}_{\text{inc}}^{\text{target}}}{\mathscr{P}_{\text{V}}^{\text{target}}}. \tag{9.43}$$

3.) Based on the assumption of a cubic SSRVE with equal edge lengths in all directions one obtains for the required edge lengths

$$L^{\text{SSRVE}} = L_x^{\text{SSRVE}} = L_y^{\text{SSRVE}} = L_z^{\text{SSRVE}} = \sqrt[3]{\frac{i\, \overline{V}_{\text{inc}}^{\text{target}}}{\mathscr{P}_{\text{V}}^{\text{target}}}}. \tag{9.44}$$

The associated unit volume of one voxel is $0.1\mu\text{m}\times0.1\mu\text{m}\times0.1\mu\text{m}$. Following the previously described method, the prescribed SSRVE's sizes, i.e. the edge length L^{SSRVE}, with $i = 1,2,3,4$ inclusions is computed using the above described method to $L_{i=1}^{\text{SSRVE}} = 3.0\mu\text{m}$ for $i = 1$, i.e. one inclusion, $L_{i=2}^{\text{SSRVE}} = 3.8\mu\text{m}$ for $i = 2$, i.e. two inclusions, $L_{i=3}^{\text{SSRVE}} = 4.3\,\mu\text{m}$ for $i = 3$, i.e. three inclusions and $L_{i=4}^{\text{SSRVE}} = 4.8\,\mu\text{m}$ for $i = 4$, thus four inclusions. Since the other statistical measures used here do not show a similar behavior regarding an interference with other measures, the computed size of SSRVEs is used in all cases for better comparability, regardless of whether the lineal-path function is incorporated in the objective function or not.

9.3.2 Coupled Micro-macro Simulations

In the following section, a brief resume of the direct two-scale micro-macro approach, also known as FE^2 method, is given. For details on the framework, the reader is referred to e.g. [Fey99], [SBM98], [MSS99], [Sch14]. The computational homogenization of micro-heterogeneous materials using a direct two-scale homogenization approach is based on a separation of scales into the macroscale and the microscale, where the microscale can be represented by a representative volume element (RVE). This RVE is attached to every gauss point in the macroscopic boundary value problem (BVP), leading to an additional BVP which has to be solved in every macroscopic integration step. On the microscale, deformations are described in terms of a microscopic deformation gradient $\boldsymbol{F}$ with $J := \det[\boldsymbol{F}(\boldsymbol{X})] > 0$, where $\boldsymbol{X}$ is the position vector to a material point in the reference configuration. The first Piola-Kirchhoff stress tensor $\boldsymbol{P}$ is the associated work-conjugated stress measure, which is related to the microscopic Cauchy stresses by $\boldsymbol{\sigma} := \frac{1}{J}\boldsymbol{P}\boldsymbol{F}^{\text{T}}$. For the macroscale, all measures are marked by an overline. Here, the deformation gradient is given by $\overline{\boldsymbol{F}}$ with $\overline{J} := \det[\overline{\boldsymbol{F}}] > 0$ and the macroscopic first Piola-Kirchhoff stress tensor $\overline{\boldsymbol{P}}$ and macroscopic Cauchy stress $\overline{\boldsymbol{\sigma}}$ can be calculated by volumetric averaging over the associated microscopic quantity, i.e.

$$\overline{\boldsymbol{P}} = \frac{1}{V}\int_{\mathscr{B}} \boldsymbol{P}\,\text{dV} \quad \rightarrow \quad \overline{\boldsymbol{\sigma}} := \frac{1}{\overline{J}}\overline{\boldsymbol{P}}\,\overline{\boldsymbol{F}}^{\text{T}}, \tag{9.45}$$

where the domain of the RVE is denoted by $\mathscr{B}$ with the associated volume V. From the macro-homogeneity condition, also known as Hill-Mandel condition, see [Hil63], boundary conditions on the microscale can be derived. Three well-known types are uniform traction boundary conditions, linear displacement boundary conditions and periodic boundary conditions. For increased efficiency in the calculations, the consistent macroscopic moduli are considered here, see [MSS99],[Sch14] and [TW08]. The material behavior of the two phases in a dual phase steel, i.e. martensite and ferrite, are modeled using an isotropic finite elasto-plasticity formulation based on the multiplicative decomposition of the deformation gradient $\boldsymbol{F} = \boldsymbol{F}^{\text{e}}\boldsymbol{F}^{\text{p}}$, with the elastic part $\boldsymbol{F}^{\text{e}}$ and the plastic part $\boldsymbol{F}^{\text{p}}$, see [Krö60], [Lee69]. In this contribution, the plastic behavior of the materials is described by an exponential hardening law, which is superimposed with linear hardening. The plastic strain energy function, cf. [Voc55], is given by

$$\psi^{\text{P}} = y_\infty\alpha - \frac{1}{\eta}(y_0 - y_\infty)\exp(-\eta\alpha) + \frac{1}{2}h\alpha^2 \tag{9.46}$$

Table 9.1 Material parameters for individual phases

	λ / MPa	μ / MPa	y_0 / MPa	y_∞/ MPa	η	h
ferrite	118,846.2	79,230.77	260.0	580.0	9.0	70.0
martensite	118,846.2	79,230.77	1,000.0	2,750.0	35.0	10.0

with the equivalent plastic strains α. The material parameters for both materials are given in Table 9.1 with the Lamé constants λ and μ, the initial yield strength y_0, plastic yield strength at initialization of linear hardening y_∞, the degree of exponential hardening η and the slope of superimposed linear hardening h. An implicit exponential update algorithm is used to integrate the flow rule, preserving plastic incompressibility, see [WA90], [Sim92], [MS92], [Kli00]. Kinematic hardening is not considered due to a lack of experimental data for the individual phases of the material.

9.3.3 SSRVEs Based on Different Sets of Statistical Measures

The following section presents SSRVEs based on different sets of statistical descriptors and compares the statistical and mechanical error compared with the real DP steel microstructure.

Table 9.2 Results of SSRVEs based on volume fraction and spectral density - values of objective function $\mathscr{E}_{\rm I}$, and individual least square functionals, number of tetrahedral elements $n_{\rm ele}$ and mechanical errors $\tilde{r}$ in % for the individual SSRVEs, compare [SBBS14].

$n_{\rm inc}$	$\mathscr{E}_{\rm I}/\ 10^{-3}$	$\mathscr{L}_{\rm V}/\ 10^{-7}$	$\mathscr{L}_{\rm SD}/\ 10^{-3}$	$n_{\rm ele}$	$\tilde{r}_x$	$\tilde{r}_z$	$\tilde{r}_{xy}$	$\tilde{r}_{yx}$	$\tilde{r}_\varnothing$
1	4.17	8.83	4.17	1638	3.3	3.7	5.0	5.0	4.32
2	2.03	1.35	2.03	8455	4.2	3.9	2.0	2.0	3.20
3	1.16	0.97	1.16	10759	2.6	1.2	3.9	3.9	3.11
4	0.74	3.38	0.74	17118	2.7	0.5	3.2	3.2	2.65

9.3.3.1 SSRVEs Based on Volume Fraction and Spectral Density

It has been shown for two-dimensional SSRVEs in [BBS14a] that SSRVEs based on volume fraction and spectral density can be improved by additionally considering the lineal-path function in view of a comparison of the mechanical error with the real target microstructure. A similar behavior is assumed in the 3D case, however the results of SSRVEs based on volume fraction and spectral density enable to observe the improvement of an additional consideration of other statistical measures. The resulting SSRVEs constructed based on a minimization of (9.31) can be found in Figure 9.6 for one, two, three and four ellipsoidal inclusions. The results of the statistical and mechanical comparison are presented in Table 9.2. A decreasing error in the objective function as well as in the overall mechanical error $\tilde{r}_{\varnothing}$ can be observed with an error of 2.65% for an SSRVE with four inclusions.

9.3.3.2 SSRVEs Based on Volume Fraction, Spectral Density and Lineal-Path Function

From the minimization of $\mathscr{E}_{II}$ given in (9.34), representations of SSRVEs with different numbers of inclusions are obtained, which are depicted in Figure 9.7. The results of the comparison of statistical measures and mechanical response are given in Table 9.3. It can be observed that a higher number of inclusions in the SSRVE is accompanied by a lower error in the objective function, which implies a closer match of the statistical measures. This behavior appears to converge with more than three inclusions in the SSRVE as the value of objective function does not decrease significantly. It can be seen that for an increased morphology complexity, i.e. adding more inclusions to the SSRVE, a higher number of finite elements is needed for a sufficient discretization. Note that mesh convergence studies have been carried out to ensure a reasonable mesh density for the SSRVEs. From the comparison of the mechanical response, i.e. $\tilde{r}_{\varnothing}$, it is observed that the SSRVEs with three and four inclusions lead to reasonably low errors, $\tilde{r}_{\varnothing} = 1.45\%$ and $\tilde{r}_{\varnothing} = 1.47\%$, respectively. Comparing the results with the ones of the SSRVEs constructed based on volume fraction and spectral density, a better approximation of the mechanical behavior is found. Based on the above results, the SSRVE with three inclusions is considered as the best representation of the real microstructure based on the statistical descriptors volume fraction, spectral density and lineal-path function, with a good degree of accurateness in the mechanical response and a moderate number of finite elements. This SSRVE is denoted by $\text{SSRVE}_{\text{LP}}^{\text{best}}$ in the following.

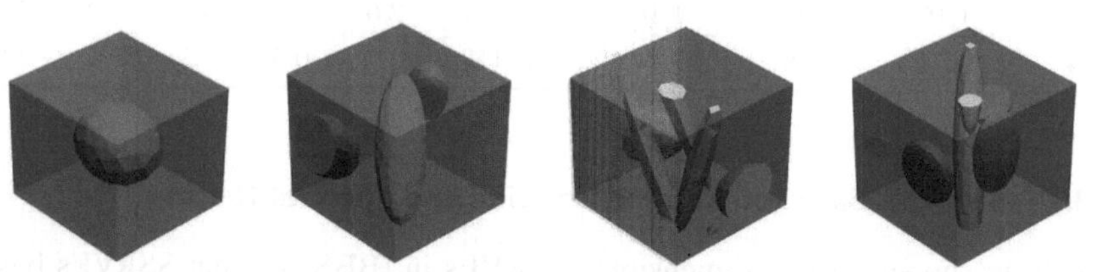

Fig. 9.6 SSRVEs based on volume fraction and spectral density, i.e. objective function $\mathscr{E}_{\text{I}}$, with different numbers of inclusions $n_{\text{inc}} = 1, 2, 3, 4$ from left to right. Taken from [SBBS14].

Table 9.3 Results of SSRVEs based on volume fraction, spectral density and lineal-path function - values of objective function $\mathscr{E}_{\text{II}}$, and individual least square functionals, number of tetrahedral elements n_{ele} and mechanical errors $\tilde{r}$ in % for the individual SSRVEs, compare [SBBS14].

n_{inc}	$\mathscr{E}_{\text{II}}/\ 10^{-2}$	$\mathscr{L}_{\text{V}}/\ 10^{-4}$	$\mathscr{L}_{\text{SD}}/\ 10^{-3}$	$\mathscr{L}_{\text{LP}}/\ 10^{-4}$	n_{ele}	$\tilde{r}_x$	$\tilde{r}_z$	$\tilde{r}_{xy}$	$\tilde{r}_{yx}$	$\tilde{r}_{\varnothing}$
1	8.432	485.29	4.5	0.31	2851	5.3	4.6	5.5	5.5	5.24
2	0.98	32.07	3.5	0.031	5015	0.5	3.6	4.4	4.4	3.60
3	0.53	3.37	3.3	0.017	15714	0.2	2.4	1.2	1.1	1.45
4	0.38	3.11	2.33	0.014	19196	1.2	2.5	0.7	0.7	1.47

9.3.3.3 SSRVEs Based on Volume Fraction, Spectral Density and Minkowski Anisotropy and Orientation Probability Density Function

The SSRVEs constructed based on a minimization of $\mathscr{E}_{III}$, cf. (9.36), are shown in Figure 9.8, the results of the statistical and mechanical comparison are summarized in Table 9.4. Again, it is observed that with a higher number of inclusions the value of objective function and mechanical error decrease. From this set, the SSRVE with four inclusion is considered as the best representation of the real microstructure, showing a mechanical error of $\tilde{r}_{\varnothing} = 1.92\%$ and requiring a number of 12133 elements for the discretization. It is denoted by $\text{SSRVE}^{\text{best}}_{\text{MA|MO}}$ is the following. A decrease of the mechanical error compared with the SSRVEs based on volume fraction and spectral density only is observable for each pair of SSRVEs with the same inclusion number in Table 9.2 and Table 9.4. From the different sets of SSRVEs, it is observed that the individual statistical errors are found to be in varying ranges from $\mathscr{E}_I$, $\mathscr{E}_{II}$ and $\mathscr{E}_{III}$. As an example, the value of least-square error for the volume fraction is found in a range of 10^{-7} for $\mathscr{E}_I$, and in a range of 10^{-4} and 10^{-5} for $\mathscr{E}_{II}$ and $\mathscr{E}_{III}$. The degree of approximation of an individual statistical measure seems to be influenced by the other measures incorporated in the objective function and the weighting factors leveling the individual errors. Furthermore, the ranges of errors of the different statistical measures are naturally located in different ranges, which is why a direct comparison of the individual statistical errors is difficult without further analysis of the range of individual statistical measures and weighting factors.

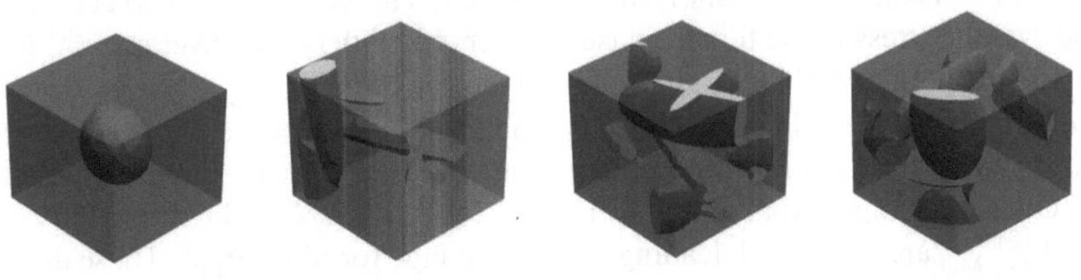

Fig. 9.7 SSRVEs based on volume fraction, spectral density and lineal-path function, i.e. objective function $\mathscr{E}_{\text{II}}$, with different numbers of inclusions $n_{\text{inc}} = 1, 2, 3, 4$ from left to right.

Fig. 9.8 SSRVEs based on volume fraction, spectral density and Minkowski measures, i.e. objective function $\mathscr{E}_{\text{III}}$, with different numbers of inclusions $n_{\text{inc}} = 1, 2, 3, 4$ from left to right. Taken from [SBBS14].

Table 9.4 Results of SSRVEs based on volume fraction, spectral density and Minkowski measures - values of objective function $\mathscr{E}_{\text{III}}$, and individual least square functionals, number of tetrahedral elements n_{ele} and mechanical errors $\tilde{r}$ in % for the individual SSRVEs, compare [SBBS14].

n_{inc}	$\mathscr{E}_{\text{III}}/\ 10^{-2}$	$\mathscr{L}_{\text{V}}/\ 10^{-4}$	$\mathscr{L}_{\text{SD}}/\ 10^{-3}$	$\mathscr{L}_{\text{MA}}/\ 10^{-2}$	$\mathscr{L}_{\text{MO}}/\ 10^{-2}$	n_{ele}	$\tilde{r}_x$	$\tilde{r}_z$	$\tilde{r}_{xy}$	$\tilde{r}_{yx}$	$\tilde{r}_{\varnothing}$
1	6.47	0.001	4.56	5.88	1.27	2519	4.0	3.7	5.4	5.4	4.69
2	4.20	0.006	2.48	1.55	2.39	5358	3.4	1.9	3.4	3.5	3.12
3	1.95	1.65	3.3	0.39	1.2	10087	2.5	0.7	3.5	3.5	2.79
4	1.41	0.19	2.01	0.38	0.82	12133	3.3	1.1	1.1	1.2	1.92

9.3.4 Comparison of Stress on Microscale

For a detailed comparison of the mechanical response of the best SSRVEs, i.e. $\text{SSRVE}_{\text{LP}}^{\text{best}}$ and $\text{SSRVE}_{\text{MA}|\text{MO}}^{\text{best}}$, with the real microstructure, the stresses on the microscale are taken into consideration. By volumetrically averaging the stresses only over the individual phases, the individual error measures r_e^{fer} and r_e^{mar} are obtained for the ferrite and martensite phase, respectively, and therein for each virtual experiment $e = x, y, xy, yx$. These error measures denote the average stress deviation in the ferrite phase and martensite phase, respectively. The results of the individual phase stress deviation for all four virtual experiments are depicted in Figure 9.9-9.12, where the average error r_e^{fer}, respectively r_e^{mar}, is plotted versus the macroscopic deformation of the complete SSRVE, i.e. $\Delta\bar{l}_e/\bar{l}_{e,0}$. The averaged errors are computed analogously to (9.39)-(9.41) and are summarized in Table 9.5. Therein, a very good accordance of the average stress in the ferrite phase is observed, with overall averages $\tilde{r}_{\varnothing}^{\text{fer}}$ lower than 1% for both SSRVEs. A slight advantage for $\text{SSRVE}_{\text{MA}|\text{MO}}^{\text{best}}$ can be detected, however both errors are in the same range. In the martensite phase, the average stress in the SSRVEs does not reproduce the average stress in the target microstructure as well. Overall average errors of $\tilde{r}_{\varnothing}^{\text{mar}} = 13.38\%$ for $\text{SSRVE}_{\text{LP}}^{\text{best}}$ and $\tilde{r}_{\varnothing}^{\text{mar}} = 17.69\%$ for $\text{SSRVE}_{\text{MA}|\text{MO}}^{\text{best}}$ are computed, leading to an advantage for $\text{SSRVE}_{\text{LP}}^{\text{best}}$. These discrepancies are not recognizable in the overall errors computed for the full macroscopic mechanical behavior, since the low martensite phase fraction compared with the ferrite phase fraction reduces the significance of the martensite microscopic response.

In general, the deficiencies in resemblance of the average stress in the martensite phase can be related to the different boundary conditions applied for the SSRVE and the target microstructure. With the application of linear displacement boundary conditions needed for the non-periodic target microstructure, a stiffening effect occurs predominantly in the stiffer phase, i.e. in the martensite. This can also be observed in Figure 9.13, where the von Mises stresses $\boldsymbol{\sigma}_{\text{vm}}$ are depicted for the target microstructure, $\text{SSRVE}_{\text{LP}}^{\text{best}}$ and $\text{SSRVE}_{\text{MA|MO}}^{\text{best}}$ in tension test in x-direction with a cut through each microstructure. This boundary stiffening effect does not occur in the SSRVE, where periodic boundary conditions are applied.

9.3.4.1 Comparison of Computational Efficiency

In the following the aspect of computational efficiency is considered. Here, not only the computation time of SSRVEs in FE^2 simulations is of interest but also the optimization time needed for the generation of an SSRVE. Both SSRVEs considered

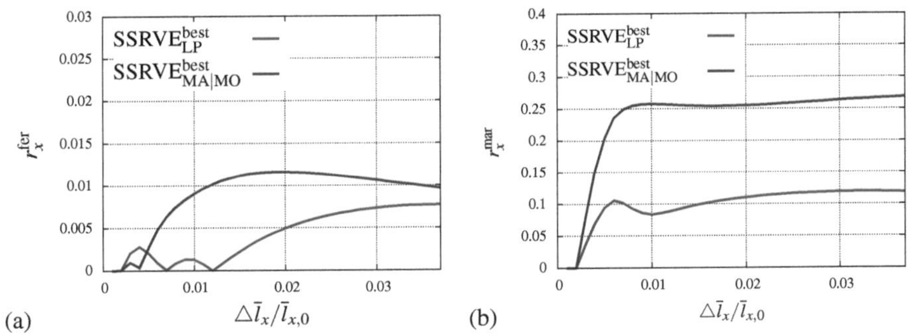

Fig. 9.9 Deviation r_x of averaged stress in x-direction of $\text{SSRVE}_{\text{LP}}^{\text{best}}$ and $\text{SSRVE}_{\text{MA|MO}}^{\text{best}}$ with (a) r_x^{fer} for ferrite and (b) r_x^{mar} for martensite. Taken from [SBBS14].

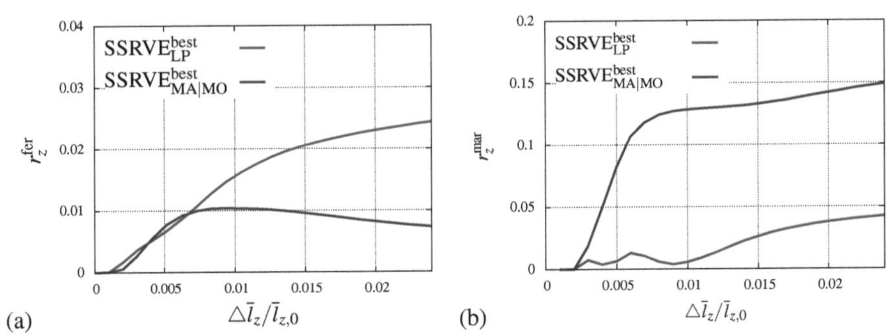

Fig. 9.10 Deviation of averaged stress in z-direction of $\text{SSRVE}_{\text{LP}}^{\text{best}}$ and $\text{SSRVE}_{\text{MA|MO}}^{\text{best}}$ with (a) r_z^{fer} for ferrite and (b) r_z^{mar} for martensite. Taken from [SBBS14].

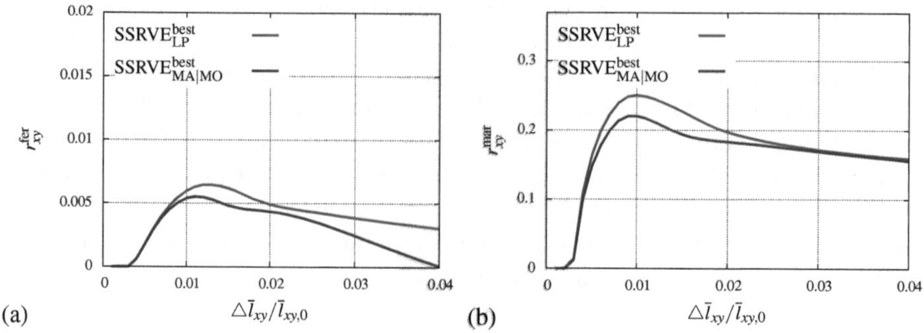

Fig. 9.11 Deviation of averaged stress in shear test xy of $SSRVE^{best}_{LP}$ and $SSRVE^{best}_{MA|MO}$ with (a) r^{fer}_{xy} for ferrite and (b) r^{mar}_{xy} for martensite. Taken from [SBBS14].

Table 9.5 Comparison of stresses on microscale for individual phases in $SSRVE^{best}_{LP}$ and $SSRVE^{best}_{MA|MO}$ evaluated for all four virtual experiments. Taken from [SBBS14].

Ferrite	$\tilde{r}^{fer}_{x}$ in %	$\tilde{r}^{fer}_{z}$ in %	$\tilde{r}^{fer}_{xy}$ in %	$\tilde{r}^{fer}_{yx}$ in %	$\tilde{r}^{fer}_{\varnothing}$ in %
$SSRVE^{best}_{LP}$	0.43	1.55	0.40	0.36	0.85
$SSRVE^{best}_{MA\|MO}$	0.90	0.75	0.29	0.30	0.62
Martensite	$\tilde{r}^{mar}_{x}$ in %	$\tilde{r}^{mar}_{z}$ in %	$\tilde{r}^{mar}_{xy}$ in %	$\tilde{r}^{mar}_{yx}$ in %	$\tilde{r}^{mar}_{\varnothing}$ in %
$SSRVE^{best}_{LP}$	9.94	2.08	17.82	17.19	13.38
$SSRVE^{best}_{MA\|MO}$	23.59	11.36	16.56	16.10	17.69

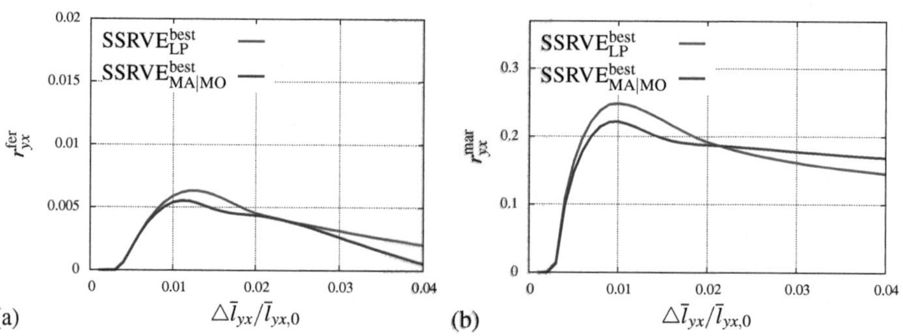

Fig. 9.12 Deviation of averaged stress in shear test yx of $SSRVE^{best}_{LP}$ and $SSRVE^{best}_{MA|MO}$ with (a) r^{fer}_{yx} for ferrite and (b) r^{mar}_{yx} for martensite. Taken from [SBBS14].

here, $SSRVE^{best}_{LP}$ and $SSRVE^{best}_{MA|MO}$, are discretized with a similar number of elements, 15714 for $SSRVE^{best}_{LP}$ and 12113 for $SSRVE^{best}_{MA|MO}$, leading to the conclusion that the efficiency of numerical two-scale calculations will also be similar, with a slight advantage for $SSRVE^{best}_{MA|MO}$. With respect to the optimization time required

(a) (b) (c)

$\langle \sigma_{vM} \rangle = 393.5$ $\langle \sigma_{vM} \rangle = 391.0$ $\langle \sigma_{vM} \rangle = 389.3$ σ_{vM}

Fig. 9.13 Von Mises stress σ_{vM} in MPa for tension test in x-direction showing the cross-section obtained from cutting through (a) the target structure (15.9 μm × 16.45 μm × 5.0 μm), (b) $\text{SSRVE}_{\text{LP}}^{\text{best}}$ (4.3 μm × 4.3 μm × 4.3 μm) and (c) $\text{SSRVE}_{\text{MA|MO}}^{\text{best}}$ (4.3 μm × 4.3 μm × 4.3 μm) and respective volume average values $\langle \sigma_{vM} \rangle$. Taken from [SBBS14].

Table 9.6 Average evaluation times t_{LP} and $t_{\text{MA|MO}}$ for the minimization of the objective functions $\mathscr{E}_{\text{II}}$ and $\mathscr{E}_{\text{III}}$ for $\text{SSRVE}_{\text{LP}}^{\text{best}}$ and $\text{SSRVE}_{\text{MA|MO}}^{\text{best}}$ in seconds. Taken from [SBBS14].

t_{LP}/s	$t_{\text{MA\|MO}}/s$	$t_{\text{LP}}/t_{\text{MA\|MO}}$
5.98	0.0031	1929.03

to solve the optimization problem and to identify the SSRVE, the two candidate objective functions $\mathscr{E}_{II}$ and $\mathscr{E}_{III}$ differ significantly. The time needed for one function evaluation of the objective function depends on the incorporated statistical measures and in some cases also on the inclusion morphology. For specific inclusion morphologies, the evaluation of the lineal-path function and Minkowski functionals is more expensive than for others, while the inclusion morphology has negligible effect on the evaluation time of spectral density and volume fraction. In the considered evolutionary algorithm, the evaluation time of one generation of candidate solutions can differ significantly, depending on the evaluated inclusion morphologies. For a decent comparison of evaluation time of the optimization processes, only the evaluation time of the objective function for the best candidate of every generation will be summed up. Dividing this value by the total number of generations to identify the SSRVE, a convincing comparative measure is obtained as an average evaluation time. These values t_{LP} and $t_{\text{MA|MO}}$ are shown in Table 9.6 as well as their ratio $t_{\text{LP}}/t_{\text{MA|MO}}$. With a factor of almost 2000, the objective function based on Minkowski functionals offers a large speedup in the optimization compared to the objective function using the lineal-path function. With regard to an efficient solution of the inner optimization problem in an automated process, measures based on Minkowski functionals show a higher efficiency.

9.3.5 Analysis of Bounds

In the previous section, SSRVEs are presented as the "best" representation of a real microstructure. As described in Section 9.2.2, the upper bound or the "worst" case scenario of a VE is constructed in the following, cf. [BSBS14]. With regard to the maximization of the inner optimization problem based on $\mathscr{E}_{II}$, i.e. using volume fraction, spectral density and lineal-path function, one would obviously obtain a microstructure only containing ferrite or martensite, yielding a maximization of the error of the volume fraction, hence trivial bounds are represented with no inherent microstructure in the RVE. Considering a minimization of the error of volume fraction together with a maximization of the error of the remaining statistical measures, a more sophisticated "worst" VE scenario can be constructed by

$$\overset{\star}{\mathscr{E}}^{\text{worst}} = \min_{\gamma_i} \left[\overset{\star}{\mathscr{E}}(\gamma_i) \right], \tag{9.47}$$

with the modified objective function given by

$$\overset{\star}{\mathscr{E}}(\gamma_i) = \omega_{\text{V}} \mathscr{L}_{\text{V}}(\gamma_i) + \left[\omega_{\text{SD}} \mathscr{L}_{\text{SD}}(\gamma_i) + \omega_{\text{LP}} \mathscr{L}_{\text{LP}}(\gamma_i)\right]^{-1}. \tag{9.48}$$

Thereby a VE is constructed with a reasonable phase fraction while the spectral density and lineal-path function, statistical measures which provide information on shape and distance of inclusions, are as dissimilar as possible compared to the measures of the target microstructure. Here, a parameterization with three ellipsoidal inclusions is considered. The parameterization of the "worst" VE is obtained by

$$\overset{\star}{\gamma}_i = \arg \left[\overset{\star}{\mathscr{E}}^{\text{worst}} \right]. \tag{9.49}$$

Figure 9.14 shows the representation of the obtained "worst" VE (top picture) and the according "best" RVE (bottom picture), which coincides with the SSRVE constructed in Section 9.3. It shall be noted that the "worst" VE here represents approximately the Voigt-bound in y- and z-direction, where a parallel arrangement of the phases occurs and the Reuss-bound in x-direction, due to the serial composition of the phases. In order to compare the results regarding a similarity of statistical

Table 9.7 Values of the objective function $\mathscr{E}_{\text{II}}$ following (9.34) and associated mechanical errors $\widetilde{r}$ for the "worst" case VE and SSRVE based on (9.34) depicted in Figure 9.14, n_{ele} denotes the number of degrees of freedom of the microscopic FE-discretization, cf. [BSBS14].

	$\mathscr{E}_{\text{II}}$ [$\times 10^{-3}$]	$\mathscr{L}_{\text{V}}$ [$\times 10^{-3}$]	$\mathscr{L}_{\text{SD}}$ [$\times 10^{-3}$]	$\mathscr{L}_{\text{LP}}$ [$\times 10^{-3}$]	n_{ele}	$\widetilde{r}_{\text{x}}$ [%]	$\widetilde{r}_{\text{z}}$ [%]	$\widetilde{r}_{\text{xy}}$ [%]	$\widetilde{r}_{\text{yx}}$ [%]	$\widetilde{r}_{\varnothing}$ [%]
worst	132.62	1.245	11.322	6.4797	4723	33.65	37.26	6.35	6.5	25.5
best	5.3	0.34	3.3	0.0017	15714	0.2	2.4	1.2	1.1	1.45

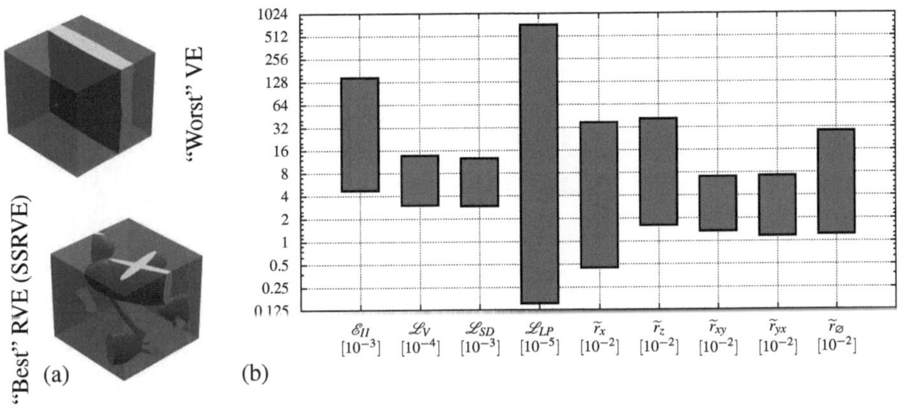

Fig. 9.14 (a) "Worst"- and "best"-case scenario of a VE or RVE with 3 inclusions and (b) intervals between bounds of values of least-square functionals and their associated mechanical errors. Taken from [BSBS14].

descriptors, the individual least-square errors of the "best" case, the SSRVE, given by $\mathscr{L}_{\text{V}}(\gamma_i)$, $\mathscr{L}_{\text{SD}}(\gamma_i)$ and $\mathscr{L}_{\text{LP}}(\gamma_i)$ and the value of objective function $\mathscr{E}_{\text{II}}(\gamma_i)$, as well as the respective values for the "worst" case, i.e. $\mathscr{L}_{\text{V}}(\overset{\star}{\gamma}_i)$, $\mathscr{L}_{\text{SD}}(\overset{\star}{\gamma}_i)$ and $\mathscr{L}_{\text{LP}}(\overset{\star}{\gamma}_i)$ and the value of objective function $\mathscr{E}_{\text{II}}(\overset{\star}{\gamma}_i)$, see Table 9.7, are used. Figure 9.14 shows the differences between the lower and the upper bound of the objective function, individual least-square errors and individual errors of mechanical tests $\tilde{r}_x$, $\tilde{r}_z$, $\tilde{r}_{xy}$ and $\tilde{r}_{yx}$ as well as the overall average mechanical error $\tilde{r}_\varnothing$ to emphasize the magnitude of the difference. Here, a logarithmic scaling on the y-coordinate is used, additionally the errors are shown in different decimal powers. It can be seen that the value of objective function differs by more than one order of magnitude, the error of the lineal-path function even varies by over three orders of magnitude. This strong deviation results in a large difference of the mechanical errors, which show a difference of up to one order of magnitude in $\tilde{r}_\varnothing$ and individual errors as high as two orders of magnitude for the tension test in x-direction. An illustration of the individual mechanical errors r_e is shown in Figure 9.15, where the error is plotted over the macroscopic strain $\triangle\bar{l}_e/\bar{l}_{e,0}$, where e denotes the respective virtual experiment, cf. Section 9.3.1. Here, a difference of the mechanical response in a tension test in x-direction of up to 50% for the "worst" VE is observed, while the SSRVE results in an error of only 1%, a similar behavior is prevalent in the tension test in z-direction. For the shear tests, the error differs not as significantly, with a range of 2% for the SSRVE and 8% for the "worst" VE. The analysis of upper and lower bounds emphasizes the importance of considering further statistical measures aside from the volume fraction. A severe error can result from disregarding further measures, which might only become visible in certain mechanical tests.

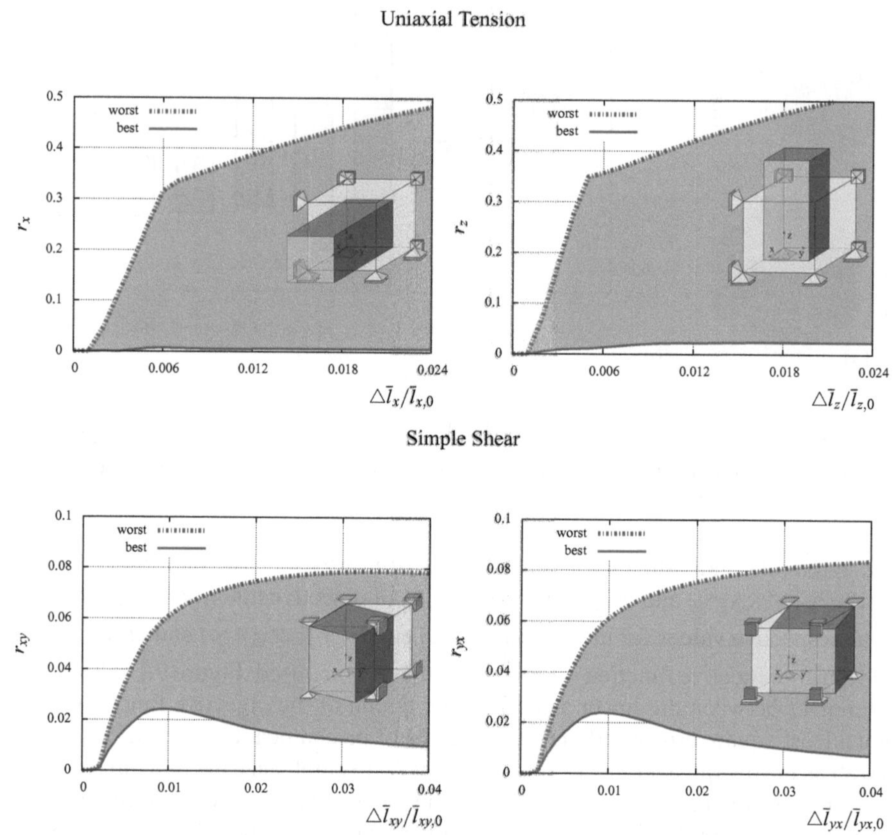

Fig. 9.15 Mechanical errors for "best" and "worst" RVE considering 3 ellipsoidal inclusions. Taken from [BSBS14].

9.4 Conclusion

This contribution details the method of constructing three-dimensional SSRVEs, which has been proposed for the two-dimensional case in [BSB09], [SBB11] and [BBS14a] and generalized to 3D SSRVEs in [BSBS14]. SSRVEs are characterized by a microstructure of reduced complexity compared with RVEs based on real random microstructures. Due to this reduction, SSRVEs can lead to advantages in direct two-scale simulations by vastly reducing computational costs. The method of construction of SSRVEs here uses a staggered optimization scheme, as proposed in [BSBS14] and extended in [SBBS14]. Therein, an outer optimization problem controls the similarity of the SSRVE with a real target microstructure regarding mechanical behavior, whereas the inner optimization problem ensures similarity with respect to statistical measures. A variation of statistical measures as well as different parameterization types for the SSRVE have been investigated. The statistical measures play an essential role in the construction of SSRVEs, since they determine the properties of the SSRVE being fitted to the ones of the real target microstructure. For an efficient construction of SSRVEs, it is advantageous to consider statistical measures which capture necessary information of the microstructure while being realizable in low computation time. Therefore, Minkowski functionals, as a base for microstructural descriptors, have been investigated here. As Minkowski functionals are defined as integrals over single geometrical objects, probability density functions have been proposed as a measure for a consideration of a microstructure built up by multiple inclusions in a matrix phase.

SSRVEs have been constructed with different numbers of inclusions considering different sets of statistical descriptors, in detail combinations of volume fraction, spectral density, lineal-path function and probability density functions based on the tensorial Minkowski functional $\mathscr{W}_1^{0,2}$ are used. It turns out that SSRVEs solely based on volume fraction and spectral density are outperformed by SSRVE additionally taking into account either the lineal-path function or probability density functions based on a Minkowski tensor regarding the comparability of mechanical behavior, regardless of the parameterization of the SSRVE, i.e. the number of inclusions. Overall, an SSRVE with three inclusions based on volume fraction, spectral density and lineal-path function performed best, with an overall error of $\tilde{r}_\varnothing = 1.45$ and a discretization using 15714 elements, whereas four inclusions are needed when considering volume fraction, spectral density and probability density functions based on the Minkowski tensor $\mathscr{W}_1^{0,2}$ instead to achieve $\tilde{r}_\varnothing = 1.92$ using a discretization of 12113 elements, cf. [SBBS14]. Hence, from the number elements in the discretizations, no strong advantage is expected for either on the SSRVEs regarding computation time. In order to take a closer look at the stress behavior in the individual phases, the average stress in each phase is calculated and compared with the average stress per phase in the target structure. Here, a very good accordance in both "best" SSRVEs mentioned above is visible in the ferrite phase, while the comparison of the average stress in the martensite phase shows some deficiencies. This difference of average stress can be ascribed to the different boundary conditions used in the virtual experiments for the real microstructure and the SSRVEs. While the SSRVEs are constructed obeying periodic extensibility and modeled using periodic boundary conditions, the real microstructure is modeled using linear displacement boundary conditions. The latter ones lead to a stiffening primarily in the stiffer phase, i.e. martensite, which does not occur under periodic boundary conditions.

In view of efficiency of the optimization procedure, a comparison of an average optimization time per generation in a differential evolution algorithm has been carried out for the "best" SSRVEs. It turns out that the measures based on the Minkowski tensor offer a speedup factor of 2000 compared to considering the lineal-path function, cf. [SBBS14]. In view of an efficient solution of the inner optimization problem, this plays an important role, especially when aiming for an automated optimization procedure of the staggered optimization problem.

The general importance of considering more sophisticated statistical measures for the construction of SSRVEs is underlined in an analysis of upper bounds of RVEs. A "worst" case RVE is constructed by a maximization of the error of statistical measures, whereas the volume fraction is matched to the according value of the real microstructure, cf. [BSBS14]. The comparison of the range of error of statistical measures and especially the range of mechanical error, which can be up to 50% for the "worst" case VE while the SSRVE leads to an error of 1%, emphasizes the need for a detailed regard of microstructure morphology, which can be described by statistical measures. This is not only valid for the construction of SSRVEs but for other types of reconstructions based on real microstructures.

Acknowledgements. The authors greatly appreciate financial funding by the German Science Foundation (Deutsche Forschungsgemeinschaft), as part of the research group "MICROPLAST" (FOR797) on *Analysis and Computation of Microstructures in Finite Plasticity*, project SCHR 570/8-2. The authors also thank Prof. D. Raabe (Max-Planck Institut für Eisenforschung, Düsseldorf, Germany) for providing the EBSD/FIB measurements of the three-dimensional microstructure for a dual-phase steel. Furthermore, assistance regarding the application of the *Mystic* optimization framework by M. McKerns is appreciated. The authors express their appreciation to the CCSS at the University Duisburg-Essen for providing computational resources on the Cray XT6m.

References

[ABRP12] Ambrozinski, M., Bzowski, K., Rauch, L., Pietrzyk, M.: Application of statistically similar representative volume elements in numerical simulations of crash box stamping. Archives of Civil and Mechanical Engineering 12, 126–132 (2012)

[AKM10] Arns, C., Knackstedt, M., Mecke, K.: 3d structural analysis: sensitivity of Minkowski functionals. Journal of Microscopy 240, 181–196 (2010)

[BBS14a] Balzani, D., Brands, D., Schröder, J.: Construction of statistically similar representative volume elements. In: Schröder, J., Hackl, K. (eds.) Plasticity and Beyond - Microstructures, Crystal-Plasticity and Phase Transitions. CISM Lecture notes, vol. 550, pp. 355–412. Springer (2014)

[BBS+14b] Brands, D., Balzani, D., Scheunemann, L., Schröder, J., Richter, H., Raabe, D.: Computational modeling of Dual-Phase steels based on representative three-dimensional microstructures obtained from EBSD data. Archive of Applied Mechanics (2014) (submitted)

[BBSC10] Balzani, D., Brands, D., Schröder, J., Carstensen, C.: Sensitivity analysis of statistical measures for the reconstruction of microstructures based on the minimization of generalized least-square functionals. Technische Mechanik 30, 297–315 (2010)

[BBSR11] Brands, D., Balzani, D., Schröder, J., Raabe, D.: Simulation of DP-steels based on statistically similar representative volume elements and 3D EBSD data. In: Computational Plasticity XI - Fundamentals and Applications, Barcelona, Spain, September 7-9, pp. 1552–1563 (2011)

[BDMW02] Beisbart, C., Dahlke, R., Mecke, K., Wagner, H.: Vector- and Tensor-valued Descriptors for Spatial Patterns. In: Mecke, K., Stoyan, D. (eds.) Lecture Notes in Physics, vol. 600 (2002)

[Ber68] Beran, M.: Statistical continuum theories. Wiley (1968)

[BMH+12] Baniassadi, M., Mortazavi, B., Hamedani, H.A., Garmestani, H., Ahzi, S., Fathi-Torbaghan, M., Ruch, D., Khaleel, M.: Three-dimensional reconstruction and homogenization of heterogeneous materials using statistical correlation functions and {FEM}. Computational Materials Science 51, 372–379 (2012)

[Bro55] Brown, W.: Solid micture permettivities. Journal of Chemical Physics 23, 1514–1517 (1955)

[BSB09] Balzani, D., Schröder, J., Brands, D.: FE^2-simulation of microheterogeneous steels based on statistically similar RVEs. In: Proceedings of the IUTAM Symposium on Variational Concepts with Application to Mechanics of Materials, Bochum, Germany, September 22-26 (2009)

[BSBS14] Balzani, D., Scheunemann, L., Brands, D., Schröder, J.: Construction of Two- and Three-Dimensional Statistically Similar RVEs for Coupled Micro-Macro Simulations. Computational Mechanics (2014)

[DW96] Drugan, W., Willis, J.: A micromechanics-based nonlocal constitutive equation and estimates of representative volume element size for elastic composites. J. Mech. Phys. Solids 44, 497–524 (1996)

[EDLR13] Eisenlohr, P., Diehl, M., Lebensohn, R., Roters, F.: A spectral method solution to crystal elasto-viscoplasticity at finite strains. International Journal of Plasticity 46, 37–53 (2013)

[EH86] Exner, H., Hougardy, H.: Einführung in die quantitative Gefügeanalyse. Deutsche Gesellschaft für Metallkunde (1986)

[FC00] Feyel, F., Chaboche, J.: FE^2 multiscale approach for modelling the elastoviscoplastic behaviour of long fibre SiC/Ti composite materials. Computer Methods in Applied Mechanics and Engineering 183, 309–330 (2000)

[Fey99] Feyel, F.: Multiscale FE^2 elastoviscoplastic analysis of composite structures. Computational Materials Science 16, 344–354 (1999)

[FS99] Fish, J., Shek, K.: Finite deformation plasticity for composite structures: computational models and adaptive strategies. Computational Mechanics in Applied Mechanics and Engineering 172, 145–174 (1999)

[GKB03] Geers, M., Kouznetsova, V., Brekelmans, W.: Multi-scale first order and second order computational homogenization of microstructures towards continua. International Journal of Multiscale Computational Engineering 1, 371–386 (2003)

[GTK97] Golanski, D., Terada, K., Kikuchi, N.: Macro and micro scale modeling of thermal residual stresses in metal matrix composite surface layers by the homogenization method. Computational Mechanics 19, 188–201 (1997)

[Hil63] Hill, R.: Elastic properties of reinforced solids: some theoretical principles. Journal of the Mechanics and Physics of Solids 11, 357–372 (1963)

[KBB01] Kouznetsova, V., Brekelmans, W., Baaijens, F.: An approach to micro-macro modeling of heterogeneous materials. Computational Mechanics 27, 37–48 (2001)

[KBC06] Kumar, H., Briant, C.L., Curtin, W.A.: Using microstructure reconstruction to model mechanical behavior in complex microstructures. Mechanics of Materials 38, 818–832 (2006)

[KFG+03] Kanit, T., Forest, S., Galliet, I., Mounoury, V., Jeulin, D.: Determination of the size of the representative volume element for random composites: statistical and numerical approach. International Journal of Solids and Structures (2003)

[Kli00] Klinkel, S.: Theorie und Numerik eines Volumen-Schalen-Elementes bei finiten elastischen und plastischen Verzerrungen. PhD thesis, Institut für Baustatik, Universität Karlsruhe (2000)

[KMS+10] Kapfer, S.C., Mickel, W., Schaller, F.M., Spanner, M., Goll, C., Nogawa, T., Ito, N., Mecke, K., Schröder-Turk, G.E.: Local Anisotropy of Fluids using Minkowski Tensors. Journal of Statistical Mechanics: Theory and Experiments (2010)

[Krö60] Kröner, E.: Allgemeine Kontinuumstheorie der Versetzung und Eigenspannung. Archive for Rational Mechanics and Analysis 4, 273–334 (1960)

[Lee69] Lee, E.: Elastic-plastic deformation at finite strains. Journal of Applied Mechanics 36, 1–6 (1969)

[LLR11] Lee, S., Lebensohn, R., Rollett, A.: Modeling the viscoplastic micromechanical response of two-phase materials using fast fourier transforms. International Journal of Plasticity 27, 707–727 (2011)

[LT92] Lu, B., Torquato, S.: Lineal-path function for random heterogeneous materials. Physical Reviews A 45, 922–929 (1992)

[MHA09] McKerns, M., Hung, P., Aivazis, M.: mystic: a simple model-independent inversion framework (2009)

[MJM08] Mantz, H., Jacobs, K., Mecke, K.: Utilizing Minkowski functionals for image analysis: a marching square algorithm. Journal of Statistical Mechanics: Theory and Experiment 12(15), 2–28 (2008)

[MS92] Miehe, C., Stein, E.: A canonical model of multiplicative elasto-plasticity formulation and aspects of the numerical implementation. European Journal of Mechanics 11, 25–43 (1992)

[MS98] Moulinec, H., Suquet, P.: A numerical method for computing the overall response of nonlinear composites with complex microstructure. Computational Mechanics in Applied Mechanics and Engineering 157, 69–94 (1998)

[MS00] Mecke, K., Stoyan, D.: Statistical Physics and Spatial Statistics: The Art of Analyzing and Modeling Spatial Structures and Pattern Formation. Springer Lecture Notes in Physics, vol. 554. Springer, Heidelberg (2000)

[MSS99] Miehe, C., Schröder, J., Schotte, J.: Computational homogenization analysis in finite plasticity. Simulation of texture development in polycrystalline materials. Computer Methods in Applied Mechanics and Engineering 171, 387–418 (1999)

[MSS+11] McKerns, M.M., Strand, L., Sullivan, T., Fang, A., Aivazis, M.A.G.: Building a framework for predictive science. In: Proceedings of the 10th Python in Science Conference (2011)

[OM00] Ohser, J., Mücklich, F.: Statistical analysis of microstructures in materials science. J. Wiley & Sons (2000)

[PBMP09] Pelissou, C., Baccou, J., Monerie, Y., Perales, F.: Determination of the size of the representative volume element for random quasi-brittle composites. International Journal of Solids and Structures (2009)

[Pov95] Povirk, G.L.: Incorporation of microstructural information into models of two-phase materials. Acta Metall. Mater. 43, 3199–3206 (1995)

[RKP14] Rauch, L., Kuziak, R., Pietrzyk, M.: From High Accuracy to High Efficiency in Simulations of Processing of Dual-Phase Steels. Metallurgical and Materials Transactions B (2014)

[RMPB12] Ramazani, A., Mukherjee, K., Prahl, U., Bleck, W.: Modelling the effect of micromechanical banding on the flow curve behaviour of dual-phase (dp) steels. Computational Materials Science (2012)

[RPBP11] Rauch, L., Pernach, M., Bzowski, K., Pietrzyk, M.: On application of shape coefficients to creation of the statistically similar representative element of DP steels. Computer Methods in Materials Science 11, 531–541 (2011)

[SBB11] Schröder, J., Balzani, D., Brands, D.: Approximation of random microstructures by periodic statistically similar representative volume elements based on lineal-path functions. Archive of Applied Mechanics 81, 975–997 (2011)

[SBBS13] Scheunemann, L., Balzani, D., Brands, D., Schröder, J.: Statistically Similar Rve Construction based on 3D dual-phase steel microstructures. In: Research and Application in Structural Engineering, Mechanics and Computation (2013)

[SBBS14] Scheunemann, L., Balzani, D., Brands, D., Schröder, J.: Design of 3d statistically similar representative volume elements based on minkowski functionals. Submitted to Mechanics of Materials (2014)

[SBM98] Smit, R., Brekelsmans, W., Meijer, H.: Prediction of the mechanical behavior of nonlinear heterogeneous systems by multi-level finite element modeling. Computer Methods in Applied Mechanics and Engineering 155, 181–192 (1998)

[Sch14] Schröder, J.: A numerical two-scale homogenization scheme: the FE^2 method. In: Schröder, K.H.J. (ed.) Plasticity and Beyond - Microstructures, Crystal-Plasticity and Phase Transitions. CISM Lecture notes, vol. 550. Springer (2014)

[SGM08] Sporer, S., Goll, C., Mecke, K.: Motion by stopping: rectifying Brownian motion of nonspherical particles. Phys. Rev. E 78, 11917 (2008)

[Sim92] Simo, J.: Algorithms for static and dynamic multiplicative plasticity that preserve the classical return mapping scheme of the infinitesimal theory. Computer Methods in Applied Mechanics and Engineering 99, 61–112 (1992)

[STKB+10] Schröder-Turk, G.E., Kapfer, S., Breidenbach, B., Beisbart, C., Mecke, K.: Tensorial minkowski functionals and anisotropy measures for planar patterns. Journal of Microscopy 238, 57–74 (2010)

[STMK+11] Schröder-Turk, G.E., Mickel, W., Kapfer, S.C., Klatt, M.A., Schaller, F.M., Hoffmann, M.J.F., Kleppmann, N., Armstrong, P., Inayat, A., Hug, D., Reichelsdorfer, M., Peukert, W., Schwieger, W., Mecke, K.: Minkowski Tensor Shape Analysis of Cellular, Granular and Porous Structures. Advanced Materials 23, 2535–2553 (2011)

[STMK+13] Schröder-Turk, G.E., Mickel, W., Kapfer, S.C., Schaller, F.M., Breidenbach, B., Hug, D., Mecke, K.: Minkowski Tensors of Anisotropic Spatial Structure. New Journal of Physics 15 (2013)

[Tor02] Torquato, S.: Random heterogeneous materials. Microstructure and macroscopic properties. Springer (2002)

[TW08] Temizer, I., Wriggers, P.: On the computation of the macroscopic tangent for multiscale volumetric homogenization problems. Computational Mechanics in Applied Mechanics and Engineering 198, 495–510 (2008)

[Voc55] Voce, E.: A practical strain hardening function. Metallurgica 51, 219–226 (1955)

[WA90] Weber, G., Anand, L.: Finite deformation constitutive equations and a time integration procedure for isotropic, hyperelastic-viscoelastic solids. Computer Methods in Applied Mechanics and Engineering 79, 173–202 (1990)

[YT98] Yeong, C.L.Y., Torquato, S.: Reconstructing random media. Physical Review E 57, 495–506 (1998)

[Zem03] Zeman, J.: Analysis of composite materials with random microstructure. PhD thesis, University of Prague (2003)

[ZRS+06] Zaafarani, N., Raabe, D., Singh, R., Roters, F., Zaefferer, S.: Three - dimensional investigation of the texture and microstructure below a nanoindent in a cu single crystal using 3D EBSD and crystal plasticity finite element simulations. Acta Materialia 54, 1863–1876 (2006)

[ZW05] Zodhi, T., Wriggers, P.: Introduction to computational micromechanics. Springer (2005)

Author Index

MIX
Papier aus verantwortungsvollen Quellen
Paper from responsible sources
FSC® C105338

If you have any concerns about our products,
you can contact us on
ProductSafety@springernature.com

In case Publisher is established outside the EU,
the EU authorized representative is:
Springer Nature Customer Service Center GmbH
Europaplatz 3, 69115 Heidelberg, Germany

Printed by Libri Plureos GmbH
in Hamburg, Germany